THE FRUIT OF THE VINE:
VITICULTURE IN ANCIENT ISRAEL

HARVARD SEMITIC MUSEUM PUBLICATIONS

Lawrence E. Stager, General Editor
Michael D. Coogan, Director of Publications

HARVARD SEMITIC MONOGRAPHS

edited by
Peter Machinist

Number 60
**THE FRUIT OF THE VINE:
VITICULTURE IN ANCIENT ISRAEL**

by
Carey Ellen Walsh

Carey Ellen Walsh

THE FRUIT OF THE VINE:
VITICULTURE IN ANCIENT ISRAEL

EISENBRAUNS
Winona Lake, Indiana
2000

THE FRUIT OF THE VINE:
VITICULTURE IN ANCIENT ISRAEL

by

Carey Ellen Walsh

Library of Congress Cataloging-in-Publication Data

Walsh, Carey, 1960–
    The fruit of the vine : viticulture in ancient Israel / Carey Ellen Walsh.
        p.   cm. — (Harvard Semitic Museum publications) (Harvard Semitic
        monographs ; no. 60.)
    Includes bibliographical references (p.).
    ISBN 1-57506-904-0 (cloth : alk. paper)
    1. Viticulture—Palestine—History.   2. Wine and wine making—
Palestine—History.   3. Viticulture in the Bible.   I. Title.   II. Series.
III. Series: Harvard Semitic monographs ; no. 60.

SB387.8.P19 W36 2000
634.8'0933—dc21

                                 00-059333

Contents

Acknowledgments

This study began as a Th.D. dissertation at Harvard University, completed in 1996, and I am grateful for the guidance and care of my committee members, Jo Ann Hackett, Peter Machinist, and Lawrence Stager. I am grateful too for Harvard's generosity, particularly in enabling me to do research in Jerusalem with the Sheldon Traveling Grant. The staff at the Albright Institute in Jerusalem was always helpful and kind. I want to thank especially the director, Sy Gitin, who provided direction and discussion on a number of occasions. I thank the ASOR committee for granting me a U.S. Information Agency Junior Research Fellowship for research at the Albright. Several people in Jerusalem made the experience fruitful and enjoyable as we explored wine together sitting in the Albright's garden. They are: Diana Edelman, Glenda Friend, Aaron Brody, and Oded Borowski. I still smile with memories from that time, and the secret knowledge that writing a dissertation was, at times, pleasurable.

Jane Geaney, my debt to you goes beyond words, but I'll stutter some anyway. God bless you for the rescue and the patience, and the comfort of staying in Portland with you. Somehow, having a friend and a witness to the exhaustion of finishing was a great solace to me. And who better than you! Kathryn Slanski and Gene McAfee too were themselves during my writing and those gifts were appreciated. We all knew each other before we wrote dissertations, and are not the same. I also am grateful to my mom, Mary Laureen Walsh for her support, a mixture of patience and concern, and pride long after my own had been depleted. Deidre too always knew just what to say and was level-headed and thoughtful.

On the Memphis leg of the writing journey, I want to thank especially Karen Smith whose support and integrity helped me and whose wisdom competes with Solomon's at times. Finally, I want to thank Michael Coogan, and Peter Machinist again. The editing process proved to be an education all its own, and yet again I had the best teachers. Peter Machinist has challenged my thinking and writing and helped to push them further than I feel I could have gone on my own. I owe a great debt of gratitude to him for this finished product, with the usual caveat that any remaining weaknesses are mine alone. But what I especially want to thank him for is that his editing was really mentoring. He taught me by example what it is to take a student seriously, long after I stopped being one of his. I hope I have the generosity to do the same with my students and their work.

Abbreviations

ABD	*Anchor Bible Dictionary*
ANET	*Ancient Near Eastern Texts Relating to the Old Testament*, 3d ed., ed. J. B. Pritchard
BA	*Biblical Archaeologist*
BAR	*Biblical Archaeology Review*
BASOR	*Bulletin of the American Schools of Oriental Research*
BCE	Before the Common Era
BDB	Brown, Driver, Briggs, *A Hebrew and Aramaic Lexicon of the Old Testament*
BHK	*Biblia Hebraica*, ed. R Kittel
CBQ	*Catholic Biblical Quarterly*
CE	Common Era
CTA	*Corpus des Tablettes en Cunéiformes Alphabétiques*, A. Herdner
DTR	Deuteronomistic historian
EI	*Eretz-Israel*
GKC	*Gesenius' Hebrew Grammar.* Edited by E. Kautzsch. Translated by A. E. Cowley. 2d. ed.
HTR	*Harvard Theological Review*
IDB	*Interpreter's Dictionary of the Bible*
IEJ	*Israel Exploration Journal*
IM	*Israel Museum*
IMJ	*Israel Museum Journal*
JAOS	*Journal of the American Oriental Society*
JB	*Jerusalem Bible*
JBL	*Journal of Biblical Literature*
JEA	*Journal of Egyptian Archaeology*
JNES	*Journal of Near Eastern Studies*
JPOS	*Journal of the Palestine Oriental Society*
JPS	*Jewish Publication Society Tanakh*
JSOT	*Journal for the Study of the Old Testament*
KB	Koehler, L., and W. Baumgartner, *Hebräisches und Aramäisches Lexikon zum Alten Testament*
KTU	*Die Keilalphabetischen Texte aus Ugarit*
LXX	Septuagint
MT	Masoretic Text
NAB	*New American Bible*
NEAEHL	*New Encyclopedia of Archaeological Excavations in the Holy Land*

NEB	*New English Bible*
NJB	*New Jerusalem Bible*
NRSV	*New Revised Standard Version of the Bible*
OAHW	*The Origins and Ancient History of Wine,* ed. P. McGovern, S. Fleming, and S. Katz
PEQ	*Palestine Exploration Quarterly*
PRU	*Le Palais Royal d'Ugarit*
REB	*Revised English Bible*
RSV	*Revised Standard Version of the Bible*
TA	*Tel Aviv*
UF	*Ugarit-Forschungen*
UL	*Ugaritic Literature*, C. Gordon
UMI	University Microfilms Incorporated
UT	*Ugaritic Textbook*, C. Gordon
VT	*Vetus Testamentum*
ZAW	*Zeitschrift für die alttestamentliche Wissenschaft*

Introduction

The biblical stories of Noah, Lot, Nabal, Uriah, Elah, and Ben-hadad include episodes of their inebriation. Drunkards incur the prophetic wrath of Isaiah, Jeremiah, Hosea, Joel, Amos, and Nahum. Biblical wisdom literature offers varying perspectives on drink, from the caution that wine will "bite" (Prov 23:32) to the simple bluntness of Qoheleth's assessment that "wine gladdens life" (Eccl 10:19).[1] And there are traditions of abstainers from wine, namely, the Nazirites (Num 6:1-21) and the Rechabites (Jer 35).

Wine use occurs frequently in the Hebrew Bible and in varying contexts—blessing, feast, procreation, wedding, cult, and military activity, among others. Jacob and Judah, both favored sons, inherit wine in blessings (Gen 27:27-29; 49:11-12). Procreation is the ostensible goal of the hosts when serving Lot and Uriah (Gen 19:30-38; 2 Sam 11:6-13): Lot's daughters want their father's seed, and David wants to obscure his patrimony of Bathsheba's child by getting her husband Uriah to sleep with her. Samson's marriage and death occur at Philistine drinking feasts (Judg 14-16). Wine is served atop Mt. Sinai in the covenant meal with Yahweh (Exod 24:11) and remains an offering in the cultic legislation of the Priestly writer (P). King Ben-hadad seems to drink his way through much of his siege of Samaria (1 Kgs 20:12, 16). Wine, in all these examples, is a constitutive facet of social encounter. What cultural attitudes towards

---

[1]Related is his refrain that "there is nothing better for mortals than to eat and drink, and find enjoyment in their toil" (Eccl 2:24; 3:13; 5:18; 8:15; 9:7).

wine are reflected in these biblical traditions? What do drinking, drunkenness, and abstention mean in these descriptions of Israelite social experience? In addition, images of vines, vineyards, and grape clusters throughout the Bible are used to convey the nature of relationships between Yahweh and his people and among humans. In Isa 5:1-7 and Ps 80:8-16, for example, Yahweh is a divine vintner and his people are first vine, then fruit. The New Testament extends this metaphor with God still as vintner, and Jesus "the true vine" (John 15:1). Jesus ostensibly literalized the imagery at the Last Supper by equating the wine with his blood, soon to be shed (Matt 26:27-29; Mark 14:23-25; Luke 22:20). In Ps 128:3, the wife of one who fears Yahweh is likened to a "fruitful vine." The fate of Pharaoh's cupbearer, which turns out to be freedom, lies in his dream of a vine (Gen 40:9-13). And, in Jotham's critique of monarchy, the vine, untempted by the chance to rule, remains committed to its noble task of "producing my wine that cheers God and mortals" (Judg 9:13). In Nebuchadrezzar's destruction of Judah, only the "vinedressers and tillers of the soil" are spared (2 Kgs 25:12). Vines are used frequently in biblical description and attain a metaphoric prominence unrivaled by any other mundane feature of Israelite life.

The grape cluster, the fruit of the vine, serves as a potent biblical metaphor as well. Its range of meaning includes: the productive potential of Canaan, when the spies return to Moses carrying a huge grape cluster (Num 13:23); sexual desire for a woman's breasts (Cant 7:8); and the remnant of people worth saving in Yahweh's harvest (Isa 65:8). Viticulture and its product, wine, then, are ubiquitous in biblical imagery, yet interpreting their meaning within the texts is incomplete, even distorted, without an historical analysis of the practical aspects of ancient viticulture. These include the cultivation and maintenance of a vineyard, the technology and tasks involved in producing wine, and the social customs surrounding the grape harvest and enjoyment of the wine. Assessing the historical context of any biblical feature is always an important first step in exegesis. It is crucial in a study on wine since alcohol often attracts the cultural and moral biases of the interpreter in a way that other features of Israelite social history, such as olive oil, dwellings, donkeys, and water, do not. Ancient Israel had its own cultural valences about wine and the vine, and these are best understood from the historical setting in which they arose.

The practice of viticulture—from planting vines to drinking wine—in Israelite culture is the focus of my investigation. I argue that viticulture was fundamental and prized in Israelite life and so remained an effective resource for biblical writers. Clearly viticulture is an inscribed feature of Israelite life in biblical descriptions.[2] It may have been a detail of daily life in ancient Israel, but its inclusion in the biblical literature marks it as culturally meaningful. Not only were vines and wine details worth mentioning by the biblical authors, but they were allowed to shape certain key descriptions. As such, viticulture is a facet of the culture represented in the Hebrew Bible. My thesis is that viticulture, no less than drinking, marked the social sphere of Israelite practitioners, and so its details were often enlisted to describe social relations in the Hebrew Bible. The biblical scenes of drinking and inebriation are often simply the fairly natural or pedestrian expressions of a society practicing viticulture, and as such connote not merely opprobrium, but rather levels of social intimacy and cohesion among the participants. The vine and vineyard are similarly enlisted to describe valued relations such as that between Naboth and his ancestors in 1 Kgs 21 and between Yahweh and the people in Isa 5:1-7.

The biblical portrayal of Israelite life is a complex nexus of theological and cultural ideologies stitched throughout with mundane features such as food and drink. These features of everyday life offer important clues for the reconstruction of Israelite social history, but they cannot be accepted off the biblical page as objective referents to everyday activity. For they have been included in, even selected for, literary representation of Israelite life by oral transmitters, authors, and redactors. This means (as it always did) that the historical quest for a biblical datum is doomed if it fails to address the literary process involved in symbolic representation. Vineyards and drinking in the Hebrew Bible are literary constructions. The interpretive task is to discern what such constructions reflect or encode of Israelite viticulture. The present study is devoted to a reconstruction of the practice of viticulture in ancient Israel through analysis of biblical and archaeological evidence. It is my contention that biblical imagery of

---

[2]That is, by virtue of its presence, viticulture shares a symbolic function in conveying the cultural values and worldview of Israelite life. It is not only an aspect of Israelite culture, but also one that has a shaping influence on that culture. Viticulture makes an indelible mark on how the Israelite cultural world is portrayed in biblical material. The biblical writers are conveying a world of symbolic meanings about community, Israel, their God, ethics, and the like, and viticulture is enlisted in that service. The notion that the writings are a cultural record of a people, and that the details share in conveying the social values of a culture, is drawn from the field of symbolic anthropology. For insightful examples of such an approach, see M. Douglas, "Deciphering a Meal," in *Implicit Meanings: Essays in Anthropology* (London: Routledge & Kegan Paul, 1975), 249-75, and V. Turner, *The Forest of Symbols: Aspects of Ndembu Ritual* (Ithaca, N.Y.: Cornell University Press, 1967). For theoretical analyses of symbolic anthropology, see M. LeCron Foster and S. Brandes, eds., *Symbol as Sense: New Approaches to the Analysis of Meaning* (New York: Academic Press, 1980), and R. J. McGee and R. L. Warms, "Symbolic and Interpretive Anthropology," in *Anthropological Theory: An Introductory History* (Mountain View, Calif.: Mayfield, 1996), 430-79.

vines, viticulture, and drinking is best illuminated through discussion of the actual processes involved in ancient vine tending and wine production. The additional thematic and theological meanings attached to such biblical representations of the vine and wine imagery are also refined by knowledge of the viticultural technology practiced by the Israelites.

Review of Biblical Scholarship on Wine

The issue of viticulture as a feature of Israelite culture has, surprisingly, received scant critical attention by biblical scholars. Wine is mentioned in commentaries and histories as an agricultural product of ancient Israel, but the technology of its production is not pursued. The best example of scholarly neglect is in the five-volume *Anchor Bible Dictionary*, published in 1992. There is quite simply no entry for "wine" or "vineyard," and "vine" receives only a paragraph description under the entry "flora."[3]

There are general descriptions of wine technology in R. J. Forbes's *Studies in Ancient Technology*, Oded Borowski's *Agriculture in Iron Age Israel*, and, most recently, Rafael Frankel's *Wine and Oil Production in Antiquity in Israel and Other Mediterranean Countries*.[4] Detailed studies on various aspects of wine technology for the ancient Near East and Mediterranean region are available in *The Origins and Ancient History of Wine*, edited by Patrick E. McGovern, Stuart J. Fleming, and Solomon H. Katz.[5] While not specific to ancient Israel, these studies provide vital information from contiguous cultures. An extensive analysis of viticulture specific to ancient Israel, however, is still lacking in biblical scholarship, to the detriment of textual interpretation. For example, one instance of neglect or oversight occurs in commentaries on Gen 27, where Jacob deceives his father, Isaac, and receives the blessing meant for Esau.

In that chapter, Jacob serves Isaac wine along with the requested game, yet neither Isaac's nor Rebekah's instructions had included it (v

---

[3] I. and W. Jacob, "Flora," in *Anchor Bible Dictionary* (ed. D. N. Freedman; 6 vols.: New York: Doubleday, 1992; hereafter *ABD*), 2:810.

[4] R. J. Forbes, *Studies in Ancient Technology*, (Leiden: Brill, 1955), 3:72-85; O. Borowski, *Agriculture in Iron Age Israel* (Winona Lake, Ind.: Eisenbrauns, 1987), 102-14; R. Frankel, *Wine and Oil Production in Antiquity in Israel and Other Mediterranean Countries* (JSOT/ASOR Monographs 10; Sheffield, Eng.: Sheffield University Press, 1999).

[5] P. McGovern, S. Fleming, and S. Katz, eds., *The Origins and Ancient History of Wine* (Luxemburg: Gordon & Breach, 1995; hereafter *OAHW*). See also L. Milano, ed., *Drinking in Ancient Societies* (History of the Ancient Near East/Studies-6; Padova: Sargon, 1994), especially 69-183 for various essays on wine and beer technology.

25). Isaac, thinking the giver is Esau, then blesses Jacob with "plenty of grain and wine." E. A. Speiser, Gerhard von Rad, and Claus Westermann do not comment on this detail, yet it is significant for understanding the action in the scene.[6]   For it means that Jacob has not simply followed the directions of his father and mother, but has improvised with a mind-altering substance. Such initiative strengthens the interpretation of Jacob as clever. Isaac's eyes are already dim (v 1). Alcohol would only further relax this old father's mental faculties and remaining senses. Wine may have been traditional at blessings or deathbeds, or a given with any meal served in ancient Israel, but since Jacob intends to deceive his father, its presence is doubly significant. It is in fact the first time that wine is explicitly mentioned within a patriarch's home.[7]   Whatever other conventions Jacob might be honoring, he has in effect slipped his father a mild sedative, whose potency we cannot know.

Speiser views Jacob as "under pressure from his strong-willed mother,"[8] and so he misses Jacob's initiative and perhaps indulges in some low-grade misogyny. He misses too a literary resonance with the story of Lot's daughters, who served their father wine in order to secure progeny (Gen 19:30-38), and perhaps with the story of Noah getting drunk and his son Ham taking advantage of him (Gen 9:20-27). This is, in fact, the Yahwist's (J) third story of a father, children, and wine. J adds a fourth when he incorporates an ancient poem now in Gen 49. There, in vv 11-12, Judah's blessing includes vines and an abundance of wine. The frequency of this topos in J of families and their wine requires analysis. In fact, the coincidence of families and wine in the Hebrew Bible is not haphazard; viticulture was practiced, as we shall see, as part of the Israelite nuclear family's farm.

Interpreters have traditionally stressed wine as a moral symbol, usually a negative one, in biblical texts.[9]   The rabbis of *Midrash Rabbah*, for example, deem Noah to be one of three agriculturalists of

---

[6]E. A. Speiser, *Genesis* (Anchor Bible 1; Garden City, N.Y.: Doubleday, 1964); G. von Rad, *Genesis* (Old Testament Library; Philadelphia: Westminster, 1972); C. Westermann, *Genesis 12-36: A Commentary* (trans. J. J. Scullion, S. J.; Minneapolis: Augsburg, 1984).

[7]Lot had plenty of wine when his daughters get him drunk (Gen 19:30-38). Abram receives wine and bread from King Melchizedek (Gen 14:18). When he later hosts three angels, no wine is present. They get instead bread, water, milk products, and a calf (Gen 18:5-8). Banquets, however, do occur and these no doubt involved wine—Abraham holds one for the weaning of Isaac (Gen 21:8); Lot tries to hold one for his guests (Gen 19:3); and Isaac makes one with Abimelech (Gen 26:30). See chapter 6, pp. 225-47.

[8]Speiser, *Genesis*, 211.

[9]J. F. Ross, "Wine," in *The Interpreter's Dictionary of the Bible*, ed. G. Buttrick et al. (Nashville: Abingdon, 1962; hereafter *IDB*), 4:851.

whom it was said, "no good was found in them."[10] For von Rad, Lot's wine-induced incest leaves his life morally "bankrupt."[11]   Magen Broshi, in his study on the wines of ancient Palestine, concurs with this dire view. For him, the deeds of Noah and Lot "were intended to serve as examples of the dangers and repulsiveness of intemperance."[12]   Such interpretations assume that drinking had a moral component to it, but they do not explicate how this came to be. Wine apparently becomes a morally laden substance because of the actions consequent to its consumption. However, one would have to demonstrate that temperance was a cultural value in ancient Israel or for the author of these tales before assuming instructional intent.[13]   A generalization can easily result about how Israelites valued the product wine divorced from its agricultural context. Morris Jastrow, for example, sees a negative thread being woven through the entire Hebrew Bible with an "opposition to viniculture expressed in Genesis—maintained by the Rechabites down to the time of the exile and implied in the Nazir's abstention."[14]

Jastrow's generalization persists in biblical studies. For example, the Rechabites abstain from wine as a part of their nomadic lifestyle (Jer 35). They do not cultivate any fruit or grains, and do not build houses. But their abstention has long been seen as a particular *moral* stance against viticulture.[15]   The reason for their abstention, however, apart from filial obedience, is left obscure in the text.

Both abstention and intoxication in the Hebrew Bible are assumed to have moral valences that reflect poorly on viticulture, the agrarian practice that produces wine. Yet this judgment is assumed by interpreters, rather than demonstrated from the texts. It ignores the fundamental importance of viticulture to the economy of ancient

---

[10]*Genesis Rabbah* (trans. H. Freedman and M. Simon; London and New York: Soncino, 1983), 1:289. Cain and Uzziah are the other two unredeemable farmers to the rabbis.

[11]von Rad, *Genesis*, 224.

[12]M. Broshi, "Wine in Ancient Palestine—Introductory Notes," *IMJ* 3 (1984): 21. Even in botanical catalogues of biblical flora, such interpretive sentiments exist without exegesis. For example, W. E. Shewell-Cooper provides this aside under the vine entry: "The Bible definitely condemns drunkenness": *Plants, Flowers, and Herbs of the Bible* (New Canaan, Conn.: Keats, 1977), 77. R. Weinhold, in his history of wine, views Noah's first vintage as "distressing": *Vivat Bacchus: A History of the Vine and Its Wine* (Watford, Eng.: Argus, 1978), 34.

[13]One would also have to exegete valences about nakedness and incest (cf. Lev 18). Counting on universal repulsion to these stories is an unacknowledged form of reader-response criticism.

[14]M. Jastrow, Jr., "Wine in the Pentateuchal Codes," *JAOS* 33 (1913): 190.

[15]See, for example, J. Bright, *A History of Israel* (3d ed.; Philadelphia/London: Westminster/ SCM, 1981), 250. See also J. Milgrom, who argues that the priestly abstention of Lev 10:8-9 is part of the overall biblical condemnation of wine, *Leviticus 1-16* (Anchor Bible 3; New York: Doubleday, 1991), 612.

Israel.[16] It also ignores the important metaphoric function viticulture has in biblical traditions of what is esteemed, such as patrimonial inheritance. Isaac, for example, gives the land of grain and wine to Jacob (Gen 27:28). Ahab is a wicked king according to the Deuteronomistic historian (henceforth Dtr) because he steals Naboth's inheritance, a vineyard. In a similar manner, Yahweh's care for his people is a theological value for Isaiah, and so he likens that care to that of a vintner in Isa 5:1-7.

Biblical scholarship, it should be remarked, is not alone in its importing of latent ideologies about alcohol into textual interpretation. Histories of food and drink often contain moralizing sentiments on the origins of wine that are not present for other technologies. Typical is this wistful, historical analysis: "It is sobering to consider that the neglected jar of fruit juice or pulp, or the half-empty honey-pot left out in the rain, set man along the road to alcoholism and the illicit still."[17] The task at present is to illuminate Israelite cultural attitudes toward wine, shorn of these anachronistic biases by analyzing what was involved in the practice of viticulture for the Israelite farmer.

In contrast to the interpretative assumptions about wine apparent in biblical studies, historical scholarship on ancient Israel has advanced our understanding of wine production and use considerably, and this is due largely to its incorporation of the relevant archaeological material. This material is, in fact, extensive, further suggesting the importance of viticulture in Israelite life. Epigraphic evidence includes the Gezer calendar of the agricultural year, the Samaria ostraca, and the *lmlk* jar handles found throughout Judah. These, along with archaeological evidence of presses, wine jugs, jar stoppers, strainers, storage facilities, and paleobotanical finds at various sites, yield a picture of significant wine production in ancient Israel. Oded Borowski and Dan Cohen offer partial reconstructions of Israelite wine production based on these archaeological finds.[18] But no study explores the influence of this wine production on the literary

---

[16]T. Unwin provides an overview of the historical and symbolic importance of wine in ancient cultures, but restricts his discussion of ancient Israelite wine to texts cautioning against excessive drinking: *Wine and the Vine: An Historical Geography of Viticulture and the Wine Trade* (London: Routledge, 1991), 83.

[17]W. J. Darby, P. Ghalioungui, and L. Grivetti, *Food: The Gift of Osiris* (London and San Francisco: Academic Press, 1977), 1:164. Similar moralizing tendencies lace *Viticulture and Brewing in the Ancient Orient* (Leipzig: J. C. Hinrichs, 1922) by H. F. Lutz; *Dionysus: A Social History of the Wine Vine* (London: Thames & Hudson, 1965) by E. Hyams; and *Wine in the Ancient World* (London: Routledge & Kegan Paul, 1957) by C. T. Seltman. For contrast, see the celebrative accounts of H. Johnson, *Vintage: The Story of Wine* (New York: Simon & Schuster, 1989) and Weinhold, *Vivat Bacchus*. For a more reasoned and descriptive history see Unwin, *Wine and the Vine*.

[18]Borowski, *Agriculture in Iron Age Israel*; D. Cohen, "On Viticulture and Wine—in Israel and the Ancient World," (Hebrew) *Beth Mikra* 37 (1991-92): 59-69; cf. Forbes, *Studies in Ancient Technology*, 3:72-85.

depictions of viticulture in the Bible. Instead, a gap currently exists between archaeological and biblical interpretation.

Methodology

My contention is that the biblical imagery of vines, viticulture, and drinking is best illuminated through discussion of the actual processes involved in ancient vine tending and wine production. Hence, it will be necessary to work with both sources of Israelite viticulture, viz., the textual and the archaeological materials for vine cultivation and wine production.

The present project seeks to bridge the current interpretive gap between biblical studies and historical study of ancient Israel and illuminate the meanings of viticulture and wine use in the culture of ancient Israel. What the textual and archaeological sources share is that they are both deposits of Israelite culture. They are the traces left behind of vine cultivation and wine use. The first task, then, is to analyze the archaeology of wine and provide a reconstruction of viticultural technology. The material culture is what remains of the realia of wine use in the Israelite world. Since this evidence has been subject to the vagaries of excavation, i.e., it represents only what has been dug and saved, it is partial. The archaeology provides the *remains* of a culture, much as the Bible does. I am necessarily working with partial data from both sources, and so these sources need to stand in dialectical tension with one another.

My focus is viticulture as a facet of daily life, and my method involves incorporating analysis of the biblical and archaeological materials into a coherent picture. To refine this picture, I draw on material from other rural agrarian cultures, both ancient and modern, for hints on precisely how agriculture influences daily life. Descriptions of pre-industrial viticulture from classical sources and ethnographic studies on Mediterranean farming provide useful parallels for reconstructing Israelite viticultural practice from the biblical descriptions of vines and wine and the archaeological materials for wine production.

Reconstruction of how viticulture operated as a facet of the agricultural strategy of Israelite farming is necessary in order to establish its value in Israelite culture. Discussions of agrarian economies by theorists working in ancient Israelite history, most notably Lawrence Stager,[19] and by comparative economic

---

[19]L. E. Stager, "The Firstfruits of Civilization," in *Palestine in the Bronze and Iron Ages* (ed. J. M. Tubb; London: Institute of Archaeology, 1985), 177-88.

anthropologists, such as Stuart Plattner,[20] have proved instrumental to my analysis of the viticultural economy. Several important features present themselves for economic analysis: the strategies of peasant farming, the practical requirements of horticulture, and the existence of industrial centers. The risk of interdisciplinary work, of course, is to end up with a hodgepodge, a pasting of various cultures onto Israel's, but the potential gain to understanding this wine culture outweighs the risk. Thomas Levy has termed such an approach "intellectual foraging,"[21] and indeed it is. He also notes that this is remarkably similar to how archaeologists work, "drawing on a wide range of disciplines to analyze and interpret their data." I view myself as a kind of excavator of biblical history, where foraging becomes, while not ideal, necessary, given the fragmented nature of the sources: a Bible, written by many different hands with many ideological bents; and archaeological remains, representing only those that can survive the centuries and that get excavated.

The biblical portrayal of viticulture is my next concern. Since vines, vineyards, and wine are evident in the Bible, authors have attributed to them symbolic import. The focus of my investigation is on what these described features tell us about viticultural practice. I take the data for wine use in ancient Israel from all the sources available—the Bible, archaeology, and epigraphy—and present as coherent a picture as possible for how Israel valued viticulture and its product wine. Such a method is social-historical to the extent that it strives to reconstruct the social world of wine. It involves as well a critical view toward the sources as literary works. An ethnographer has always to interpret the stories of multiple informants, as those stories are themselves social constructs. Through them, she or he can begin to rule out the idiosyncrasies of one teller and begin to construct a picture of the custom generally described by the pool of informants. With the Bible, the "informants" are literary creators and often have political and theological agendas. One must continually ask what part is likely to be factual, that is, reflective of an actual custom, what part rhetorical or metaphoric. These two, the factual and metaphoric, are never separate whenever language is used. To ask what viticulture meant to a people is a complicated question involving economy and social customs. To ask it of texts and archaeological remains is perhaps even more complicated.

---

[20]S. Plattner, ed., *Economic Anthropology* (Stanford: Stanford University Press, 1989).

[21]T. E. Levy, ed., *The Archaeology of Society in the Holy Land* (New York: Facts on File, 1995), 2.

The question of how vineyards and wine were esteemed in ancient Israel is investigated through remains of its culture—both material and literary. It should be stressed at the outset, however, that the reconstructive and exegetical tasks are not separate, since biblical traditions are themselves a source for historical reconstruction. Rather, these tasks stand in dialectical relation, honing and refining the picture of Israelite viticulture. A study of Israelite viticulture provides insight into an agricultural adaptation and technology practiced by Israelites. It makes a contribution to the study of the social history of ancient Israel. It also makes a contribution to biblical interpretation. Knowledge of the details involved in cultivating a vineyard, pruning, harvesting, treading grapes, storing the wine in jars, and then drinking it, illuminates the meanings of scenes and poetic metaphors in the biblical text where vines and wine are present.

Chapter 1 begins with a summary of the history of viticulture in Syria-Palestine, and a brief discussion of the role of alcohol in ancient Egypt and Mesopotamia. Next follows a discussion of the geographic conditions and agricultural calendar of ancient Israel. Chapter 2 is a discussion of the sociology of the vintner in ancient Israel. It includes analysis of the archaeology of the family farm, the labor pool involved, and the custom of patrimonial inheritance. Chapter 3 is concerned with the cultivation of the grape vine, from starting a vineyard, through tilling the soil, planting and training the vines, and finally the measures taken to protect the vineyard. Chapter 4 is a discussion of the two primary installations of the vineyard, viz., the tower and the winepress. Chapter 5 is a reconstruction of the elements of the Israelite grape harvest and of wine production. Chapter 6 is a study of wine consumption, from daily sustenance to the important role of the banquet in family life.

The cultivation of the vineyard, production of wine, and the resulting commensality and feasting all involve and help to define social interaction. The use of an intoxicant, with its ability to loosen inhibitions and depress the central nervous system, can both solidify and potentially threaten these social encounters. Israelite cultural values toward the vine and its eventual wine, then, have significance beyond the agricultural product. They reflect social relations. Hence, investigation of the meanings of viticulture and wine use in ancient Israel and the Hebrew Bible is focal to understanding Israelite culture.

Chapter 1

Viticulture as a Vital Facet of Israelite Culture

> Talk about vines and you are talking about society,
> political power, an exceptional labor process, in fact
> an entire civilization. If wheat is the prose of our
> long history, wine is its more recently born poetry,
> illuminating and ennobling the landscape.[1]
>
> Fernand Braudel

Though Braudel is writing of France here, his insight applies to other civilizations where wine played its part. Ancient Israel is one such civilization, yet histories of viticulture often include only a short excursus on Noah's stupor to indicate ancient Israel's negative stance towards viticulture.[2] Yet wine did far more for ancient Israel than humiliate the occasional father such as Noah. It shaped its very culture. In general, horticulture, the cultivation of vines and fruit trees, is an important supplement to grain production, as it provides additional foodstuffs for the household and flourishes during the summer months, when there is no grain farming activity. Horticulture also indicates settlement, since trees and vines take years to come to initial fruition. It also adds sources of energy (and diversity) to the diet, such as figs, dates, and grapes, and in olive oil provides both fat and fuel for lamps.

The production of wine marks a further significant shift in the human control of food through cultivation of a plant. For with the advent of viticulture, agriculture turned toward producing a product not necessary for biological subsistence, but still important on a *social*

---

[1]F. Braudel, *The Identity of France* (trans. Sian Reynolds; New York: Harper & Row, 1988-1990), 2:317.
[2]*Genesis Rabbah*, 289; Broshi, "Wine in Ancient Palestine," 21-22; Darby et al., *Food: The Gift of Osiris*, 1:164; Jastrow, "Wine in the Pentateuchal Codes," 190.

level, in that wine brought enjoyment.[3] Wine was a liquid resource, brought levity, and was storable. It was thus valuable for trade and contributed to the economic stability of ancient farmers. Desired by others, wine could act as a cash crop to exchange for other crops or tools as the need arose.

I. History of Viticulture in Syria-Palestine

The origins of viticulture in Syria-Palestine go back well before the Iron Age to the Early Bronze I period (3500-3000 BCE).[4] Domesticated grape pips have been found for this period in Hama in Syria;[5] and in Jericho, Lachish, and Arad in Palestine.[6] Evidence of

---

[3]For an insightful argument for the role enjoyment can play as an incentive for historical change in societies, see V. G. Childe "The Neolithic Revolution" in *Man Makes Himself* (London: Watts, 1951), 67-72.

[4]Grape pips have been reported in the Caucasus mountain region from EB as well: R. Gorney, "Viniculture and Ancient Anatolia," in *OAHW*, 172; G. N. Lisitsina, "The Caucasus: A Centre of Ancient Farming in Eurasia," in *Plants and Ancient Man, Studies in Palaeoethnobotany* (ed. W. van Zeist and W. A. Casparie; Rotterdam/Boston: A. A. Balkema, 1984), 285-92; D. Zohary, "The Domestication of the Grapevine *Vitis Vinifera* L. in the Near East," in *OAHW*, 28; H. P. Olmo, "Grapes," in *Evolution of Crop Plants* (ed. N. W. Simmonds; Essex, Eng.: Longman Scientific & Technical, 1976), 295; Unwin, *Wine and the Vine*, 63; A. Goor, "The History of the Grape-vine," *Economic Botany* 20 (1966): 46; M. Zohary and M. Hopf, *Domestication of Plants in the Old World: The Origin and Spread of Cultivated Plants in West Asia, Europe, and the Nile Valley* (Oxford: Clarendon, 1988), 134; J. Einset and B. H. Barritt, "The Inheritance of Three Major Fruit Colors in Grapes," *Journal of the American Society of Horticultural Science* 94 (1969): 87-89. Stager adds that horticulture is evident earlier, as four fruits—olive, date, fig, and pomegranate—begin to appear in Syria-Palestine by the late Chalcolithic; the grape, however, is not attested until EB I: "Firstfruits," 173; see also Broshi, "Wine in Ancient Palestine," 22.

[5]H. Helbaek, "Late Cypriot Vegetable Diet at Apliki," in *Opuscula Atheniensia* 4 (Lund: CWK Gleerup, 1962), 181; H. Helbaek, "Les empreintes de céréals," Appendix I, in *Hama, Fouilles et recherches de la fondation Carlsberg II, 3* (ed. P. J. Riis; Copenhagen: Gyldendalske Boghandel, 1948), 205.

[6]For Jericho, see M. Hopf, "Jericho Plant Remains," in *Excavations at Jericho* (ed. K. M. Kenyon and T. A. Holland; Jerusalem: British School of Archaeology in Jerusalem, 1983), 5:587. For Arad, see M. Hopf, "Plant Remains, Strata V-I," in *Early Arad* (ed. R. Amiran; Jerusalem: Israel Exploration Society, 1978), 73-74. Hopf notes that the pips had blisters equally spread over their surface, an indication that they were probably dried before being charred, rather than fresh berries which would have been spotted irregularly with larger blisters or splits. Hence, they are evidence of raisin production. She adds that only a small number were found at Arad and so is doubtful that wine was being produced this early at the site. For Lachish, see H. Helbaek, "Plant Economy in Ancient Lachish," Appendix A, in *Lachish 4* (ed. O. Tufnell; London: Oxford University Press, 1958), 309-17. A small number

wine dates back to the Early Bronze period as well, though so far in regions outside of Syria-Palestine. Jar shards with red deposits that contained tartaric acid, an acid from wine, were discovered at Godin Tepe in Iran, dating from 3100-2900 BCE (Late Period V).[7] Descriptions of wine exist from Egypt from the First Dynasty (3100-2890 BCE).[8] The wild ancestor of the grape vine cultivated for wine, *Vitis vinifera*,[9] is not native to Syria-Palestine.[10] Hence, vine remains discovered there, i.e., in areas beyond the vine's natural distribution zone, are considered to be the most convincing evidence of viticulture in those areas.[11] Botanists place the origins of the wild *Vitis vinifera* in the region between the Caspian and Black Seas, in the Armenian hills.[12] At present, early finds are too sporadic to establish a precise location of origin.[13] It is possible that Gen 8:4 reflects an ancient tradition that recalls the origin of the vine, when Noah's ark rests on

---

of pips from the Early Bronze Age was found. They were also raisins rather than grapes as they show signs of having been dried before burning, 310; D. Zohary and P. Spiegel-Roy, "Beginning of Fruit Growing in the Old World," *Science* 187 (1975): 321; Helbaek, "Late Cypriote Vegetable Diet at Apliki," 181.

[7]V. R. Badler, P. McGovern, and R. H. Michel, "Drink and Be Merry! Infrared Spectroscopy and Ancient Near Eastern Wine," *MASCA Research Papers in Science and Archaeology* 7 (1990): 27-28; see also V. R. Badler, "The Archaeological Evidence for Winemaking, Distribution and Consumption at Proto-Historic Godin Tepe, Iran," in *OAHW*, 45-65.

[8]A. Lucas and J. R. Harris, *Ancient Egyptian Materials and Industries* (London: Edward Arnold, 1962), 17; L. Lesko, *King Tut's Wine Cellar* (Berkeley: B. C. Scribe, 1977), 11, 22-36.

[9]The species is likely to have been *Vitis vinifera silvestris*, and if cultivated, *Vitis vinifera sativa*. Its subgenus is *euvitis* for Eurasia: H. J. de Blij, *Wine: A Geographic Appreciation* (Totowa, N.J.: Rowman & Allanheld, 1983), 9, 11, 40; A. J. Winkler, J. A. Cook, M. M. Kliewer, and L. A. Lider, *General Viticulture* (2d ed.; Berkeley: University of California Press, 1974), 18. Ninety percent of the world's grapes today are still produced from *Vitis vinifera*: de Blij, *Wine: A Geographic Appreciation*, 19.

[10]Helbaek, "Late Cypriot Vegetable Diet at Apliki," 181; Goor, "The History of the Grape-vine," 46. Olmo, "Grapes," 244. This makes the vine an apt metaphor for transplanted, cultivated people, e.g., Ps 80:8.

[11]Hopf, "Jericho Plant Remains," 587.

[12]de Blij, *Wine: A Geographic Appreciation*, 19; H. N. Moldenke and A. L. Moldenke, *Plants of the Bible* (New York: Dover Publications, 1952), 242; Broshi, "Wine in Ancient Palestine," 22; Goor, "The History of the Grape-vine," 46; Stager, "Firstfruits," 173; Winkler et al., *General Viticulture*, 17; Lutz, *Viticulture and Brewing in the Ancient Orient*, 1; Helbaek, "Late Cypriot Vegetable Diet in Apliki," 181; Olmo, "Grapes," 295.

[13]Helbaek, "Late Cypriot Vegetable Diet at Apliki," 181. Godin Tepe is south of the Armenian Mountains and has evidence of wine dating from 3100-2900 BCE: Badler, "The Archaeological Evidence for Winemaking," in *OAHW*, 45-65; Gorney, "Viticulture and Ancient Anatolia," in *OAHW*, 133-173. Gorney notes the presence of Early Bronze Age plastered basins at Titriş Höyük with tartaric acid residues, acids primarily found in wine, 162.

the mountains of Ararat (ancient Urartu, modern Armenia). For Noah's first act after disembarking from the ark is to plant a vineyard.[14] The Mesopotamian flood parallels, namely the Sumerian Deluge, Gilgamesh 11, and Atrahasis, do not contain this viticultural tale.[15] The Yahwist's account of the origins of viticulture with the figure of Noah may well contain a memory of the region from which the first vines came to his land.

Given its geographic location, Syria-Palestine undoubtedly played a significant role in the history of wine through export to Egypt from EB I and later to other lands around the Mediterranean.[16] Stager suggests that domestication first began in southern Anatolia, came to Syria-Palestine and then to the Egyptian Delta via an old EB I trade route along the north Sinai coast.[17] Canaanite wine is one of the goods imported by the pharaohs of Egypt as early as the Middle Bronze Age (2000-1550 BCE) as the presence of the "Canaanite jar" in Egyptian tombs and paintings indicates.[18] The Egyptian Tale of Sinuhe, also from the Middle Bronze Age, praises the land of Syria-Palestine for its horticulture and the abundance of its wine:

> It was a good land, named Yaa.
> Figs were in it, and grapes.
> It had more wine than water.[19]

---

[14]Gen 9:18-28 is considered to be from the Yahwist source, while 8:4 is considered Priestly. P could, then, be adding an historical flourish to J's story—a nice one too, since Noah would have to get rootstock for planting his vineyard from somewhere. Only animals were aboard the ark; there is no mention of provisions.

[15]"The Deluge," translated by S. N. Kramer, in *Ancient Near Eastern Texts Relating to the Old Testament* (3d ed.; ed. J. B. Pritchard; Princeton: Princeton University Press, 1969; hereafter *ANET*), 42-44; "Atrahasis," translated by E. A. Speiser (*ANET*, 104-6); "Gilgamesh," translated by S. Dalley, in *Myths From Mesopotamia* (Oxford: Oxford University Press, 1989), 33, 133, n.135. Utnapishtum does, however, land on a mountain—see E. A. Speiser, "Southern Kurdistan in the Annals of Ashurnasirpal and Today," *Annual of the American Schools of Oriental Research* 8 (1926-27): 18.

[16]de Blij, *Wine: A Geographic Appreciation*, 11. *Vitis vinifera* thrives in moderate climates, in two latitudinal zones, namely, 30-50 degrees north latitude and 30-50 degrees south latitude. It originally diffused from its Eurasian source around the world and to the Southern Hemisphere by human intervention. Olmo, "Grapes," 295.

[17]Stager, "Firstfruits," 173; Olmo, "Grapes," 295.

[18]A. Zemer, *Storage Jars in Ancient Sea Trade* (Haifa: National Maritime Museum, 1978), 7; for a slightly later example, from 1450 BCE, see Unwin, *Wine and the Vine*, 70.

[19]"The Story of Si-nuhe," translated by J. A. Wilson (*ANET*, 19).

Two plastered basins possibly served as a winepress at Taanach in MB II C (1650-1550 BCE).[20] Forty-five grape pips were found in Shiloh from the same period (Stratum VII).[21] Wine production continued in Canaan into the Late Bronze period (1500-1200 BCE) as evidenced at Ugarit on the Syrian coast [22] and Aphek in Palestine.[23] Biblical and archaeological materials discussed throughout the present work demonstrate that wine production continued in Palestine during the Iron Age (1200-586 BCE).

During the Iron Age, the westward movement of the vine probably fanned out from Asia Minor and Greece, following the Phoenician sea routes.[24] Then later, in the Late Roman or Early Byzantine period, the vine's spread was associated with the spread of Christianity, as wine was a necessary ingredient in the ritual of the Eucharist.[25] While the transmission of viticulture with religion's role is itself an interesting issue, it lies beyond the scope of the present project. What is not beyond its scope is the economic and symbolic place of vines and wine in Israelite life. Israelite culture reflects the considerable impact of this agricultural product.

The Iron Age farmers of Palestine undoubtedly followed in the tradition of viticulture and viniculture practiced by the Canaanites (later the Phoenicians). Whether they did so from the beginning of the Iron Age (1200 BCE), or some time later in the Iron Age is hard to discern given the limitations of the nature of the evidence; wine would have been drunk or evaporated; vines would naturally have decayed; and winepresses, since they were often built on bedrock, lack stratigraphy and are therefore difficult to date. Moreover, the study of ancient agricultural fields and paleobotanical remains has only recently become an important component of archaeological research.

---

[20]P. Lapp, "The 1963 Excavation at Ta'annek," *BASOR* 173 (1964): 18, figs. 7, 19.

[21]M. Kislev, "Food Remains," in *Shiloh: The Archaeology of a Biblical Site* (ed. I. Finkelstein, S. Bunimovitz, and Z. Lederman; Tel Aviv: Tel Aviv University Press, 1993), 355.

[22]From the third millennium on vineyards played a large economic role in Ugarit as in most of the coastlands of the eastern Mediterranean: M. Heltzer, *The Rural Community in Ancient Ugarit* (Wiesbaden: Dr. Ludwig Reichert, 1976), 40; M. Heltzer, "Vineyards and Wine in Ugarit (Property and Distribution)," *UF* 22 (1990): 119-35.

[23]M. Kochavi, "The History and Archaeology of Aphek-Antipatris," *BA* 44 (1981): 75-86.

[24]Stager suggests that the Phoenicians and Greeks introduced grapes and olive oil into the central Mediterranean as imports into colonies by the eighth century BCE: "Firstfruits," 181; de Blij, *Wine: A Geographic Appreciation*, 42.

[25]Broshi, "Wine in Ancient Palestine," 22; de Blij, *Wine: A Geographic Appreciation*, 46-47; Olmo, "Grapes," 295.

A hoard of grape pips discovered from Iron I Shiloh (Stratum V) does, however, likely indicate the cultivation of vines in the area.[26]

Thus far, all our epigraphic notations of wine from the Iron Age come from Iron II,[27] but then so too does the bulk of all evidence for writing. In other words, it may be writing, rather than wine, which was scarce in Iron I. However, the Gezer calendar, dated to the latter half of the tenth century, the very beginning of Iron II, does mention olive gathering, pruning, and a fruit harvest for an individual farmer.[28] These horticultural endeavors are listed in sequence with other farm activities, as if they are routine, rather than innovative, parts of a farming regimen. I infer from this calendar that viticulture was a facet of the agricultural calendar for some time before it was written.[29] Otherwise, we might expect that any new farming strategies and products either would not be mentioned in a calendar or would be remarked upon or highlighted in some fashion.

The ceramic repertoire from the Iron I period includes storage jars, jugs, and chalices from sites throughout Israel. At Taanach, Bethel, Tell Qiri, and ʿIzbet Ṣartah, for example, numerous storage jars, pilgrim flasks, juglets, and chalices were discovered from the

---

[26]Kislev, "Food Remains," 356; Hans Kjaer discovered a collared rim store jar with a horizontal streak inside, which he suggested was from wine, whose chemical influence left its trace. He dated this jar to the Iron I period. "The Excavation of Shiloh, Preliminary Report," *JPOS* 10 (1930): 99; A. Kempinski, "Shiloh," *The New Encyclopedia of Archaeological Excavations in the Holy Land* (ed. E. Stern; Jerusalem: Israel Exploration Society and Carta, 1993; hereafter *NEAEHL*) 4:1364-66.

[27]The Samaria ostraca (chapter 2, pp. 51-59); Gibeon jar handles (chapter 4, pp. 161-62); Ashkelon ostracon (chapter 5, pp. 200-202); Hazor jug and Judean wine jug (chapter 5, pp. 194, 205-6).

[28]See pp. 33-39 below.

[29]Baruch Rosen contends that there was not substantial horticulture in the Iron I period: "Subsistence Economy in Iron Age I," in *From Nomadism to Monarchy: Archaeological and Historical Aspects of Early Israel* (ed. I. Finkelstein and N. Naʾaman; Jerusalem: Israel Exploration Society, 1994), 345-47. In contrast, I. Finkelstein argues for a transition to a horticulture-based economy during the latter half of the Iron I period in Ephraim, first in the northern slopes, then in the southern slopes: *The Archaeology of the Israelite Settlement* (Jerusalem: Israel Exploration Society, 1988), 187, 199-200. However, his theory is largely based on his interpretation that collared-rim store jars were used primarily for oil and wine storage. The jars could have been used for water as well, and their presence need not indicate substantial horticulture.

Iron I period.[30]   The pottery from Tell Qasile, Stratum X (late eleventh-early tenth century) also contained many jugs, juglets, chalices, and flasks.   One room next to two vats contained a number of storage jars, which the excavator, Benjamin Mazar, interpreted as a storeroom for wine and oil vessels.[31] The presence of oil presses can sometimes suggest at least the possibility that wine technology could also be evident, since rock-cut presses are similar for both oil and wine production.   At Yoqne'am, for example, an olive oil installation stood in a courtyard from Iron I, with many jars, some stone weights, and a large quantity of olive pits.[32] This discovery demonstrates that horticultural production, namely of olive oil, was practiced in the Iron I period, but without any additional evidence in the form of chalices and cups, as at Tell Qasile, it does not indicate that wine production also occurred.   In addition, even with the assemblage of storage jars, juglets, and even flasks and chalices, we cannot be sure whether their contents were wine, water, or olive oil. The contents were either used up or have long since evaporated, and spectroscopic analysis of clay shards for traces of contents has so far only been done on Roman amphorae, Egyptian wine jars, and jars from Godin Tepe, Iran, the latter dating to 3100-2900 BCE.[33] Storage jars, of course, could be used for all foodstuffs, such as grain, olive oil, and wine. Jugs and chalices could be used both for water and for wine. Jugs no doubt were used for water transport from cisterns and wells, for farm use and meal service, but wine could have been present as well. Since

---

[30]W. Rast, *Taanach 1: Studies in Iron Age Pottery* (Cambridge, Mass.: American Schools of Oriental Research, 1978): medium-sized jars 68, 70, 82-84, 94; storage jars and jugs 74, 86, 94; and chalices from Period 1 (ca. 1200 BCE) 9; W. F. Albright, and J. L. Kelso, *The Excavation of Bethel: 1934-1960* (Cambridge, Mass.: American Schools of Oriental Research, 1968), 100-103; M. Hunt, "The Pottery," in *Tell Qiri: A Village in the Jezreel Valley: Report of the Archaeological Excavations 1975-1977* (ed. A. Ben-Tor and Y. Portugali; *Qedem* 24; Jerusalem: The Institute of Archaeology, The Hebrew University of Jerusalem, 1987), 88, 92-97, 103; I. Finkelstein, *'Izbet Ṣartah: An Early Iron Age Site near Rosh Ha'ayin, Israel* (Oxford: B.A.R. International Series, 1986), 44-45.
[31]B. Maisler [Mazar], "The Excavations at Tell Qâsile: Preliminary Report," *IEJ* 1 (1950-51): 130.
[32]A. Ben-Tor and R. Rosenthal, "The First Season of Excavations at Tel Yoqne'am, 1977," *IEJ* 28 (1978): 57-82.
[33]See, for example, the discussion of how infrared spectroscopy is done on wine jars from Godin Tepe, Iran, Period V. Badler et al., "Drink and Be Merry! Infrared Spectroscopy and Ancient Near Eastern Wine," 25-36.

the Iron I period is one of relative poverty[34] and little evidence of international trade,[35] it is reasonable to assume that these jars were primarily for home use.

While most installations for olive oil and wine production are from Iron II, three sites likely have winepresses from Iron I. The installations for olive oil and wine atop Samaria before Omri contain pottery dated from the eleventh to the ninth centuries BCE.[36] One winepress at Tell en-Nasbeh is dated to Stratum IV (twelfth century).[37] In Stratum X (late eleventh-early tenth century) at Tell Qasile, two square vats with a hole connecting them may have been a winepress.[38] Storage jars and jugs were found nearby, and, as we shall see in chapter 4, winepresses were typically rectangular, rather than circular like olive presses.

Finally, two biblical poems, Gen 49 and Deut 33, which may be dated to the eleventh century on the grounds of orthography and stylistic features close to Ugaritic poetry,[39] mention wine in the land (Gen 49:11-12 and Deut 33:28). These instances could well be anachronistic insertions, of course, especially in Gen 49:11-12, since its thematic intent, the glorification of Judah, may reflect that region's

---

[34]Small, unwalled villages dot the highlands, instead of the large, fortified cities of the Late Bronze Age, and the ceramic repertoire is "surprisingly limited": J. M. Miller and J. Hayes, *A History of Ancient Israel and Judah* (Philadelphia: Westminster, 1986), 83; see also A. Mazar, *Archaeology of the Land of the Bible* (New York: Doubleday, 1990), 345-46.

[35]"No pottery from either Mycenae or Cyprus is found in Israelite settlements, in contrast to its presence in earlier settlements": T. Dothan, "In the Days When the Judges Ruled—Research on the Period of the Settlement and the Judges," in *Recent Archaeology in the Land of Israel* (ed. H. Shanks and B. Mazar; Washington, D.C.: Biblical Archaeology Society, 1984), 35-41. In a study of pottery found in the sea off the coast of Israel, while Iron Age pottery is evident, none was from the twelfth-ninth centuries BCE: Zemer, *Storage Jars in Ancient Sea Trade*, 101.

[36]L. E. Stager, "Shemer's Estate," *BASOR* 277/278 (1990): 93-107.

[37]J. Zorn, "Tell en-Nasbeh," *NEAEHL*, 3:1099. In addition, the two Late Bronze winepresses at Aphek-Antipatris might have continued in use in Iron I: M. Kochavi, "The First Two Seasons of Excavations at Aphek-Antipatris," *Tel Aviv* 2 (1975): 17-42.

[38]Maisler [Mazar], "The Excavations at Tell Qâsile: Preliminary Report," 130.

[39]Albright, on the basis of repetitive parallelism and paronomasia, that is, word play: *Yahweh and the Gods of Canaan* (London: Athlone, 1968), 9-18; D. N. Freedman, on the basis of divine names and epithets: *Pottery, Poetry, and Prophecy: Studies in Early Hebrew Poetry* (Winona Lake: Eisenbrauns, 1980), 85-88, 90-92; F. M. Cross, using both divine epithets and prosodic form: *Canaanite Myth and Hebrew Epic* (Cambridge: Harvard University Press, 1973), 123; More generally, Cross and Freedman date the blessings of Gen 49 and Deut 33 to some time toward the end of the period of the judges. They add that blessings likely had a long oral history before their completion as literary pieces. F. M. Cross and D. N. Freedman, *Studies in Ancient Yahwistic Poetry* (Missoula, Mont.: Scholars Press, 1975), 7, 69.

supremacy in Iron II.[40] Still, Judah's glorification lies in the splashing abundance of his wine, not the mere presence of it. These collected pieces of disparate evidence, tenuous by themselves, suggest to me the presence of wine in the Iron I period, and, since this was a period marked by little trade, that it was being produced locally.

The textual and archaeological evidence for Iron II wine production is extensive and the focus of the present project. As a preview of the more extensive discussion to come, let me note that one iconographic representation of Israelite vines exists in the reliefs of Sennacherib's seige of Lachish. In it, grape clusters on vines dot the background, on hills.[41] Archaeological remains and biblical literature, then, indicate that wine was a part of Israelite life.[42] These materials suggest that wine rather than beer was the prevailing alcoholic beverage of ancient Israel. Why was this so? Why did the Israelite farmer go to the trouble of incorporating a vineyard into his agricultural regimen?

Beer, in many ways, would have been easier to make, for it came from barley, which the farmer was likely to be growing already.[43] Wheat was his principal grain crop, yet he harvested some barley to supplement this yield and to feed his farm animals.[44] To make beer, a farmer need only have increased his barley yield somewhat and apportion a percentage of the crop for brewing, since beer came from bread making. The beer brewer, often a woman in Egyptian clay models and Mesopotamian texts,[45] would moisten some of the barley and allow it to germinate. She would then roll the dough into loaves and put them in jars of water to produce beer.[46]

---

[40]Other scholars consider the poems to be earlier than the Yahwist (tenth-ninth c BCE), but do not pinpoint it further: C. Westermann, *Genesis 37-50: A Commentary* (trans. J. J. Scullion, S. J.; Minneapolis: Augsburg, 1986), 232; von Rad, *Genesis*, 424-25; Speiser, *Genesis*, 371.

[41]J. M. Russell, *Sennacherib's Palace without Rival at Nineveh* (Chicago: University of Chicago Press, 1991), photos 203-7; 210-11, 220.

[42]J. F. Ross, "Wine," in *The Interpreter's Dictionary of the Bible* (ed. G. Buttrick; Nashville: Abingdon, 1962), 4:849.

[43]Rosen, "Subsistence Economy in Iron Age I," 339-51; Gezer calendar, line 4, see section IV below; e.g., Exod 9:31; Lev 27:16; Num 5:15; Deut 8:8; Judg 7:13; Ruth 1:22; 2:17, 23; 3:2, 17; 2 Sam 14:30; 17:23; 21:9; Job 31:40; Isa 28:25; Jer 41:8.

[44]Rosen, "Subsistence Economy in Iron Age I," 339-51.

[45]For representative examples of women brewers in Egyptian clay models, see H. Wilson, *Egyptian Food and Drink* (Aylesbury, Eng.: Shire, 1988), 18; and M. Dayagi-Mendels, *Drink and Be Merry: Wine and Beer in Ancient Times* (Jerusalem: Israel Museum, 1999), 120; for mention of a woman beer brewer in Mesopotamian texts, see "The Code of Hammurabi," translated by T. Meek (*ANET*, 108-109, 111[laws]).

[46]Lucas and Harris, *Ancient Egyptian Materials and Industries*, 11.

Since barley was a yearly crop, the farmer could afford to experiment with yield and brewing technique. Simplest of all, beer could be brewed as needed throughout the year by taking grain from storage. In this way, the farmer never risked a shortage of food because of miscalculation.

Wine production, by contrast, required a longer-term investment strategy of the farmer. He would first devote a portion of his land or acquire additional property to cultivate vines. Then, he would purchase vine stocks and wait 4 to 5 years for a first meager harvest.[47] It would be ten years until a full harvest.[48] While it may be well worth the wait for today's vintner, the average life expectancy for a male during the Iron Age was 30 to 45 years.[49] His wait for the harvest, then, could take as much as a sixth of his life. Once the vines were in the ground, however, the lead time would be absorbed by the first-time vintner. Hence, subsequent generations would not have to wait so long for their wines. Vine and tree crops indicate two features of the society. They demonstrate, first, a sedentary life for these farmers, one where initial planting is worth their wait and, second, an investment and a devotion to or consideration of the next generation. The farmer who cultivated a vineyard likely had his son(s) in mind to carry on the farm and its vines after he was gone (1 Kgs 21; 2 Kgs 9:26). Planting vines, as I shall argue below,[50] was at once agricultural strategy and patrimonial legacy. Ancient viticulture, because it entailed sedentary settlement in an area and traditional land inheritance through a family, presumed a social nexus. The social consequences of wine, then, begin not with a first sip, but right when

---

[47]D. C. Hopkins adds that it could well be a decade before "significant production" could result from newly-planted vines: *The Highlands of Canaan: Agricultural Life in the Early Iron Age* (Sheffield, Eng.: Almond, 1985), 227.

[48]S. Aschenbrenner, "A Contemporary Community," in *The Minnesota Messenia Expedition: Reconstructing a Bronze Age Regional Environment* (ed. W. A. McDonald and G. R. Rapp, Jr.; Minneapolis: University of Minnesota Press, 1972), 55.

[49]R. Jones, "Paleopathology," *Anchor Bible Dictionary* (ed. D. N. Freedman; 6 vols.; New York: Doubleday, 1992), 5:67; Peter Garnsey suggests 20-30 years as a reasonable estimate in ancient societies as mortality rates were high and life-expectancy at birth low: *Famine and Food Supply in the Graeco-Roman World* (Cambridge: Cambridge University Press, 1988), 64; K. Hopkins estimates 25 years: "On the Probable Age Structure of the Roman Population," *Population Studies* 20 (1966): 245-264; S. Dixon estimates the median life expectancy of a female in Roman times to be 27, the male 37: *The Roman Family* (Baltimore: Johns Hopkins University Press, 1992), 149.

[50]Chapter 2, pp. 63-85.

the vine is planted in the ground.   Already, botany shapes the social life of the vintner.[51]

Additional tasks would be required of the farmer throughout the year, which, while not hard, took time.   He would need to weed and train the vines, prune them, maintain any vineyard or terrace walls, and provide protection from pests and intruders.   At harvest time, he would need a pressing installation and vat(s), jars for decanting and storage, and a labor pool to pick and tread the grapes. None of these additional farm chores exists for the beer maker, for beer came from bread making.

Viticulture, then, demanded a considerable investment of lead time, labor, and initial capital from the Israelite farmer.   Since he was, for the most part, a poor, independent farmer,[52] those resources would probably be stretched.   Hence, the question remains, why did the Israelite farmer make wine more than or instead of beer?   Before addressing this question in detail, let us take a cursory look at the beer cultures of Egypt and Mesopotamia.

II. Egypt and Mesopotamia

Egypt, as mentioned above, did import and praise Canaanite wine.[53]   It also ran royal vineyards primarily in the Delta.   The best-known viticultural areas were the oases of Khargah and Dakhlah.[54] Wines are recorded on jar seals from near Lake Mareotis, Tanis, and Memphis from the third millennium BCE.[55]   Papyrus Harris, dated from the reign of Ramses III (1197-1165 BCE), lists 513 vineyards

---

[51]"Surely there are few human pursuits that generate as close a relationship between people and the land they cultivate as does viticulture": de Blij, *Wine: A Geographic Appreciation*, 5; C. Eyre notes more generally that all food production is a social endeavor: "The Agricultural Cycle, Farming, and Water Management in the Ancient Near East," in *Civilizations of the Ancient Near East* (ed. J. M. Sasson; New York: Scribner's, 1995), 1:175.

[52]See chapter 2.

[53]A. Gardiner, *Ancient Egyptian Onomastica* (London: Oxford University Press, 1947), 1:96; 2:216-35.

[54]H. Kees, *Ancient Egypt* (London: Faber & Faber, 1961),   81. For a thorough discussion of the role of wine in ancient Egypt's court life and religious practice, see M.-C. Poo, *Wine and Wine Offering in the Religion of Ancient Egypt* (London: Kegan Paul International, 1995).

[55]Unwin, *Wine and the Vine*, 68; Kees, *Ancient Egypt*, 81; Herodotus mentions a journey to Bubastis, where everyone is served more wine on one day than they get for the rest of the year: *Histories* 2.60.

under temple control.[56]   There are many tomb paintings depicting vines and wine use, and jars from tombs whose notations list their product as wine.[57]   But these are the tombs of the ruling elite. Their tomb contents represent what the wealthy wanted for the afterlife. Vineyards and their wine remained under state control in Egypt and were not for the poor farmer. Wine was served for the pharaohs and their officials and in mortuary offerings for their afterlife.[58]   It remained throughout Egypt's history primarily a luxury product.[59]

The problem here, then, is one of source remains in two senses. First, tomb paintings may not represent Egyptian life, but rather constitute an artistic wish or bon voyage for the afterlife.[60] No wine vats, for example, have been excavated in Egypt so far; they exist only in tomb paintings.[61]   Second, these tombs are of the elite; their remains are rich and in no way represent the average Egyptian. The daily life of a peasant farmer in Egypt is largely a matter of inference. As Christopher Eyre notes for the ancient Near East in general, "nobody had any reason to document bare subsistence and poverty."[62] To surmise, then, what the majority of the population drank, one has to look to the agricultural foundations of Egypt.

Agrarian life in Egypt was of course built around the yearly inundation of the Nile for the water and alluvium it supplied. Its flooding was mostly regular and insured a fairly stable crop yield for the Egyptian farmer.[63] Rainfall is no more than 100-200 mm a year in the Delta,[64] which is below the 300 mm threshold for agriculture. The Nile valley fairs only slightly better, since its rainfall averages hover right at that minimum threshold, but is never guaranteed.[65] The Egyptian farmer depended on the inundation of the Nile between July and October and managed this water supply to harvest his main

---

[56]Unwin, *Wine and the Vine*, 71.
[57]Three dozen wine jars were discovered in the tomb of Tutankhamon. Darby et al., *Food: The Gift of Osiris*, 2:568; Lesko, *King Tut's Wine Cellar*, 7; Many wine jars were found at the Ramesseum at Thebes and at the royal city of Amarna, Kees, *Ancient Egypt*, 82.
[58]Kees, *Ancient Egypt*, 82.
[59]J. Baines and J. Malék, *Atlas of Ancient Egypt* (Oxford: Phaidon, 1984), 17; J. G. Wilkinson, *The Manners and Customs of the Ancient Egyptians* (London: John Murray, 1842), 2:170.
[60]Kees, *Ancient Egypt*, 82.
[61]Darby et al., *Food: The Gift of Osiris*, 2:557.
[62]Eyre, "The Agricultural Cycle," 186.
[63]Kees, *Ancient Egypt*, 47; J. Goody, *The Oriental, the Ancient, and the Primitive* (Cambridge, Eng.: Cambridge University Press, 1990), 319.
[64]Baines and Malék, *Atlas of Ancient Egypt*, 14.
[65]Cairo averages 1-1.5 inches of rainfall (254-381 mm) a year: Kees, *Ancient Egypt*, 47.

crops, emmer and barley.[66]   Yearly inundation made the soils in which he planted mainly silt.[67]   Barley can grow in such soils, but vines prefer a mixture of silt, clay, and sand.   A mainly silt soil would not offer the support and drainage that vines need. The Nile's yearly inundation meant that the farmer could rely on it when calculating his grain needs for food and beer.   The Nile gave the Egyptian farmer his rich soil, his water, and a dependable agricultural cycle. Since life depended on that consistent pattern of the river, Egypt truly was, as Herodotus claimed, the gift of that Nile.

The small farmer would eke out a meager existence largely of crop cultivation and animal husbandry, with dues paid to the temple or state.[68]   From his grain, he made a coarse beer.[69]   Herodotus is aware that the Egyptians "use a drink made of barley."[70]   In the same line, he further (over) generalizes that there are no vines in Egypt.[71]   Beer is attested in Egyptian literature as a greeting, "bread and beer," and the combined hieroglyphs for "bread and beer" formed the generic determinative for food.[72] Beer occurs in many Egyptian tales, whereas wine appears only in the Tale of Sinuhe, a story about an Egyptian official's trip to Syria-Palestine.[73]   These textual clues suggest that beer was rather commonplace in Egypt.   Beer was certainly cheaper and easier to make than wine for the poor farmer, given the dependable barley harvest and abundance of water. Egypt's river shaped and defined her agriculture and led to a social distinction in alcoholic beverage. A similar development arose in Mesopotamia as well, with agriculture there defined by the Tigris and Euphrates.

---

[66]Baines and Malék, *Atlas of Ancient Egypt,* 16; Darby et al., *Food: The Gift of Osiris,* 2:479, 483, 487.

[67]Baines and Malék, *Atlas of Ancient Egypt*, 15; Darby et al., *Food: The Gift of Osiris,* 2:453.

[68]B. G. Trigger, B. J. Kemp, D. O'Connor, A. B. Lloyd, *Ancient Egypt: A Social History* (Cambridge, Eng.: Cambridge University Press, 1983), 227, 326; Eyre, "The Agricultural Cycle," 187.

[69]Darby et al., *Food: The Gift of Osiris,* 2:531; D. and P. Brothwell, *Food in Antiquity* (London: Thames & Hudson, 1969), 166; Baines and Malék, *Atlas of Ancient Egypt,* 17; Forbes, *Studies in Ancient Technology,* 3:71; Unwin, *Wine and the Vine,* 67; Johnson, *Vintage,* 33; Lutz, *Viticulture and Brewing in the Ancient Orient,* 75; Lesko, *King Tut's Wine Cellar,* 13.

[70]Herodotus *Histories* 2.77. See also Athenaeus *Deipnosophistae* 1.34b; Strabo *Geography* 17.2.5.

[71]Herodotus *Histories* 2.77.

[72]Bread and beer were, then, apparently staples and could be made at the same time. Darby et al., *Food: The Gift of Osiris,* 2:533; W. C. Hayes, "Daily Life in Ancient Egypt,"*National Geographic* 80 (1941): 455.

[73]Darby et al., *Food: The Gift of Osiris,* 2:533.

In Mesopotamia, wine was likewise restricted primarily to the royal domain.[74]   Viticulture was known in Mesopotamia from an early period. Sumerian texts from the third millennium BCE note the presence of grapes and the vine with the term *geštin*, but wine itself remained a rare commodity, always much more difficult to produce and therefore more costly than the barley beer.[75] Vineyards seem to be a cause of monarchic pride in Assyrian annals. Assurnasirpal II (883-859 BCE) is credited with building vineyards at Calah.[76] Sennacherib's palace reliefs are filled with vines.[77] The Nimrud wine lists give extensive listings of wine rations for the male and female court officials in eighth-century BCE Calah.[78] The production and use of wine in both Assyria and Babylonia remained centered around the king. Beer was apparently much more widespread throughout these lands and a part of household economies. Barley beer—Sumerian *kaš*, Akkadian *šikāru*—is attested frequently in texts

[74]M. Powell, "Wine and the Vine in Ancient Mesopotamia: The Cuneiform Evidence," in *OAHW*, 101; Darby et al., *Food: The Gift of Osiris*, 2:532; J. Bottéro, "Boisson, banquet et vie sociale en Mésopotamie," in *Drinking in Ancient Societies* (ed. L. Milano; History of the Ancient Near East/ Studies–6; Padova: Sargon, 1994), 3-13.

[75]Powell, "Wine and the Vine in Ancient Mesopotamia," in *OAHW*, 97-121; Unwin, "Wine and the Vine," 64: R. Zettler and N. Miller, "Searching for Wine in the Archaeological Record of Ancient Mesopotamia of the Third and Second Millennia B.C.," in *OAHW*, 123-31; L. Milano, "Vino e birra in Oriente. Confini geografici e confini culturali," in *Drinking in Ancient Societies* (ed. L. Milano; History of the Ancient Near East/ Studies–6; Padova: Sargon, 1994), 421-40.

[76]D. Wiseman, "Mesopotamian Gardens" *Anatolian Studies* 33 (1983): 137-144; P. Albenda, "Grapevines in Ashurbanipal's Garden," *BASOR* 215 (1974): 5-17; F. Pinnock "Considerations on the 'Banquet Theme' in the Figurative Art of Mesopotamia and Syria," in *Drinking in Ancient Societies* (ed. L. Milano; History of the Ancient Near East/ Studies–6; Padova: Sargon, 1994), 15-26.

[77]E. Bleibtreu, *Die Flora der neuassyrischen Reliefs* (Vienna: Verlag des Instituts für Orientalistik der Universität Wien, 1980), 131-39; Russell, *Sennacherib's Palace Without Rival at Nineveh*, photos 26-27; 29; 70; 101; 106-7; 109; 132-33; 203-7; 210-11; 220.

[78]J. V. Kinnier Wilson, *The Nimrud Wine Lists: A Study of Men and Administration at the Assyrian Capital in the Eighth Century B.C.* (Cuneiform Texts from Nimrud, vol. 1; London: British School of Archaeology in Iraq, 1972), text nos. 127-56; F. M. Fales, "A Fresh Look at the Nimrud Wine Lists," in *Drinking in Ancient Societies* (ed. L. Milano; History of the Ancient Near East/ Studies–6; Padova: Sargon, 1994), 361-80.

on beer production and drinking.[79]   Many cuneiform texts describe
the brewing process, and the Laws of Hammurapi detail laws
concerning alewives and their pubs.[80]

In Mesopotamia, as in Egypt, the small farmer practiced grain
farming and animal husbandry and paid taxes to a central government.
He too could rely on water and flooding from the Tigris and
Euphrates for his annual needs. The rivers' waters insured him some
reliability, though not to the same degree as the Nile. And, it yielded
another liquid, beer, from the barley grown in the year. The flooding
of the rivers here also meant that the soils were largely silt and
alluvium, whose poor drainage is not supportive of vine growth.[81]
The Mesopotamian farmer's choice of drink, thus, was also beer[82]
and, as in Egypt, determined by his geography.

In both Egypt and Mesopotamia, in sum, wine was a prestige
drink—saved for the king and his officials, kept from the commoner.
Prestige was bought either in the achievement of royal vineyard
production or in the state's ability to import wine from afar. In either
case, wine required the imperial power to obtain it. Little would have
trickled down to the lowly farmer. He presumably made do with
beer, as the simple by-product of his grain farming. The elite of these
societies, on the other hand, would have feasted using both wine and
beer.

Both these ancient Near Eastern cultures had centralized
governments.[83] Both are alluvial cultures, their agriculture based on
the rivers flowing through their lands. The soils were rich for grain
farming, but not for the hardy needs of the vine. What such riverain
societies offered the individual farmer was a relative dependability of
crop yield, barring invasion or some unforeseen disaster. The flood
patterns of the Tigris and the Euphrates were not nearly as predictable

---

[79]See especially A. L. Oppenheim and L. Hartman, *On Beer and Brewing
Techniques in Ancient Mesopotamia, JAOS* Supplement 10 (Baltimore, Md.:
American Oriental Society, 1950); Unwin, *Wine and the Vine*, 64; A. L. Oppenheim,
Ancient Mesopotamia: Portrait of a Dead Civilization (rev. and ed. Erica Reiner;
Chicago: University of Chicago Press, 1977), 44; Also, W. Röllig, *Das Bier im
Alten Mesopotamien* (Berlin: Gesellschaft für die Geschichte und Bibliographie des
Brauwesens, 1970); and various essays in L. Milano, ed., *Drinking in Ancient
Societies.*
[80]Forbes, *Studies in Ancient Technology*, 3:68.
[81]Powell, "Wine and the Vine," 104; Lutz, *Viticulture and Brewing in the Ancient
Orient*, 37; T. Jacobsen, *Salinity and Irrigation in Antiquity* (Malibu, Calif.: Undena,
1982), 5; J. Cox, *From Vines to Wines* (New York: Harper & Row, 1985), 40.
[82]Lutz, *Viticulture and Brewing in the Ancient Orient*, 41.
[83]In ancient Egypt, the monarch was divine and sole owner of the land, and there
was a state-run system of production and distribution of foodstuffs: Kees, *Ancient
Egypt*, 61.

as those of the Nile in Egypt. Still, the agricultural cycle was far more predictable for the farmers in Egypt and, to a lesser extent, in Mesopotamia, than in ancient Israel. These farmers may have merely subsisted on grain farming and animal husbandry, surrendering a portion of their yield to the government, but the rivers afforded them a stability not shared by the Israelite farmer. The centralized states of these two societies provided relative security from invasion, and the rivers gave water. The river valleys and river-based agriculture of these two areas made barley beer the common drink of these lands.[84]

Abundant and reliable water is precisely the commodity that the Israelite farmer lacked. In this, he differed from his ancient Near Eastern neighbors in a way that made itself felt in aspects of his culture and lifestyle, among them his drink. This fact set him apart from the farmers of Egypt and Mesopotamia (cf. Deut 11:10-11). His livelihood was dry farming without a predictable amount of rainfall in each year. The average annual rainfall for Israel varies due to the microclimates. Jerusalem, for example, averages 550 mm a year, while Mount Carmel can get as much as 800-900 mm a year. The highest rainfall for Israel occurs in central Upper Galilee with annual averages of 1,000-1,100 mm, while the lowest rainfall occurs in the south, at Eilat, which averages only 15 mm.[85] The variability of rainfall within each year and from year to year can be as much as 30 percent.[86] That variability, with few perennial rivers, was enough to threaten the Israelite farmer.

Drought and famine gave a vicious edge to Israelite farming. Wine became the farmer's drink because he could afford neither the water for beer brewing nor the risk of concentrating all his efforts on one kind of crop, grain. Beer production took one liquid, water, to make another. It did not add to the valued liquid provisions for the household. Wine, by contrast, gave a juice without tapping into precious water supplies. In fact, it even required little water for growth on the vine. And vines allowed the Israelite farmer to diversify his yield, so that if one crop failed, another might survive. He was then in a position to barter or exchange for grain, in other words, to subsist. So-called monoculture farming, basing livelihood on one main crop, was too risky for the semi-arid climate of ancient Israel. The ancient Israelite practiced viticulture as part of his

---

[84]Broshi adds that had Jesus been an Egyptian, the Christian Eucharist would include beer, not wine. If so, the symbolism of Jesus' blood would have been significantly obscured: "Wine in Ancient Palestine," 22.

[85]E. Orni and E. Efrat, *Geography of Israel* (2d ed.; Jerusalem: Israel Program for Scientific Translations, 1966), 112, 114.

[86]Hopkins, *The Highlands of Canaan*, 91.

polycultural farming of grain, olives, grapes, and figs in addition to stockbreeding. Wine, in other words, was the result of an adaptive strategy to the environment of Israel that differed from those of Egypt and Mesopotamia, and it became a cultural marker distinguishing Israel from these neighbors in the ancient Near East. How the Israelite farmer made this agricultural adaptation is the subject of the next section. That he did led to his greater security and enjoyment.

III. Geographic Conditions

The land of Israel lies in the Mediterranean zone whose "outstanding characteristics [are] . . . winter rains and summer drought."[87] Hence, its agricultural regimen fits these characteristics: grain, legumes, and vegetables grow in the winter season and are harvested in the spring; vines and fruit trees can withstand the dry summers and come to fruition in the fall. Ecological conditions, most notably water resources and soil composition, were largely responsible for the adaptation of viticulture into the agricultural regimen of the Israelite farmer. Viticulture was practiced throughout Israelite history quite simply because it both thrived in the environment and offered worthwhile social and economic benefits to the society.

The climate and topography of the land Israel were conducive to the cultivation of vines and other tree crops such as the olive and fig. Israel is a semi-arid country located in the warm temperate zone within latitudes 31-33 degrees north, from the southern rim of the Dead Sea to Dan. The best areas in the world for viticulture lie between 30 and 48 degrees north.[88] Hence, Israel inhabits a zone with Greece, Italy, France and California, all eminent viticultural domains. While the microclimates of these wine-producing regions vary, they

---

[87]Orni and Efrat, *Geography of Israel*, 57, 105; D. Baly, *The Geography of the Bible* (2d ed.; New York: Harper & Row, 1979), 80. M. Zohary, *Plant Life of Palestine* (New York: Ronald, 1962), 20. Hence, the largest producers of wine circle the Mediterranean, with Italy, France and Spain in the lead: Olmo, "Grapes," 294. For this reason, Hyams terms the vine a plant of "Mediterranean man," *Dionysius*, 11.

[88]H. S. E. Fisher, "Wine: The Geographic Elements. Climate, Soil and Geology are the Crucial Catalysts," *Geographical Magazine* 51 (1978): 86; Unwin, *Wine and the Vine*, 34; Hyams, *Dionysius*, 9. Viticulture thrives in southern regions of the same latitudes as well, so Chile, Peru, and Australia all enjoy heady and renowned wine production today.

all share temperatures averaging from 10 to 30 degrees C and benefit
from moist breezes brought in from bodies of water to their west.
The Atlantic Bay of Biscay provides this temperate effect for the
region of Bordeaux, France, and the Pablo Bay does the same for the
Napa Valley in California. Israel, Greece, and Italy together enjoy
the moderating effect of the Mediterranean. The night winds blow
from land to the sea, and in the day, sea winds blow in over the land.[89]

Rainfall provided the primary source of water for agriculture
in ancient Israel (Deut 11:11). Irrigation farming was not viable to
the extent that it was in Egypt and Mesopotamia because there were
too few perennial streams. The Keziv and Ga'aton in the north and
the Hadra and Yarkon in the central hill country, along with the
Jordan, are the only notable perennial rivers and even these can
reduce to a trickle in parts.[90] Water for daily household needs came
from natural springs, but these too were dependent on rainfall.
Grain, legume, and vegetable crops also depended on rainfall, and so
their growth cycles occurred entirely within the winter or rainy
season, from approximately mid-October to the beginning of May.
Dry farming is cultivation relying on the natural cycle of rainfall
without the human manipulation in irrigation. Israel, then, practiced
dry farming, in contrast to the primarily irrigation agriculture of
Egypt and Mesopotamia. It relied on rainfall, not rivers, for the
growth of its crops.

The average annual rainfall for Jerusalem in the twentieth
century is 550 mm.[91] Mount Carmel, because of its proximity to the
sea and steep slopes, will average 800-900 mm; the Jezreel valley,
though surrounded by hills, gets rain-bearing winds through the
southwest "corridor" in the Menashe region and can average up to 700
mm.[92] Viticulture requires between 400 mm and 800 mm,[93] and not
much of any sort of agriculture can survive below 300 mm.[94] By way

---

[89]The sea breeze reaches the central hills around noon: Orni and Efrat, *Geography of Israel*, 109.

[90]Hopkins, *Highlands of Canaan*, 95; Orni and Efrat, *Geography of Israel*, 443.

[91]Orni and Efrat, *Geography of Israel*, 114: this is roughly what London averages in a year, with the difference being in distribution. The rain falls on average in fifty days in Jerusalem, while the same amount falls in three hundred days in London; Baly cites 601 mm for the average, *Geography of the Bible*, 56; Hopkins cites 550 mm, *Highlands of Canaan*, 86.

[92]Orni and Efrat, *Geography of Israel*, 113-14. Not incidentally, the Jezreel Valley is where the contested dispute over Naboth's vineyard occurs, and Mount Carmel is remembered in Elijah's contest where he drenches the offering until water floods the surrounding area.

[93]Unwin, *Wine and the Vine*, 43.

[94]Hopkins, *Highlands of Canaan*, 85.

of comparison, the equatorial lakes that serve as the source of the Nile average 1016 mm to 2032 mm of rain a year, which the White Nile then carries down to Egypt's Nile River.[95]

Two aspects of Israel's rainfall, however, significantly affected agricultural activity in antiquity. First, whatever rainfall there was for any given year fell between mid-October and the beginning of April.[96] No rain typically falls in the five-month-long summer in Israel today—none. Almost half of the year is spent dry, an astonishing obstacle, agriculturally speaking. Dew in the summers provided a limited amount of moisture and was important for the summer crops.[97] Second, annual rainfall varied significantly within any sequence of years.

Using data collected for about a century—from 1846 to 1953—for rainfall in modern Palestine,[98] David Hopkins has estimated that roughly 30 percent of a decade will be well below the average rainfall. So, taking Jerusalem as an example, with its annual average, which he cited as 550 mm, he calculated that for three out of ten years, rainfall would be only 489 mm, with one of those years falling even below 418 mm.[99] Rainfall in the amount of 400 mm year, as was noted above, is the base amount for successful vine cultivation. Such a drop in annual rainfall, then, threatened viticulture and drew perilously close to the 300 mm base minimum for agriculture itself. If such a drop in annual rainfall did occur during roughly 30 percent of any decade in ancient Israel, then it is quite likely that drought, with its menace of famine, was experienced by each generation at least once. As Yohanan Aharoni noted: "Years of drought and famine run like a scarlet thread through the ancient history of Palestine."[100]

Temperatures further complicate this picture, since the sun affects evaporation of water and soil moisture during both the summer and the rainy seasons.[101] Between December and March the

---

[95]E. C. Semple, *The Geography of the Mediterranean: Its Relation to Ancient History* (New York: Henry Holt, 1931), 158.

[96]Orni and Efrat add that 70 percent of this rain falls between November and February, with very little rain in the second half of October and the first half of April: *Geography of Israel*, 114; Hopkins, *Highlands of Canaan*, 84; G. Hamel, *Poverty and Charity in Roman Palestine, First Three Centuries C.E.* (Berkeley: University of California Press, 1990), 102.

[97]Orni and Efrat, *Geography of Israel*, 115.

[98]J. Neumann, "On the Incidence of Dry and Wet Years," *IEJ* 6 (1956): 58-63.

[99]Hopkins, *Highlands of Canaan*, 86-91.

[100]Y. Aharoni, *The Land of the Bible* (trans. A. F. Rainey; 2d ed.; London: Burns & Oates, 1979), 14.

[101]Orni and Efrat estimate as much as 50-60 percent of rainfall can be lost to evaporation: *Geography of Israel*, 148.

average daily temperature does not exceed 11 degrees C (52 degrees F). Hill temperatures are generally lower than those of the rest of the country. Mean temperatures for Jerusalem are between 9.7 degrees C (49.5 degrees F) in January, and 25 degrees C (77 degrees F) in August, the hottest month.[102] The coldest period in winter is short, with the temperatures rarely reaching the freezing point of 0 degrees C. Hence, frost, though it occurs, is uncommon. Vines themselves can withstand short periods of freezing temperatures in the winter without damage.[103] They enter a period of dormancy in the winter months. Then they stay dormant until the mean daily temperature reaches 10 degrees C (50 degrees F).[104] Without this period of dormancy, in fact, vines yield small crops of poor-quality grapes, as is generally true of viticultural attempts in the tropics.[105] The temperatures of the summer months combined with the lack of rainfall leach the soil of moisture and dry out vegetation. The vine is one of the few plants that prefer hot, dry summers, when their fruit matures, and the dormancy offered by wet, cool winters.[106] Vines "have very extensive root systems and can withstand more drought than many other crops."[107] In fact, the "absence of summer rains is often a favorable characteristic for a wine-growing area" because then there is no risk of mildew.[108] Viticulture, then, was well suited to the Israelite climate.

Settlements in the Iron I period (1200-1000 BCE) were primarily in the central highland ridge, away from the Sharon and Shephelah valley plains and the desert fringes of the southern Judean and Negev regions. Expansion into the plains occurred in Iron II (1000-586 BCE), but settlement remained dense in the central highlands. Israelite agriculture was, then, a kind of hill country farming that made use of the varied terrains of slopes and intermontane valleys. Within the central hill country, *terra rossa* is the predominant soil type. It is high in clay content, shallow, and, of course, red.[109]

---

[102]Hopkins, *Highlands of Canaan*, 81; Orni and Efrat, *Geography of Israel*, 138.

[103]Unwin, *Wine and the Vine*, 34; Baly, *Geography of the Bible*, 58.

[104]Unwin, *Wine and the Vine*, 33.

[105]Unwin, *Wine and the Vine*, 34.

[106]Baly asserts the vine was a "miracle" to ancient Israel for precisely its environmental "fit": *Geography of the Bible*, 85.

[107]M. A. Amerine and V. L. Singleton, *Wine: An Introduction* (2d ed.; Berkeley: University of California Press, 1965), 43.

[108]Amerine and Singleton, *Wine: An Introduction*, 43.

[109]L. E. Stager, "The Archaeology of the Family in Ancient Israel," *BASOR* 260 (1985): 4.

Terra rossa is the most characteristic soil type in the mountain regions of Israel and other Mediterranean locales such as Greece and Italy,[110] though it poses certain demands on the farmer. Its clay content, for instance, means that it retains moisture well, but is poor in permeability and susceptible to erosion.[111] *Terra rossa* is low in humus,[112] so necessary for the nitrogen it provides to plants. Its high iron and low humus accounts for its red color.[113] Also, *terra rossa* is slightly alkaline and hinders the absorption of other nutrients. Specifically, its pH balance is 6.8-7.8, and is slightly above the neutral balance beneficial to most plants.[114] Vines, however, prefer this soil of low humus and slightly alkaline character because too much nitrogen results in weak plant growth.[115] In fact, as Jeff Cox advises the novice Californian vintner, the soil must have a pH balance between six and eight for a successful vineyard.[116] *Terra rossa* of Israel and the Mediterranean falls precisely within that range.

Vines today, as in antiquity, do best in loam soils of a mixture of clay with silt, sand, stones, and organic material.[117] The limestone and clay soils found on slopes in the Mediterranean basin have always been excellent for vineyards because of this natural mixture.[118] The extensive root systems have guarded against the erosion characteristic of clayey soil, and rocks offset poor permeability.

Rocky soils are best for vines, as they are for fig and olive trees, because they encourage the roots to extend deep into the soils to reach the moister earth beneath.[119] Gravelly, loam soils are not of much use for grain crops, but ideal for the vine. Château Beaucaillou ("beautiful pebbles") in Bordeaux offers modern testimony to the value of rocky soils for viticulture.[120]

---

[110]M. Zohary, *Plant Life of Palestine*, 11; Semple, *The Geography of the Mediterranean*, 380; Hopkins, *Highlands of Canaan*, 127.

[111]Stager, "Archaeology of the Family in Ancient Israel," 4; Hopkins, *Highlands of Canaan*, 127; Baly adds that *terra rossa* is suitable for all typical Mediterranean crops, but often shallow, low in humus, and easily washed off hillsides: *Geography of the Bible*, 79.

[112]Hopkins, *Highlands of Canaan*, 125.

[113]A. Reifenberg, *The Soils of Palestine* (trans. C. L. Whittles; London: Thomas Murby & Co., 1947), 74.

[114]Hopkins, *Highlands of Canaan*, 128.

[115]Cox, *From Vines to Wines*, 41.

[116]Cox, *From Vines to Wines*, 40.

[117]Cox, *From Vines to Wines*, 40.

[118]Semple, *The Geography of the Mediterranean*, 381.

[119]Semple, *The Geography of the Mediterranean*, 381.

[120]Cox, *From Vines to Wines*, 38.

Slopes still tend to be the best area for planting a vineyard.[121] The main benefit of hillsides to vineyards is that air movement moderates the temperature. The cool night air nearest the soil is denser than the air mass above and will slide downhill. The coolest air will thus collect in pools on valley floors, and it is there, if temperatures are near freezing, that frost will threaten.[122] In addition, vines on an incline are apt to receive sunlight on the entire plant. This distribution helps with the sugar and acid content of the fruit; sunlight on exposed soil affects the acid level of grapes, while sunlight on the leaves affects the sugar levels.[123] Valley floors in Israel and the Mediterranean were reserved for grain farming, so hillside cultivation in no way detracted from the availability of arable land. In addition, the village was typically atop a hill, so vineyards on slopes would allow shorter travel time from home and make it thus easier to keep watch.

IV. The Israelite Agricultural Calendar

Viticulture, as we have seen, was remarkably well suited to the geographic and ecological conditions of ancient Israel. It fit as well into the yearly agricultural cycle of the Israelite farmer. Grapes, olives, and figs ripened during the dry, hot summer months, when not much else would. Deuteronomy 11:14 reflects this fit by placing the harvests, in sequence, after the last rains of the spring:

> Rainfall for your land in its season,
> the early and latter rain (יוֹרֶה וּמַלְקוֹשׁ),
> and you will gather your grain, wine, and oil.[124]
> (Deut 11:4)

The agricultural year began, as the passage indicates, with the onset of the first rains, יוֹרֶה, approximately mid-October.[125] They softened the soil from the previous summer drought.[126] Grapes thrive

---

[121]Cox, *From Vines to Wines*, 35.
[122]Cox, *From Vines to Wines*, 36.
[123]Cox, *From Vines to Wines*, 36.
[124]The translation of all biblical texts from the Hebrew is my own.
[125]Hopkins, *Highlands of Canaan*, 84.
[126]Baly, *Geography of the Bible*, 52.

in the hot, dry summer months, yet benefit from the only moisture of
that season, that of the nightly dew. Hopkins argues that the quantity
of dew does not contribute in any agriculturally significant way to
Israelite crops, yet it does.[127] The one crop that does benefit from just
this much moisture and no more is the grape. It cannot endure a rain
shower because too much moisture would cause mold to grow on the
skin. But a trace amount of precipitation, such as that of dew, helps
the skin expand and enhances the fruitfulness. Today in the Napa
Valley of California, vineyards continue to benefit from morning
dews from the bay, but have rain covers readied onsite should it rain.

With this appreciation of the dew's usefulness for the grape, the
particularity of Isaac's blessing to his mistaken son becomes indeed
quite keen:

> May God give you of the dew of
> heaven, and of the fat
> of the earth,
> and plenty of grain and wine.
> (Gen 27:28)

Again, in another deathbed blessing, Moses describes Israel's dwelling

> in a land of grain and wine,
> where the heavens drop down dew.
> (Deut 33:28)

Isaiah 18:4, Hos 14:5-6, and Zech 8:12 also associate dew with the
vine and its wine. In addition, a Ugaritic text associates dew with
grapes: "the dew which drops on the grapes."[128]   Dew, then, is not
incidentally present in these biblical and extrabibilical contexts
involving the vine. It enhances the grape in its ripening during the
summer months, and this important role is recalled in the texts above.

Given the rainfall pattern of Israel, farming no doubt adhered
to an agricultural calendar such as those found in Exod 23:10-19,
34:18-23, Lev 23:1-44, and Deut 16:1-17, which detail seasons for
barley, wheat, and fruit harvests. An epigraphic representation of an

---

[127] Hopkins, *Highlands of Canaan*, 98.
[128] *KTU* 1.19, lines 40-41; M. D. Coogan, trans., *Stories from Ancient Canaan*
(Philadelphia: Westminster, 1978), 41.

agricultural calendar also exists for ancient Palestine. It was unearthed at Gezer and is dated to the tenth century.[129] Scholarly opinion varies on the precise nature of the inscription, that is, on whether it is an official document,[130] or that of a farmer,[131] or schoolboy,[132] but there is wide consensus that the contents are agricultural and that they span somehow the period of a year.[133] This calendar, like the biblical calendars, reflects a mixed-farming regimen with grain sowing, barley, and fruit cultivation:

1. ירחו אספ ירחו
2. זרע ירחו לקש
3. ירח עצד פשת
4. ירח קצר שערם
5. ירח קצר וכל
6. ירחו זמר
7. ירח קץ[134]

---

[129]W. F. Albright, "The Gezer Calendar," *BASOR* 92 (1943): 18; Borowski, *Agriculture in Iron Age Israel*, 32; S. Talmon, "The Gezer Calendar and the Seasonal Cycle of Ancient Canaan," *JAOS* 83 (1963): 177.

[130]For Talmon, it is an official document from Solomon's reign (961-922 BCE), probably for tax purposes, listing the main farming sequence of the region: "The Gezer Calendar and the Seasonal Cycle of Ancient Canaan," 177; for S. Ronzevalle, it is an official document and not the work of "an idle peasant": "The Gezer Hebrew Inscription," *Palestine Exploration Fund Quarterly Statement* 41 (1909): 111. We ignore for the moment the question of just what could be considered idle about peasantry, or, for that matter, inscribing in stone.

[131]M. Lidzbarski, "An Old Hebrew Calendar-Inscription from Gezer," *Palestine Exploration Fund Quarterly Statement* 41 (1909): 26.

[132]This is Albright's proposal, which is the most persuasive, because it accounts for several features, viz., form, content, and orthography. The form has three double months, followed by three single months, then one double, one single, forming a kind of "poetic ditty." The content, he argues, would be too well known to require mnemonic devices for a farmer: "For a grown peasant, to whom the succession of agricultural activities was as familiar as the use of his senses," rhythmic enumeration has no value, but for a child it could be as useful as our "thirty days hath September." Finally, the handwriting is unskilled and varied: Albright, "The Gezer Calendar," 25.

[133]Rosen notes that the calendar does not mention herding at all: "Subsistence Economy in Iron Age I," 349. It would seem, then, that the Gezer calendar is strictly agricultural and so does not reflect all of a typical Mediterranean economy.

[134]Borowski presents a clear drawing of the calendar and compares the tasks listed to the modern agricultural cycles in Israel and the West Bank: *Agriculture in Iron Age Israel*, 34-35.

1. his[135] two months of ingathering[136]/ his two months
2. of sowing/ his two months of later planting[137]
3. his month of ʿṣd pš/pśt[138]
4. his month of barley harvest
5. his month of harvest ...[139]
6. his two months of pruning[140]
7. his month of summer fruit.[141]

---

[135]See chapter 2, p. 47. Albright, and F. M. Cross and D. N. Freedman identify the *waw* as reflecting a dual noun plus possessive suffix: Albright, "The Gezer Calendar," 24; F. M. Cross and D. N. Freedman, *Early Hebrew Orthography* (New Haven, Conn.: American Oriental Society, 1952), 46-47. Borowski's translation does not take into account the *waws* after ירח: *Agriculture in Iron Age Israel*, 38; For Ronzevalle, the *waw* signals "and," and so line 1, for example, is "a month and gathering": Ronzevalle, "The Gezer Hebrew Inscription," 109; also G. B. Gray, "The Gezer Inscription," *Palestine Exploration Fund Quarterly Statement* 41 (1909): 192.
[136]See chapter 5, pp. 167-71. For Albright, it is specifically the olive harvest whose importance is indicated by two months, "The Gezer Calendar," 22.
[137]Albright, "The Gezer Calendar," 22; Borowski, *Agriculture in Iron Age Israel*, 34.
[138]For reconstructions with פשת, "flax," see Albright, "The Gezer Calendar," 22; Lidzbarski, "An Old Hebrew Calendar-Inscription from Gezer," 28; G. B. Gray, "An Old Hebrew Calendar-Inscription from Gezer, 2," *Palestine Exploration Fund Quarterly Statement* 41 (1909): 31; and U. Cassuto, "The Gezer Calendar and its Historical-Religious Value," in *Biblical and Oriental Studies* (trans. I. Abrahams; Jerusalem: Magnes, 1975), 2:214. For reconstructions with פשת, as "hoeing" weeds of some sort, see Borowski, *Agriculture in Iron Age Israel*, 38; Talmon, "The Gezer Calendar and the Seasonal Cycle of Ancient Canaan," 178.
[139]The consonants are unclear with the last term, except for a *lamed*. Various options for reconstruction have been proposed, and most scholars distinguish two letters before the *lamed*. Borowski suggests *kwl* for "and measuring (grain)": *Agriculture in Iron Age Israel*, 36 ; *kol*, "all," is suggested by others: Lidzbarski, "An Old Hebrew Calendar-Inscription from Gezer," 28, photo, fig. 1; G. Dalman, "Notes on the Old Hebrew Calendar-Inscription from Gezer," *Palestine Exploration Fund Quarterly Statement* 41 (1909): 118. Albright reconstructs *wgl*, "festivity": "The Gezer Calendar," 23. Cassuto reconstructs *klh* and renders it "finishing" harvest: "The Gezer Calendar and its Historical-Religious Value," 215. The grapheme looks more like a *kaf* than a *gimel* with a scratch over it, so *kol* or *klh* seem to be the best suggestions though they remain tentative.
[140]Several scholars render *zmr* as "grape harvesting": Borowski, *Agriculture in Iron Age Israel*, 36; Gray, "The Gezer Inscription," 192; A. Lemaire, "*Zāmir* dans la tablette de Gezer et le Cantique des Cantiques," *VT* 25 (1975): 15-26; Cassuto, "The Gezer Calendar and its Historical-Religious Value," 216- 17.
[141]So, Borowski, *Agriculture in Iron Age Israel*, 32. The other possible translation is "end," Ronzevalle, "The Gezer Hebrew Inscription," 111. The contraction of diphthong indicates a North Israelite spelling rather than Judahite, as do the suffixed forms and perhaps the *kl* in line 5. Hence, "summer fruit" is likely on orthographic grounds and offers a completion to the list of agricultural products. See Cross and Freedman, *Early Hebrew Orthography*, 47.

Fruit of the Vine

The Gezer calendar is important for our analysis of Israelite viticulture because it is evidence of grape cultivation by individual farmers as early as the tenth century BCE. It indicates that vine tending was part of the yearly cycle along with other agricultural endeavors. Lucian Turkowski's study of peasant agriculture in the Judean hills from 1943 to 1947 is instructive at this point, for it describes the major agricultural tasks of the small landowners who practiced mixed farming there, and compared it with the Gezer calendar.[142] The farmer's year began in the fall with two months of the "ingathering." אסף/אסיף in the Gezer calendar was the fruit harvest and included olives and grapes (September-October).[143] This period of ingathering was, Turkowski notes, the most enjoyable period for the Judean farmers, as it constituted the "crowning of their efforts and the gathering of grapes and other fruits."[144] Next followed two months of sowing, of barley and then wheat.[145] This period corresponds roughly to late November-December, but is dependent on when the first rains fall.[146]

Broadcasting seed was done in the fall, so that grain could grow during the rainy season of the winter months. Since ingathering and sowing are listed in lines 1-2, the calendar likely begins with the fall. The peasants plowed twice for cereals some time after the rainy season had begun. Typically, a shallow plowing occurred first to break up the ground and prevent weeds, followed by a second plowing before sowing.[147] A period roughly corresponding to January-February followed with the planting of legumes and vegetables (line 2: לקש).[148] Then followed a month (roughly mid-March/mid-April) of 'ṣd pš/pśt (line 3: פשת), which in antiquity was probably either

[142]L.Turkowski, "Peasant Agriculture in the Judaean Hills," *PEQ* 101 (1969): 21-33, 101-12.
[143]Borowski, *Agriculture in Iron Age Israel*, 34.
[144]Turkowski specifically mentions only September, but he is detailing the peak time period of the ingathering harvest only, "Peasant Agriculture in the Judaean Hills," 27. For the small farmers of Vasilika, in Boeotia, too, the grape harvest is also the height of their year: E. Friedl, *Vasilika: A Village in Modern Greece* (New York: Holt, Rinehart & Winston, 1962), 18.
[145]Turkowski, "Peasant Agriculture in the Judaean Hills," 25.
[146]Borowski, *Agriculture in Iron Age Israel*, 34.
[147]Turkowski, "Peasant Agriculture in the Judaean Hills," 28.
[148]Albright, "The Gezer Calendar," 22; Borowski, *Agriculture in Iron Age Israel*, 34; Turkowski, "Peasant Agriculture in the Judaean Hills," 28.

flax[149] or the "spread" of weeds and other wild grasses.[150]   Gustaf Dalman records having seen extensive weeding in twentieth-century Palestine during the months of February and March.[151]   Turkowski notes that during these two months the Judean peasants worked in the olive groves and vineyards, though he is not specific.   It is possible that they were hoeing for weeds.   This would have to have been done at various points throughout the year (Isa 5:6; 7:25),[152] but perhaps a lull in other agricultural activities allowed for concentrated weeding at this point in the late winter/early spring.   Hoeing vineyards and orchards during the rainy season would have been a wise agricultural strategy, as it would have helped to keep the soils moist.[153]   During April, the Judean farmers spent much of their time working their vegetable gardens.[154]

In late April-May, the barley harvest occurred,[155] followed by the wheat harvest about a month later, in late May-early June.[156]   Line 4 clearly mentions the barley harvest (קצר שערם), which we know from ethnographic parallels was harvested in April,[157] and we can infer that the harvest of line 5, though unspecified, was the other major grain crop, wheat.   With this view, then, lines 2-5 likely incorporate the agricultural tasks from mid-October through mid-May,[158] i.e., the rainy season.   The tasks in lines 1, 6, and 7, then, occur in the period of drought of the summer and early fall period. These lines constitute the evidence for horticulture.

During July and August, Turkowski describes the two primary activities for the Judean farmers: the grain was threshed and winnowed; and "for the children and old people it is a rather pleasant season for looking after vineyards."[159]   Unfortunately, Turkowski

---

[149]Turkowski, "Peasant Agriculture in the Judaean Hills," 27; Albright, "The Gezer Calendar," 22; Lidzbarski, "An Old Hebrew Calendar-Inscription From Gezer," 28; cf. Exod 40:3; Deut 22:11; Judg 14:4; Prov 31:13.

[150]Talmon, "The Gezer Calendar and the Seasonal Cycle of Ancient Canaan," 178; Borowski, Agriculture in Iron Age Israel, 38; G. Dalman, Arbeit und Sitte in Palästina (7 vols.; Gütersloh: Bertelsmann, 1928-39), 2:216-17; פשה cf. Lev 13:7.

[151]Dalman, Arbeit und Sitte in Palästina, 2:216-17.

[152]See chapter 3, pp. 97-98.

[153]Hopkins, Highlands of Canaan, 227.

[154]Turkowski, "Peasant Agriculture in the Judaean Hills," 27.

[155]Turkowski, "Peasant Agriculture in the Judaean Hills," 27; Baly, Geography of the Bible, 85.

[156]Borowski, Agriculture in Iron Age Israel, 37; Turkowski, "Peasant Agriculture in the Judaean Hills," 27; Baly, Geography of the Bible, 85.

[157]Turkowski, "Peasant Agriculture in the Judaean Hills," 27; Baly, Geography of the Bible, 85.

[158]Hopkins, Highlands of Canaan, 84; Baly, Geography of the Bible, 52.

[159]Turkowski, "Peasant Agriculture in the Judaean Hills," 27.

does not indicate precisely what they did, but presumably pruning was part of it. The calendar devoted two months to pruning (line 6: זמר).
In biblical Hebrew, this root (זמר) denotes the pruning of grape vines, as well as singing. The context here argues for pruning. Vine pruning and the grape harvest both require cutting with a tool, the biblical מְזַמְּרוֹת.[160] Pruning enhances the fruit growth over wood or leaf growth.[161] Cutting back on the wood growth in branches diverts the plant's nutrients into the fruit.[162] There are typically two pruning periods for the vine, one in the winter and then a second one sometime in the summer.[163] In the summer pruning, the leaves as well as branches are pruned so that the sun has greater access to the fruit in its last stage of growth to ripening in the fall. Line six of the Gezer calendar probably indicates this second vine pruning,[164] rather than the grape harvest itself,[165] because the task occurs in the list between the spring wheat harvest and the harvest of the end/summer fruits (line 7). If *zmr* in line 6 does indicate a second pruning, the grape harvest is not thereby omitted by the calendar, but is instead contained in the ingathering (אסף) of line 1. אסף/אסיף in biblical calendars denotes the collection of fruit in the fall harvest (Exod 23:16; 34:22; Deut 16:13; Lev 23:39; cf. Exod 23:10-11).

The last line likely signals the end (line 7: קץ) of the agricultural year.[166] Most scholars contend that the line denotes the gathering of "summer fruit."[167] Cross and Freedman argue for "summer fruit" (קיץ), noting, as already indicated, that the contraction of the diphthong reflects a northern dialect. Their translation is the most persuasive on orthographic grounds, as the diphthong contraction is evident elsewhere in the calendar with the *waw* endings on ירחו reflecting a dual noun plus possessive suffix.[168] On the grounds

---

[160]See chapters 3, pp. 119-22, and 5, pp. 171-74.

[161]See chapter 3, pp. 119-22.

[162]Semple, *The Geography of the Mediterranean*, 392; Olmo, "Grapes," 294; Weinhold, *Vivat Bacchus*, 83. Columella provides detailed descriptions of ancient pruning techniques: *De re rustica* 4.7-11; 4.23-24.

[163]R. Lyons, *Vine to Wine* (Napa, Calif.: Stonecrest, 1985), 9; Dalman, "Notes on the Old Hebrew Calendar-Inscription from Gezer," 119.

[164]It could refer to the grape harvest itself, but this takes place in the fall, in September and October. See chapter 5, pp. 167-71.

[165]For the latter view, see n. 140.

[166]Ronzevalle, "The Gezer Hebrew Inscription," 111.

[167]Albright, "The Gezer Calendar," 23; Gray, "An Old Hebrew Calendar-Inscription from Gezer, 2," 31; Borowski, *Agriculture in Iron Age Israel*, 32. Cassuto suggests "summer fruit" or "final month": "The Gezer Calendar and its Historical-Religious Value," 216-17.

[168]See n. 135 and chapter 2, p. 47.

of content as well, "summer fruit" is apt as it completes the list of agricultural products.   Summer fruit is the last harvest in the chronological progression of the agricultural year, and so it logically comes at the end of this calendar. The root קץ, then, easily could have signaled both the end of the agricultural year and its final harvest, viz., of summer fruits, as it does in Amos 8:1-2.   After it was harvested, the new year's cycle would begin again for the farmer with the grape and olive gathering of the fall.

Olives and grapes come due at the same time in the fall, yet they differ in their harvests. Olives can stay on the trees even through the first rains of the new year and still be good.[169] This meant that the olive harvest could be spread out over the two months from the fruit's ripening through the first rains, roughly late September to November.[170] The grape harvest was much different. Grapes ripened and peaked in a short period of time and had to be harvested during that time.[171]   Otherwise, they would drop off the vine, shrivel, or catch the eager attention of pests.

Ideally, the farmer would give priority to the grape harvest and pick olives when the demands of the grape harvest were not as urgent. Also, grapes differed from olives in that they had to come off the vine before the first rains of the new year began. The window of time for the grape harvest was then much shorter than that of the olive harvest, even though both fruits ripened in the same season. This difference may explain why the grape harvest could be included in the ingathering of the Gezer calendar, rather than receiving a separate line. Both olives and grapes, then, were harvested during the two months of ingathering, but with differing periods of intensity within that time period.   Two months would easily be given over to the harvesting of these most valued crops of the fall harvest.

In conclusion, the grape harvest fit into the individual farmer's yearly cycle and was part of the fall labor period along with the harvest of olives. It occurred before the early or first rains (יוֹרֶה) of the new agricultural year.   Rain at that point in the grape's development could easily ruin the crop. Hence, while the farmer might hope for good early rains to initiate his agricultural year, he was nevertheless particular about the timing of these showers; namely, that they (or his deity) hold off until *after* the grape harvest.

Wheat, barley, legumes, and vegetables were all sown and harvested in the year between late November to early June. Had these

---

[169]Hopkins, *Highlands of Canaan*, 230.
[170]Baly, *Geography of the Bible*, 101.
[171]Hopkins, *Highlands of Canaan*, 229.

been the Israelite farmer's only crops, his food supply for a year would have been dependent on 6-6.5 months of labor. A small farmer would not likely have taken such a risk or vacation, particularly in a land of scarce water resources. Vine cultivation complements, rather than competes with the other crop cycles and thrives in the environmental conditions of the Israelite land. Vines thrive in soils and on slopes that are not useful for the other farming ventures of grain farming and sheep-goat pastoralism.[172] Vines thrive in the very regular summer drought period, unlike grain, legumes, and vegetables.[173]

Horticulture enabled the Israelite farmer to make use of the summer months for growth, and increase his yields considerably in the form of grapes, olives, and figs. Even so, the Bible stories of famine are unnervingly frequent, dramatic, and poignant, e.g., Gen 12:10; 26:1; 41:56; 47:4, 13; Ruth 1:1; 2 Sam 21:1; 1 Kgs 18:2; 2 Kgs 6:25; 25:3. Much more than fanciful trope, they gave narrative voice to a pitiful threat looming for the Israelite farmer.

A combination of grain farming, stockbreeding, and the cultivation of vine and tree crops characterized the agrarian basis of Israelite life.[174] This threefold combination typified Mediterranean economies in antiquity, as regions around that sea practiced similar mixed farming. The Israelite farmer, like the Greek and Roman farmers, produced and lived off his varied yield: grain, in the form of wheat and to a lesser extent barley; milk, meat, and skins of sheep and goats; and olive oil, figs, and wine from his orchard and vineyard. Grains were harvested in the spring, fruit trees and vines in the fall. This crop regimen fit the Mediterranean zone with its characteristic winter rainy season and summer drought. Vines and fruit trees became then important and useful complements to grain farming and contributed to the diet. So Deut 8:8 offers agricultural testimony of a land that is specific of mixed farming:

---

[172]The herds would have needed to graze, of course, but this need not have been done on the hill slopes used for viticulture. There was plenty of shrub area for the farmer even if he had to walk his herds for a few kilometers.

[173]Hopkins, *Highlands of Canaan*, 196. Legumes offered the farmer the chance to practice green fallow by planting after a grain harvest in the same field. It was a way to refurbish the soil without letting it go vacant. Roman farming included green fallowing with legumes: K. D. White, *Roman Farming* (London: Thames & Hudson, 1970).

[174]L. E. Stager, "Agriculture," in *The Interpreter's Dictionary of the Bible Supplementary Volume* (ed. K. Crim; Nashville: Abingdon, 1976), 12-13; Borowski, *Agriculture in Iron Age Israel*, 7; Hopkins, *Highlands of Canaan*, 27.

> a land of wheat and barley,
> of vines and fig trees and pomegranates,
> a land of olive trees and honey.

A mixture of grain farming and horticulture often describes the land's capability in general, and it was farming strategy. The triad of wine, olive oil, and grain is a constant motif throughout the Bible and reflects how central these products were to agriculture.[175] The next chapter demonstrates how this "Mediterranean" mixed farming was the regimen of the Israelite small farmer.

---

[175]Gen 27:28, 37; Deut 7:13; 8:8; 11:14; 12:17; 14:23; 18:4; 28:51; 33:28; 2 Kgs 18:32; 2 Chr 31:5; 32:28; Neh 5:11; 10:39; 13:5, 12; Ps 4:7; Isa 36:17; Lam 2:12; Hos 2:10 [ET 2:8]; 8:22; Joel 1:10; 2:17; Hag 1:11; Sir 39:26.

Chapter 2

The Sociology of the Vintner

Who Was a Vintner?

The question posed in this chapter is sociological, namely, who was a typical Israelite vintner? So far, we have reviewed the climatic and geographical conditions that made viticulture a useful, even essential, adaptation for Israelite agriculture. The discussion now turns to the social component of who ran the farm, harvested the grapes, and trod them into wine. For behind the labor and care involved in bringing a vineyard to its first and successive harvests were the lives of Israelite vintners. In the present chapter biblical and archaeological materials, in comparison with evidence drawn from the classical world and from modern peasant societies, are worked into a profile of the Israelite vintner.

The biblical examples of vintners, besides guest appearances in metaphor by the deity (Ps 80:8-9; Isa 5:1-7, 63:2-3; Jer 2:21; Lam 1:15), are rare or muted. Noah (Gen 9:20-27), the Timnites (Judg 14:5), Naboth (1 Kgs 21), the Solomon of Song of Songs and his lover, and, by inference, Isaac (Gen 27:28, 37), Judah (Gen 49:11-12), and perhaps Nabal (1 Sam 25) are marked as vintners. In addition, Kings David (1 Chr 27:27) and Uzziah (2 Chr 26:10) oversee or support vineyards in their kingdoms. Vintner activities are used in a variety of prophecies (Isa 16:10; 37:30; 65:21; Jer 31:5; 48:32; Ezek 28:26; Amos 4:9; 5:11; 9:14; Zeph 1:13), and we can infer that much of the intended audience of these prophecies either practiced them or was familiar with their meaning.

A social history of the Israelite vintner necessarily entails some theoretical assumptions about the economic matrix of which he was a part. My primary assumption is that the agrarian economy of ancient Israel was inseparable from its society, and so an investigation into viticulture illuminates something important, not incidental, about

Israelite daily life. Farmers were the majority and backbone of the Israelite population.[1] Trade and larger-scale operations undoubtedly were a part of Israelite life. There are, for example, wine amphorae from Israel found in eastern Mediterranean sites.[2] And there is some evidence for vineyards having been run under the auspices of the crown. In 1 Chr 27:27, David has an official for the vineyards, and one for the wine cellars. In 2 Chr 26:10, King Uzziah has vintners and farmers for whom he builds towers and cisterns. Several of the Samaria ostraca—the eighth-century BCE records of wine and oil deliveries—lack individual names or town names, which might well indicate that they stem from royal vineyards.[3] Jar handles with the inscription *lmlk*, "to/of the king," dating to the late eighth-early seventh century BCE, number over 1500 and were likely all made from the same pottery house.[4] These jar handles seem to represent the royal distribution of provisions, if not the production of agricultural materials as well.[5] An Iron II winery at the Philistine site of Ashkelon was housed in a monumental building in the center of the city. It had large winepresses, capable of considerable wine production, and was probably a royal winery.[6] Nevertheless, the focus of the present investigation is on the daily life of the average Israelite vintner. A living had to be made by the populace, and viticulture contributed to that endeavor.

---

[1]Aharoni, *The Land of the Bible*, 13; P. J. King, *Amos, Hosea, Micah: An Archaeological Commentary* (Philadelphia: Westminster, 1988), 108; M. Zohary, *Plants of the Bible* (Cambridge, Eng.: Cambridge University Press, 1982), 36; Borowski, *Agriculture in Iron Age Israel*, xix; Hopkins, *Highlands of Canaan*, 15.

[2]Zemer, *Storage Jars in Ancient Sea Trade*, nos. 5-17 (pp. 11-23), which represent storage jars recovered off the coast of Syria-Palestine. Nos. 5-6, 11 (pp. 11, 17) represent types commonly found in southern sites of ancient Israel and the coast; nos. 9-10 (p. 14) represent types common to the northern sites; nos. 12-14 (pp. 18-20) were of types found in a few sites in Israel, near Akko, yet were common in Cyprus; no.15 (pp. 18-20) also was rare for Israel but very common in Cyprus. All these are examples of a flourishing sea trade in wine for Iron II.

[3]See below, p. 56. Ostraca nos. 53, 54 wine; 55 oil; 58, 61, no commodity mentioned. Aharoni suggests that these might be royal vineyards, *Land of the Bible*, 364. Most of the ostraca mention place names, individuals, and commodities.

[4]F. M. Cross, "Judean Stamps," *EI* 9 (1969): 22.

[5]N. Na'aman, "Hezekiah's Fortified Cities," *BASOR* 261 (1986): 5-21; D. Ussishkin, *The Conquest of Lachish by Sennacherib* (Tel Aviv: Tel Aviv University Press, 1982); Ussishkin, "Excavations at Tel Lachish 1978-1983: Second Preliminary Report," *Tel Aviv* 10 (1983): 97-177; Cross, "Judean Stamps," 20-27; A. D. Tushingham, "New Evidence Bearing on the Two-Winged LMLK Stamp," *BASOR* 287 (1992): 61-65; A. D. Tushingham, "A Royal Israelite Seal (?) and the Royal Jar Handle Stamps (Part One)," *BASOR* 200 (1970): 71-78; A. D. Tushingham, "A Royal Israelite Seal (?) and the Royal Jar Handle Stamps (Part Two)," *BASOR* 201 (1971): 23-35; H. Darrell Lance, "The Royal Stamps and the Kingdom of Josiah," *HTR* 64 (1971): 315-32.

[6]L. E. Stager, "The Fury of Babylon: Ashkelon and the Archaeology of Destruction," *BAR* 22 (1996): 56-69, 76-77.

Most Israelites farmed for their livelihood, and this activity shaped their social relations. The written record of ancient Israel, contained in the Bible primarily, is infused with agriculture because farming was the livelihood of its society.[7] The sheer dominance of agriculture in the Bible bespeaks its importance, even preoccupation, in Israelite society. The archaeology for Iron Age Israel as well indicates an agrarian society through domestic architecture, storerooms, farm implements, and olive and winepresses. Hence, the historical reconstruction of the vintner ought to open a fairly large window into the social world of this agrarian society. Through analysis of the details of viticulture—of vine tending, wine production, and drinking—a vital aspect of Israelite life comes into relief.

Diversification into various crops, as we saw in chapter 1, was an adaptive strategy to the environment.[8] The climate was semi-arid and the lands rocky and hilly. Summers of potential drought followed the cold, wet winters. Hill slopes and rocky soils proved poor contexts for grain farming. The strategic importance of mixed farming to such environments is underscored by the classical scholar Peter Garnsey:

> The Mediterranean small farmer has traditionally
> practised mixed farming, the polycropping of arable
> and trees on the same land with the addition of a little
> livestock. The goal is self-sufficiency but also the
> minimalisation of risk: since the growth requirements
> of the various products differ, the possibility that the
> farmer will be left with nothing is reduced.[9]

The Israelite farmer, like his Bronze Age predecessors in the Levant, met the environmental challenge laudably with his inclusion of horticulture. He too practiced this mixed farming so characteristic of Mediterranean lands, viz., vine and tree crop cultivation, grain

---

[7]M. Zohary, *Plants of the Bible*, 36; King, *Amos, Hosea, Micah*, 108; Aharoni, *Land of the Bible*, 13; Borowski, *Agriculture in Iron Age Israel*, xix.
[8]Stager, "Agriculture," 12-13; see chapter 1, pp. 27-41.
[9]Garnsey, *Famine and Food Supply in the Graeco-Roman World*, 54.

farming,[10] and sheep and goat pastoralism.   Vines, as I noted in chapter 1, were a useful inclusion in the agricultural calendar because they could thrive in the soil and rainfall patterns of Israel without detracting from grain production.[11]

Several sources of evidence indicate that Israelite agriculture was characterized by mixed farming by households. Since vines were a facet of that mixed farming, discussion of this evidence is necessary in order to provide the context for Israelite viticulture.   The first section of this chapter is an examination of the archaeological evidence for Israelite farming inclusive of horticulture.[12]   This evidence consists of the Gezer calendar, domestic architecture, and the Samaria ostraca.   Section two contains a discussion of the primary labor pool of the farm, that is, the family, through analysis of biblical texts and ethnographic materials. A third section focuses on Israelite inheritance, since this practice enabled vineyards, and farming in general, to continue.

I.   Archaeological Evidence for Family Farms

a. Gezer Calendar

The Gezer calendar, as we saw in chapter 1, apparently lists the major agricultural activities of the year for ancient Israel, and these included the harvests of grapes, grain, and olives.[13] Its importance to the present discussion is that this calendar also provides evidence that

---

[10]Grain remains the primary crop in mixed farming regimens today: Hopkins, *Highlands of Canaan*, 214. In the Arab village of Baytin (biblical Bethel), grain remains the most important crop: "Bread is literally the staff of life in Baytin." It is consumed at every meal and wheat is the principal crop of each farmer: R. T. Antoun, *Arab Village: A Social Structural Study of a Transjordanian Peasant Community* (Bloomington: Indiana University Press, 1972), 112. So too for the rural farming families of Greek Ambeli. They resist market breads, even as they are readily available. The most common explanation is voiced by one villager from du Boulay's study: "Why should I buy my bread when I can grow it myself?": J. du Boulay, *Portrait of a Greek Mountain Village* (Oxford: Clarendon, 1974), 35.

[11]Baly, *Geography of the Bible*, 85; Zohary, *Plants of the Bible*, 11; Semple, *The Geography of the Mediterranean*, 380-81; Hopkins, *Highlands of Canaan*, 120, 196; Stager, "Archaeology of the Family," 4; Cox, *From Vines to Wines*, 38, 40-41.

[12]For archaeological evidence of farms in regions too dry to have been able to sustain significant horticulture, see Y. Aharoni, M. Evenari, L. Shanan, and N. H. Tadmor, "The Ancient Desert Agriculture of the Negev: An Israelite Agricultural Settlement at Ramat Matred," *IEJ* 10 (1960): 23-36, 97-111; A. Mazar, "Giloh: An Early Israelite Settlement Site near Jerusalem," *IEJ* 31 (1981): 1-36; M. Haman, "The Iron Age II Sites of the Western Negev Highlands," *IEJ* 44 (1994): 36-61.

[13]See chapter 1, pp. 34-41.

the mixed farming of grain and fruits was carried on by *individual farmers*. Four of the months listed in the calendar have a vertical mark after the *ḥet*. That is, four of them are written ירחו instead of ירח. Albright and Cross and Freedman correctly identify this mark as the letter *waw*.[14] This *waw* cannot represent the vowel letter "o" for the masculine singular suffix ("his month"), since this would be represented in preexilic Hebrew by the letter *he*. The retention of the *waw*, then, as Cross and Freedman argue, signals the presence of a dual noun form, "two months," plus the masculine singular suffix. The form *-êw* ("his two months") results when the contraction of the diphthong *-ay-* preceded syncope of intervocalic *he*.[15] Had the *he* been syncopated first, the *waw* would not be evident in the orthography. The presence of the *waw*, in other words, signifies the masculine singular suffix and a dual form of the noun ירח. The Gezer calendar, then, is written with the individual farmer in mind.

Four of the months listed in the calendar have the *waw* indicating the masculine singular suffix and dual form, and therefore represent two months each, for a total of eight months. The remaining four months do not. The total number of months noted in the Gezer calendar, then, is twelve. The presence of the letter *waw* after ירח in the calendar is important on two counts: it provides further evidence that the calendar represents a yearly agricultural cycle; and, it shows that an individual, male farmer is indicated.[16] The calendar becomes, then, a kind of almanac for the ancient individual farmer, listing the tasks of each of 'his months.'

b. Domestic Architecture

The architecture of domestic structures throughout sites from Iron I and Iron II Israel bears a striking uniformity. The so-called

---

[14]Albright, "The Gezer Calendar," 24; Cross and Freedman, *Early Hebrew Orthography*, 46-47.

[15]Cross and Freedman, *Early Hebrew Orthography*, 46-47.

[16]The remaining months indicated in lines 3-5, 7, Cross and Freedman argue, would, on the grounds of parallelism, have the masculine suffix on a singular noun, though not represented with a *he*. In addition, the contraction of the diphthong here, as with קץ in line 7, is characteristic of a northern Israelite dialect: *Early Hebrew Orthography*, 46-47. Borowski does not incorporate Cross and Freedman's important analysis in his reading of the calendar: *Agriculture in Iron Age Israel*, 34-35; Ronzevalle interprets the *waw* as a separate word "and" repeating throughout the calendar: "The Gezer Hebrew Inscription," 109; Gray interprets the *waw* to mean "as" to specify the primary task of a month: "The Gezer Inscription," 192.

four-room pillared house is found at many Israelite sites.[17]   The
number of rooms in each house actually varies between three and
four, with a few examples containing only two rooms. Nevertheless,
the presence of multiple rooms, with a broad room and pillars, marks
the dwellings as similar. Examples of such homes occur in Tell Beit
Mirsim A, Tell en-Nasbeh, Beth-Shemesh II, Tell Qasile X-VIII, Tell
el-Farah III-II, 'Ai, Hazor VIII-V, and Beersheba.[18]   Up through
1987, 155 such dwellings had been found in Israel,[19] with more
unearthed in subsequent archaeological seasons. The broad room of
this structure extended the length of the house and contained pillars,
either of tooled stone or with flat, large bases that perhaps supported
wooden or thatched roofs.[20] Typically the larger four-room house
would have two rows of pillars, while the smaller three-room house
would have one row.[21] One to three smaller rooms then flanked the
broad room. A courtyard often joined the house in front of the broad
room. The courtyard often served as a work area where agricultural
installations are found.[22] They served as outdoor pens for herds as
well.

       Lawrence Stager has reconstructed facets of Iron I family life
using the domestic structures from 'Ai and Raddana.[23] In them were
found abundant agricultural installations such as storage jars known as
collared-rim store jars, large stone saddle querns, mortars and pestles,
and numerous flint segments from sickles used to harvest cereals. In
addition, faunal remains of sheep and goat were found in every

---

[17]Pillars and a fourfold division of space are typical of Israelite public buildings as
well, such as stables and storehouses. J. S. Holladay, Jr., "House, Israelite" in *ABD*
3:310; Y. Shiloh, "The Four-Room House: Its Situation and Function in the Israelite
City," *IEJ* 20 (1970): 180-90; J. S. Holladay, Jr., "Houses, Syro-Palestinian," in
Oxford Encyclopedia of Archaeology in the Near East, ed. E. M. Meyers (New
York: Oxford University Press, 1997) 3:94-114; R. David, "La maison à piliers dans
l'argumentation concernent l'émergence d'Israel en Palestine à l'époque du fer I," in
Où demeures-tu? (ed. J.-C. Petit; Saint-Laurent, Québec, Canada: Éditions Fides,
1994), 53-69; F. Braemer, *L'architecture domestique du Levant à l'age du fer* (Paris:
Éditions Recherche sur les civilisations, 1982).
[18]Shiloh, "The Four-Room House: Its Situation and Function in the Israelite City,"
180-90. In fact, for Shiloh and Holladay, it marks them as even distinctively
Israelite, though some have been unearthed at Philistine Tell esh-Sharia Stratum VIII,
and Sahab and Tell es Sa'idiyeh in Jordan; Stager, "Archaeology of the Family," 17.
[19]Holladay, "House, Israelite," 308.
[20]Holladay, "House, Israelite," 309. The structures at Tell Qasile and Tell es-
Sa'idiyeh were made from mudbrick.
[21]Holladay, "House, Israelite," 308-9.
[22]Mazar, *Archaeology of the Land of the Bible* (1990), 392; G. Barkay, "The Iron
Age II-III," in *The Archaeology of Ancient Israel* (ed. A. Ben-Tor; New Haven: Yale
University Press, 1992), 332.
[23]Stager, "Archaeology of the Family," 11-12.

quarter.[24]   Several of the smaller side rooms would have been for storage, given the amount of store jar shards in some of them.

Other examples of Iron II pillared houses with evidence of agricultural installations or storage were found at Gibeon, Tell Qasile, and farmsteads at Tel Michal and in Western Samaria. At Gibeon, pillared four-room houses were unearthed for the eighth-seventh centuries BCE, with evidence of agricultural production such as hand mills for grain on the floor, storage jars, and a flint tool.[25]   Botanical remains, of 50 olive pits, were found in one of the houses.[26]   At Tell Qasile, a four-room house from the tenth century BCE (Stratum IX) had a plastered wine vat in the wall of the court, and an oven, millstone and store jars.[27] Another house from the same period, house K, also had a winepress in the courtyard and three storage jars within the house, one of which was filled with vetch seed.[28]

A tenth-century BCE farmstead house (structure 1522) at Tel Michal contained storage jars, bowls, kraters (for wine use) and a silo where "large quantities of charred grape seeds" were found.[29]   In other houses, olive pits were discovered for the same period.[30] In a survey of Western Samaria, Shimon Dar found small farmsteads with winepresses, e.g., Farm 1 in Nahal Beit Arif between modern Lod and Rosh Ha'ain in a hilly area.[31]   It comprised rooms along a wall, a farmyard of 20 m x 20 m that contained a winepress and circular basins measuring 36 cm x 46 cm in diameter, which likely had an agricultural function.[32] Overall, his survey of a 35 km[2] area yielded sixty farmsteads dating from the Iron II through Hellenistic periods. All of the farmsteads had evidence of domestic agricultural work, and many had oil and wine installations.[33]

The significance of these finds is remarkable, for they indicate that much of the layout of these domestic buildings seems to have been devoted to agricultural storage or activity. Barkay concludes that the

---

[24]Stager, "Archaeology of the Family," 11-12.

[25]J. B. Pritchard, *Gibeon: Where the Sun Stood Still: The Discovery of the Biblical City* (Princeton: Princeton University Press, 1962), 107.

[26]Pritchard, *Gibeon: Where the Sun Stood Still*, 107.

[27]Maisler [Mazar], "The Excavations at Tell Qâsile: Preliminary Report," 137; for other examples of presses and analysis of the features of a winepress, see chapter 4, pp. 142-57.

[28]Maisler [Mazar], "The Excavations at Tell Qâsile: Preliminary Report," 137.

[29]Locus 1529: Z. Herzog, G. Rapp, Jr., and O. Negbi, *Excavations at Tel Michal, Israel* (Minneapolis: University of Minnesota Press and Tel Aviv: Sonia and Marco Nadler Institute of Archaeology, Tel Aviv University, 1989), 64.

[30]Herzog et al., *Excavations at Tel Michal, Israel*, 219.

[31]A thorough analysis of how to identify agricultural installations as winepresses follows in chapter 4, pp. 142-57.

[32]S. Dar, *Landscape and Pattern: An Archaeological Survey of Samaria 800 BCE–636 CE* (2 vols.; Oxford: B.A.R. International Series, 1986), 1:4.

[33]Dar, *Land Use and Pattern*, 1:4-6.

architecture of domestic buildings demonstrates that the first floor was given over to agricultural and household work, for family labor.[34] In fact, so much of the layout is for economic activity that both Stager and Holladay postulate a second story for sleeping. Since archaeology can glimpse only the remains of a house, often the foundation, with little left of the walls and nothing of a ceiling, ethnographic studies of Middle Eastern peasant dwellings help to discern house function.

One study, by Carol Kramer, focuses on domestic architecture in a small Kurdish village, and throughout forecasts what of the village is likely to remain in the archaeological record. The villagers practiced mixed farming on a subsistence level of wheat, barley, alfalfa, lentils, and sheep and goat pastoralism.[35] Each house had several rooms, one larger than the others, and a courtyard.[36] The small rooms were devoted to the storage of grain, legumes, fodder, tools, and dung, which was the chief fuel source for the family. Each house had an indoor stable and courtyard.[37]

Holladay, using Watson's study on another Iranian village, estimates that half of the roofed, domestic space of the Israelite house was devoted to economic activities, such as stabling and storage.[38] He then took Watson's estimate of food needs to discern how much storage space an Israelite family might have needed. Watson estimated the food needs for a year for an impoverished family of five and their livestock to be about 1800 kg of wheat, 1080 kg of barley.[39] Holladay uses the Judean *lmlk* jars, whose capacity has been studied,[40] to calculate how many jars would have been needed for this amount of grain. With an average capacity of 45 liters,[41] 55 such jars would be needed for wheat, and 36 for barley. If stacked two jars high, they would fill a 12 m^2 room. The Israelite store rooms average 12.4 $m^{2.}$[42] In other words, grain would take up one average storage room.

---

[34]Barkay, "The Iron Age II-III," in *The Archaeology of Ancient Israel*, 332.

[35]C. Kramer, "An Archaeological View of a Contemporary Kurdish Village: Domestic Architecture, Household Size, and Wealth," in *Ethnoarchaeology: Implications of Ethnography for Archaeology* (ed. C. Kramer; New York: Columbia University Press, 1979), 141.

[36]Kramer, "An Archaeological View of a Contemporary Kurdish Village," 147.

[37]Kramer, "An Archaeological View of a Contemporary Kurdish Village," 147.

[38]Holladay, "House, Israelite," 312.

[39]P. J. Watson, *Archaeological Ethnography in Western Iran*, Viking Fund Publications in Anthropology 57 (Tucson: University of Arizona Press, 1979), 292-3; J. S. Holladay Jr., "The Kingdoms of Israel and Judah: Political and Economic Centralization in the Iron IIA-B (ca. 1000-750 BCE)," in *The Archaeology of Society in the Holy Land* (ed. T. E. Levy; New York: Facts on File, 1995), 387.

[40]Ussishkin, "Excavations at Lachish 1978-1983: Second Preliminary Report," 97-177.

[41]Ussishkin, "Excavations at Lachish 1978-1983: Second Preliminary Report," 162.

[42]Holladay, "House, Israelite," 314.

For ancient Israel, wine and oil would have to be factored in as well, since these Iranian communities have no olive orchards or grape vines. In chapter 6, I estimate the wine yield from a two-dunam vineyard to be approximately 694 liters.[43] This amount would fill 15.4 *lmlk* size jars, which could easily be stored in one of the other store rooms.

c. Samaria Ostraca

The Samaria ostraca, from the eighth century BCE, are evidence of wine sent from villages and clan districts to the northern capital, Samaria, and they detail the personnel involved.[44] Jars of wine and oil are recorded along with personal names, towns, clans, and regnal years. Almost half of the 63 ostraca, 29 in all, detail the commodity specifically. Wine and oil are the only commodities noted on these ostraca. The ostraca that do not mention a specific commodity note simply a number one or two, presumably to indicate the number of jars, or they denote nothing about the commodity. These records of shipments detail something of who was producing the wine, i.e., who was filling the jars sent to Samaria.

The ostraca were unearthed in corridor fill on the east side of a building complex on the Samaria royal acropolis during Reisner's 1910 excavation season.[45] A mid-eighth century BCE date is fairly well established for the ostraca on the basis of stratigraphy, pottery analysis, and paleography. The ostraca were found in a 10-40 cm fill layer under a courtyard floor in use before the 721 BCE destruction of Samaria.[46] Since they were found in the fill and not on the floor

---

[43]Chapter 3, pp. 111-12 for plot estimates; chapter 6, pp. 211-12, 218-19 for wine estimates.

[44]Aharoni, *The Land of the Bible,* 356, 367; Y. Yadin, "Recipients or Owners, A Note on the Samaria Ostraca," *IEJ* 9 (1959): 184-87; A. S. Rainey, "Administration in Ugarit and the Samaria Ostraca," *IEJ* 12 (1962): 62-63; A. S. Rainey, "Wine From the Royal Vineyards," *BASOR* 245 (1982): 57-62; A. S. Rainey, "The Samaria Ostraca in the Light of Fresh Evidence," *PEQ* 99 (1970): 32-41; I. T. Kaufman, "The Samaria Ostraca: A Study in Ancient Hebrew Palaeography. Text and Plates," (Th.D. diss., Harvard University, 1966): 156, 159; Kaufman, "The Samaria Ostraca: An Early Witness to Hebrew Writing," *BA* 45 (1982): 229-39; Kaufman, "Samaria (Ostraca)," in *ABD,* 5:921-26; B. Maisler, [Mazar], "The Historical Background of the Samaria Ostraca," *JPOS* 21 (1948): 117-33; G. A. Reisner, C. S. Fisher, and D. G. Lyon, *Harvard Excavations at Samaria 1908-1910,* vol. 1 (Cambridge, Mass.: Harvard University Press, 1924); A. Lemaire, *Inscriptions hébraïques,* vol. 1, Les Ostraca (Paris: Cerf, 1977); W. H. Shea, "The Date and Significance of the Samaria Ostraca," *IEJ* 27 (1977): 16-27.

[45]Kaufman, "The Samaria Ostraca: A Study in Ancient Hebrew Palaeography," 101.

[46]Reisner et al., *Harvard Excavations at Samaria,* 62.

above, the ostraca have to be earlier than the floor. Pottery analysis
of the shards established the ostraca to be from the second and third
quarter of the eighth century BCE.[47] The paleography of the ostraca
also establishes a mid-eighth century BCE dating.[48]

The ostraca shards were not taken from the jars that held the
commodities themselves.[49] Ivan Kaufman drew this conclusion from
his analysis of both the paleography and the pottery of the ostraca. He
notes, for example, that ostraca nos. 43-44 detail the different districts
of Hoglah and Shechem, yet the shards are from the same bowl.[50]
Also, he argues, nos. 17, 19, and 21 have writing so similar in style
that they are almost certainly from the same scribe, yet they come
from three different places.[51] This evidence—viz., of the same
handwriting on three ostraca with three different place
names—suggests that the ostraca as a whole probably originated in
Samaria, rather than in the places named in the ostraca. Kaufman
views the ostraca as accounts recorded at the destination point of the
commodities. In this, he differs from George Reisner, the excavator,
and Benjamin Mazar who considered the ostraca to have been invoices
sent along with the commodities.[52] Yohanan Aharoni's view accords
with Kaufman's.[53] He termed the ostraca "scratch-pad notations" of
scribes at Samaria, from which he postulated a later more formal
ledger done on papyrus.[54]

The data on the ostraca are year, place name by town or clan
district, commodity, and personal names, some with an affixed *lamed*,
the *l*-men of scholarly literature, and some without a *lamed*, deemed
non-*l*-men[55] or secondary men.[56] The terminology in the literature
itself betrays where scholarly interest lay, viz., in the identity of the *l*-
men. However, for our purposes, the non-*l*-men are just as
important.[57] For the relation of both sets of names to the jars of wine
offers us a glimpse into the Israelite wine economy at least for the
eighth century BCE in the north.

---

[47]Kaufman, "The Samaria Ostraca: An Early Witness to Hebrew Writing," 232-33.
[48]Kaufman, "The Samaria Ostraca: An Early Witness to Hebrew Writing," 233-34;
Yadin, "Recipients or Owners, A Note on the Samaria Ostraca," 184-87.
[49]Kaufman, "The Samaria Ostraca: An Early Witness to Hebrew Writing," 237.
[50]Kaufman, "The Samaria Ostraca: A Study in Ancient Hebrew Palaeography," 155.
[51]Kaufman, "The Samaria Ostraca: A Study in Ancient Hebrew Palaeography," 155.
[52]Reisner, et al., *Harvard Excavations at Samaria 1908-1910* (1924), 1: 232; Maisler
[Mazar], "The Historical Background of the Samaria Ostraca" (1948), 117.
[53]Aharoni, *Land of the Bible*, 363; see also Shea, "The Date and Significance of the
Samaria Ostraca," 16-27.
[54]Aharoni, *Land of the Bible*, 363.
[55]Aharoni, *Land of the Bible*, 363.
[56]This is Kaufman's term, "The Samaria Ostraca: A Study in Ancient Hebrew
Palaeography," 153.
[57]For the sake of convenience, I will refer to the names without *lameds* as "non-*l*-
men," this being less pejorative than the label, "secondary men."

The ostraca vary in their detail. In ostraca nos. 1-2, 4-19, and 21, town names are listed with jars of either wine or oil.[58] These all contain one *l*-man each. Only nos. 1-2 have non-*l*-men. Nos. 22-29 list both an *l*-man and a non-*l*-man, a district name, and a town name. But in these ostraca, no commodity or jar numbers are given. Nos. 30-34, 36-39, 42, 44, 45, 47-50, and 57 have no town names, but provide district names along with both *l*-men (except no. 36) and non-*l*-men. This group does not provide the commodity either, except for nos. 36 and 44. The variation in the formulae continues to fuel scholarly debate on the ostraca and may shroud the ostraca in mystery for years to come. However, some patterns can be detected for historical reconstruction.

First of all, probable identification of the towns has been made by Aharoni,[59] Albright,[60] and Martin Noth,[61] and this has then yielded the general locations of the districts associated with the towns. The importance of these identifications for biblical studies and the historical study of ancient Israel is supreme. For remarkably, the place names in the ostraca correlate with the clan names associated with the tribe of Manasseh in Josh 17. The land is distributed in Josh 13-19 according to the tribes and their clans. The place names given as clan names for Manasseh are: Abiezer, Helek, Asriel, Shechem, Hepher, and Shemida (Josh 17:2). All but Hepher are listed in the Samaria Ostraca. Abiezer occurs in nos. 13 and 28; Helek occurs in nos. 22-27,[62] Asriel is noted in nos. 42 and 48, Shechem in no. 44 only, and Shemida is mentioned most often, in sixteen ostraca: nos. 3, 29-40, 57, 62, and 63.[63]

While Hepher is not represented in the ostraca, two of five places given to his granddaughters (Zelophehad's daughters, Josh 17:3), are represented, namely Noah in no. 50 and Hoglah in nos. 45 and 47.[64] What this tends to indicate is that the region of Hepher is indicated in the ostraca, but it is going by more specific place names

---

[58]For a helpful chart, see Aharoni, *Land of the Bible*, 358-62; Kaufman, "The Samaria Ostraca: A Study in Ancient Hebrew Palaeography," 141-44.

[59]Aharoni, *Land of the Bible*, 367.

[60]W. F. Albright, "The Site of Tirzah and the Topography of Western Manasseh," *JPOS* 11 (1931): 248.

[61]M. Noth, "Der Beitrag der samarischen Ostraka zur Lösung topographischen Fragen," in *Palästinajahrbuch* (1932), 54-67.

[62]Nos. 25 and 26, however, have only the *kaf* represented. They are still thought to be Helek ostraca since both the place name and *l*-man name are the same as in the other Helek ostraca.

[63]Ostraca no. 34 is reconstructed as Shemida, though the name is missing. This is based on the *l*-man name being the same as for the Shemida ostraca (Helez [son of] Gaddiyau). No. 35 has only the *shin* present, but bears as well the same *l*-man name. See Aharoni, *Land of the Bible*, 361; Kaufman, "The Samaria Ostraca: A Study in Ancient Hebrew Palaeography," 143.

[64]Kaufman, "The Samaria Ostraca: An Early Witness to Hebrew Writing," 230.

given in Josh 17:3. In sum, six of the seven clan names for Manasseh in Josh 17 are place names in the Samaria ostraca, and two of the five place names of Zelophehad's daughters are listed as well. In all, this shows a remarkable corroboration between the biblical description of the Manasseh region with that represented in the Samaria ostraca. The places listed in the ostraca are located 4-8 miles from Samaria, except for *Yšb* in Asriel, which is 12 miles south of Samaria.[65]   Hence, biblical clan regions can be mapped and are evidence of lineage grouping around Samaria in the eighth century BCE.   A distance of 4-8 miles for transport of the jars into the capital would have been fairly easy.[66]

Now we turn to the personnel noted in the ostraca, beginning with the *l*-men. Comparison of the *l*-men to the non-*l*-men yields three differences that might have significance for determining the relation of the personal names to the commodities. First, the *l*-men names recur regularly on several ostraca, while it happens only rarely (probably once) with the non-*l*-man names (see below). Gaddiyau, for example, serves as *l*-man on eight ostraca (nos. 2; 4-7; 16-18) and is listed, perhaps as a patronymic, in four other ostraca (nos. 30; 33-35) with Helez as the preceding name. Ahinoam is the *l*-man on five ostraca (nos. 8-11; 19), while Shemaryau is listed as the *l*-man for four ostraca (nos. 1; 13-14; 21). Asa is the *l*-man on eight ostraca (nos. 22-29; no. 25 is partially reconstructed), while Helez is the *l*-man on six ostraca (nos. 30-35).[67]

While the *l*-men recur on multiple ostraca, only one non-*l*-man recurs with the same place and clan names, that of Helez in nos. 22 and 23.[68]   A non-*l*-man named Baala is listed in three ostraca, but these have three different place and clan names and so probably indicate three different persons (nos. 27, 28, 31). Otherwise, the non-*l*-men names are not repeated on ostraca the way the *l*-men names are.

---

[65] Kaufman, "The Samaria Ostraca: An Early Witness to Hebrew Writing," 230.

[66] A question arises at this point as to why other, more distant towns are not mentioned in the ostraca. Several explanations are possible. Perhaps, the ostraca find—in floor fill, it will be recalled—excluded distant sites by chance, with the *Yšb* ostraca providing the only hint of towns further out. Or the number of Samaria satellite villages could have been reduced at this point to the environs represented by the ostraca town names. Another possibility is that a second site or building in Samaria, still unexcavated, could have been the collection point for jars from other towns. Finally, the ostraca may detail only the horticultural take for the capital, with grain, if collected, presumably being stored elsewhere, or exacted from other parts of the kingdom, like the valleys.

[67] Helez may occur another time, on no. 49, according to Aharoni's and Kaufman's reconstruction: Aharoni, *Land of the Bible*, 362; Kaufman, "The Samaria Ostraca: A Study in Ancient Hebrew Palaeography," 144.

[68] Aharoni adds no. 26, though the name is not evident, *Land of the Bible*, 360; cf. Kaufman, "The Samaria Ostraca: A Study in Ancient Hebrew Palaeography," 142.

The non-*l*-men also never appear with more than one *l*-man.[69] For example, Helez's *l*-man is only Asa, while Asa is listed for other non-*l*-men too. Ostraca nos. 1 and 2 are unique in that they list multiple non-*l*-men to one *l*-man. No. 1 lists five non-*l*-men with one town to one *l*-man, Shamaryau. No. 2 lists four non-*l*-men with one town to one *l*-man, Gaddiyau. The *l*-men are linked to several districts and several towns, but the non-*l*-men are not.[70] The repetition of *l*-men names for various districts and towns, without the same repetition for the non-*l*-men, suggests that the *l*-men are the centralized figures in Samaria, the recipients of district commodities, while the non-*l*-men are the senders who are associated with only one town.

Second, the name formations in the two groups bear a striking difference, which Albright first noticed.[71] Names with Yahwistic elements are more frequent than those with Baal, and tend to be *l*-men. Names with Yahwistic elements occur in nos. 1 (twice); 2; 4-7; 13-14; 16-18; 21; 30; 33-35; 41; 42 (twice); 45; 48; 50; 52; and 58. Names with Baal occur in ostraca nos. 1; 2 (twice); 3; 12; 27-28; 31, and 37.[72] While the names with Yahwistic elements occur for both *l*-men and non-*l*-men, the majority are *l*-men. Twenty of the 28 Yahwistic names, a full 71 percent, occur as *l*-men. More striking still is that the Baal names, with the exception of nos. 3 and 12, occur *only* as non-*l*-men.

The Baal names occur, for the most part, as non-*l*-men, while most of the Yahwistic names occur as *l*-men. If the *l*-men were working for the government in the capital of Samaria, we might expect them to be more likely than non-government workers to participate in or at least originate from the traditional and state sanctioned religion of Yahwism, and therefore to have Yahwistic names. Practitioners in the Baal religion, no doubt, were also present in Samaria as well and in its environs, but these people might have been less likely to gain successful employment as tax collectors for the state. The names, in other words, might reflect a subtle discrimination policy on the part of the Samaria palace, whereby more Yahwists, or at least individuals from Yahwist backgrounds, worked as *l*-men than did Baalists. This remains conjecture, but the preponderance of names with Yahwistic elements as *l*-men supports the notion that the latter were based in Samaria, with Samaria officially sanctioning Yahwism as the state religion.

---

[69]Kaufman, "The Samaria Ostraca: A Study in Ancient Hebrew Palaeography," 151.
[70]The Baala, mentioned above, has a patronymic in no. 27 and a different patronymic in no. 31, and none in no. 28, and so these are likely three different persons.
[71]W. F. Albright, *Archaeology and the Religion of Israel* (5th ed.; Garden City, N.Y.: Anchor Books, 1969), 160.
[72]Albright argued that this difference in names might reflect a social context of waning Baalism in the north: *Archaeology and the Religion of Israel*, 160.

Third, though the ostraca vary in formula, the *l*-men are rarely omitted.[73] They are listed with a regularity that underscores their significance. The several ostraca that do omit the *l*-men are sent by *kerem*-sites, i.e., place names with the word כרם in them (nos. 20; 53-55; 60-61).[74] These ostraca note only a vineyard and a commodity,[75] and have no personal names attached to them. These *kerem*-sites, viz., כרם התל and כרם יחועלי, may be evidence of royal vineyards. Two fragments as well detail only כרם התל and a jar of wine (nos. 3892-3893).[76] No mention is made either of town or district, presumably because their location would be known to the scribes. And none of these *kerem* ostraca mention personal names. It may be that the personal names of non-*l*-men would be superfluous, since no one person owned and operated a royal vineyard. It is also possible that, with royal vineyards, even the names of the *l*-men would have become unnecessary.[77] For, if these ostraca evince basically the state contributing its own share, then the record-keeping need not have been as meticulous as it was with individual farmers.[78] If the shipments from royal vineyards were simply assumed to be regular and dependable, then there might have been no need to record the *l*-men who received them.

The above three factors about the *l*-men names suggest that, over against the non-*l*-men, these are the more likely to be situated in Samaria, serving in some sort of official capacity. *Mems*, denoting "from," are affixed to the town and district names and the *lameds*, "to," are affixed to the names of recipients, i.e., the *l*-men. The non-*l*-men, then, are from the towns and districts mentioned. In no. 27 this is explicit as the patronymic name matches the town name listed.[79] There is a scholarly consensus that the wine and oil noted in the ostraca come from the regions noted. From a record of shipment it is reasonable to infer locus of production. The ostraca that mention wine, then, constitute evidence of wine production from the individuals named. The ostraca that do not mention a commodity might also represent a wine or oil shipment and production that is assumed, rather than stated. It then becomes a matter of discerning why *l*-men in Samaria received commodities.

There is general consensus that the ostraca denote some sort of accounting system, and one that was based in Samaria, since several of them have the same handwriting, and several are from shards from

---

[73]Only nos. 36, 51-52 list no *l*-man.
[74]Ostracon 58 is an exception. It notes a *kerem* site with an *l*-man.
[75]Not always wine. Oil is noted for nos. 20 and 55.
[76]Aharoni, *Land of the Bible*, 362.
[77]Aharoni, *Land of the Bible*, 322.
[78]See below for a discussion of the purpose of the Samaria ostraca.
[79]"Baala Baalmeonite" and Baalmeon.

the same piece of pottery.[80]   There are two primary suggestions concerning the deliveries noted in the ostraca: (1) they reflect taxes in kind to the capital from the surrounding regions;[81] or (2) they reflect provisions delivered to officials stationed in Samaria from their own farms.[82] The second interpretation is improbable for several reasons. First, if the *l*-men owned the farms producing the wine and oil, why would scribes bother to mention non-*l*-men as senders from those estates?   Second, if the *l*-men of the capital supplied their own provisions from their farms, why bother to transport these through the central royal storehouse?   It would have been simpler to transport the jars directly to the *l*-man's urban dwelling and not involve scribal bookkeeping and storage.   And third, only wine and oil are noted. The *l*-men would have needed other provisions for daily living, unless grain was provided in some other way, or was the commodity of the ostraca that do not explicitly mention one.

The overall patterning of the *l*-men suggests to me that the *l*-men were not the owners of the farms listed, but were instead recipients of taxes in kind from the surrounding regions of Samaria, option (1) above.   Yadin rejects this notion that the *l*-men were recipients of taxes because, he says, they are associated with enough different towns so as to constitute administrative mayhem.[83] However, the area represented by the ostraca is small, with a radius of approximately 4-8 miles.[84]   I do not see how such a defined area could result in mayhem, especially since it is unlikely that the *l*-men themselves would have collected the taxes. It seems more probable that delivery of the commodity into the capital would have been part of the fiscal responsibility of the taxed farmer.   The scribes' task then was simply to note the arrival of commodities in the capital.   For Yadin, though, the *l*-men were the owners of the farms who sent the commodities to the city.[85]   I propose, by contrast, that the *l*-men were not associated at all with the regions denoted in the ostraca.   Rather, the *l*-men as tax recipients could be responsible for different tax times or seasons, whether they be annual or from each harvest.   In this view, the scribal notation of the *l*-men reflects which official was in charge of a certain season's shipments—Gaddiyau, say, for an autumn tax, Shamaryau for a spring one.   In this case, the regions named by

---

[80]Kaufman, "The Samaria Ostraca: An Early Witness to Hebrew Writing," 235.

[81]Yadin, "Recipients or Owners, A Note on the Samaria Ostraca," 184-87; Kaufman, "The Samaria Ostraca: A Study in Ancient Hebrew Palaeography," 151-60.

[82]Rainey, "Administration in Ugarit and the Samaria Ostraca," 62; Rainey, "The Samaria Ostraca in the Light of Fresh Evidence." 32-41; Lemaire, *Inscriptions hébraïques*, 76.

[83]Yadin, "Recipients or Owners, A Note on the Samaria Ostraca," 184-87.

[84]Not counting the site of *Yṣb*, which was 12 miles from the capital.

[85]Yadin, "Recipients or Owners, A Note on the Samaria Ostraca," 186.

the ostraca would not yield an organized pattern, because they have no bearing on which *l*-man receives the shipment.

The commodities themselves suggest that the shipments were taxes. The number of jars indicated in the ostraca is one, except for no. 1, where two jars of wine are listed for one non-*l*-man. Only wine and oil are noted, but their precise description signifies a value. The oil is always "washed oil" (שמנ רחצ),[86] and the wine is "old wine" (ינ ישנ). Stager has convincingly argued that this oil was the finest produced.[87] It was produced by crushing the olives, immersing them in water, and then skimming the "washed oil" off the top.[88]   As for the "old wine,"[89] this occurs on ostraca nos. 1; 3; 5-10; 12-14; and 36.[90] ינ without the qualifier occurs only twice, in nos. 11 and 44. By analogy, we can reasonably infer that "old wine" was the finer quality wine than that noted without the qualifier ישנ. The tax then seems to be a single jar, but it is of the most valued yield.

It is, then, the non-*l*-men from their towns and districts who owed a tax in kind to the capital and would bring in a jar of their best wine or oil for the scribes to record and then store. The *l*-men were the officials overseeing the shipments. Personal names and towns were recorded because these identified the individual family vineyards and farms. Five different men come from Poraim in ostracon no.1 because these are the individual farmers of that town who contributed to the tax period reflected by this ostracon. The scribes have noted these additional personal names because they are the owners of the farms involved. Pega, for example, bears the distinction of being the farmer in ostracon no. 1 who brought two jars of old wine.

The data from ostraca nos. 3-19 seem to argue against this view, for while they have town names and *l*-men, no non-*l*-men are given. Several of the town names recur on ostraca marked with a different year.[91] Ostraca nos. 3-19 are marked with year nine or ten, while ostraca nos. 22-63[92] are marked with year fifteen. It is possible that the lack of non-*l*-men on nos. 3-19 reflects a less detailed notation

---

[86]In nos. 16-21, 53-55, 59 (ן illegible in 59). שמנ, unlike ינ, does not occur without its qualifying term. Kaufman, "The Samaria Ostraca: A Study in Ancient Hebrew Palaeography," 141-44; Aharoni, *Land of the Bible*, 358-63.

[87]L. E. Stager, "The Finest Oil in Samaria," *JSS* 28 (1983): 242.

[88]For a discussion of the production of this oil, see Stager, "The Finest Oil in Samaria," 241-45.

[89]See chapter 5, pp. 206-7.

[90]For Nos. 3, 6, 8, and 36, one of the three graphemes for ישנ is not discernible. For no. 8, two are not legible. Kaufman, "The Samaria Ostraca: A Study in Ancient Hebrew Palaeography," 141-44; Aharoni, *Land of the Bible*, 358-63. See further discussion of the ostraca in chapter 5, pp. 206-7.

[91]For discussions of the two different regnal years indicated, see the literature cited in n. 44 above.

[92]Except no. 51.

system from the earlier period. In ostraca nos. 3-19, several town names repeat. Kozoh, for example, occurs on three ostraca (nos. 4-6), Yazith is named five times (nos. 9, 10, 19, 45, 47), and Hazeroth is noted twice (nos. 15, 18) and then again on the later ostraca nos. 22-26.[93] Perhaps these towns were small enough that their taxes were noted simply by town name when they arrived into Samaria.[94] The *l*-man would then know, for example, to expect three jars from Kozoh and Yazith. Later, in year fifteen, the ostraca included mention of non-*l*-men in addition to the town names. While the *l*-men names have proven to be fascinating evidence for reconstructing Samaria administrative activity, the non-*l*-men are no less so. They may give us the actual names of small farmers who lived out their daily life in the Iron II period, and, because their farms were small and family operated, their personal names get preserved forever in the ostraca.

In sum, the evidence of the Samaria ostraca, though complicated, suggests small farm production of wine and oil. This accords with the domestic architecture for the Iron Age, which shows clear signs of having been used for agricultural production. In addition, too, the Gezer calendar lists agricultural tasks for a year for the individual farmer. Together, these archaeological materials reflect the domestic locus of agricultural production of which viticulture was a part. The next issue, then, is to discern how the small Israelite farm operated.

II. Labor Pool

The first point is that the individual farmer assumed in the Gezer calendar and named in the Samaria ostraca did not work alone. He and his family were the farmers of the family land. The Israelite vintner was then the individual male farmer and his often unnamed family. Pragmatically, it would make sense that the Israelite vintner would have called upon his wife, sons, daughters, and any other relations or trusted friends to help with the harvest, even as he remained the primary agent in the various duties of farming his land. The patriarchal organization of Israel meant that the father was in charge, that he remained head of the household. It did not mean,

---

[93]Aharoni, *Land of the Bible*, 359; Kaufman, "The Samaria Ostraca: A Study in Ancient Hebrew Palaeography," 143.

[94]The wine and oil taxes from villages outside Ugarit were recorded by village names, rather than by individual farms: see Heltzer, *The Rural Community in Ancient Ugarit*, 18; M. Heltzer, "Vineyards and Wine in Ugarit (Property and Distribution)," *UF* 22 (1990): 127; J. Nougayrol, ed., *Le Palais Royal d'Ugarit* (Paris: Imprimerie Nationale, 1955; hereafter *PRU*), III. 10. 044.

however, that the women were rendered inactive or consigned to the indoors. Relegating women to the hearth is partly an expression of the leisure afforded industrialized economies. In the ethnographic examples used below, women and children are active in small, family-run vineyards and farms of this century. Hence, we can reconstruct their presence in the Israelite vintner farm through a combination of these parallels from societies practicing traditional agriculture and biblical texts. I seek in this section on the labor pool not to belabor the obvious, but simply to provide it.

In traditional farming communities of the Middle East and Mediterranean today, families work together to bring in their seasonal harvests. In Baytin (biblical Bethel), for example, the harvesting of grain is carried out by husband and wife, their children and any spouses of the children.[95] In Iran, the wife collects the wheat after threshing the grain. She will also wet down the chaff to keep it from blowing away during winnowing.[96] The women typically do not plow in these communities, but they do winter hoe work.[97] In Vasilika, a modern Greek village, the less well-off farmers will ask their wives to do the deep hoeing: "Although such a procedure is frowned upon in Vasilika, a farmer considers the practical necessities of his situation as more important than any possible loss of prestige."[98] Cooperation characterizes Vasilika farming because family members are working to maintain or improve their common lot. In Messenia, Greece, women and children actively participate in all the harvests throughout the year, except for plowing and heavy transport.[99] Among the Jebaliyah Bedouin of southern Sinai, a division of labor is noticeable as the wife and female children are put in charge of goat and sheep raising, while the men cultivate the orchards.[100]

Women are mentioned in fields only rarely in the Hebrew Bible (Judg 13:9; Ruth 2:23), but it is probable that they helped in the chores for processing grain. The daughters of Zelophehad win the right to inherit land when there are no male heirs (Num 27:1-11; cf. Job 42:15). The ruling implies that they are able to work and live off that land.

---

[95]Antoun, *Arab Village*, 12.
[96]Watson, *Archaeological Ethnography in Western Iran*, 80.
[97]Antoun, *Arab Village*, 9.
[98]Friedl, *Vasilika: A Village in Modern Greece*, 20.
[99]Aschenbrenner, "A Contemporary Community," 54.
[100]A similar division of labor might be hinted at in the stories of Rebekah and Zipporah, who both become betrothed when they are seen watering the animals (Gen 24:15-22; Exod 2:16). The Jebaliyah women help with the orchards in the summer months: A. Perevolotsky, "Orchard Agriculture in the High Mountain Region of Southern Sinai," *Human Ecology* 9 (1981): 343.

Women are present in gardens and vineyards in biblical texts. The "mighty woman" (אֵשֶׁת חַיִל) of Prov 31:16 is praised and desired for her viticultural participation, even control:

> She considers a field and buys it;
> with the fruit of her hands she plants a vineyard.
> She girds herself with strength,
> and makes her arms strong.

As a laborer, she plants vines and acquires strong arms from doing so. Her value to the Israelite farmer is evident from this poem.

A vineyard is the setting for the Song of Songs, a book of love and its consummation. A vineyard's association with coupling may derive in part from the presence of women during the festive grape harvest.[101] Joyful, dancing daughters and wives (Judg 21:20-22) would have made the grape harvest a natural setting for love imagery. So too would the fruit's own qualities at the peak of its ripening. For when ripe, the darkened grapes hang heavily on the vine, engorged with juice, and in triangular clusters. A ripe vineyard is stupefyingly erotic, if subliminally. This being so, the lament in Lam 1:15 then becomes excruciating: "Yahweh has trodden in a winepress the virgin daughter Judah."[102] Here, one of the joyful, dancing virgin daughters has been thrown into the press and crushed. Such biblical images of horror are by no means a pleasant way to reconstruct the social history of women in Israelite viticulture, but such shunted glimmers are often the only textual remnants we have. Women likely danced and worked at the grape harvest.

In the Song of Songs, the woman owns her own vineyard, and it is more valuable than all of King Solomon's grape harvests. In 1:6, she laments that she has not worked her own vineyard because her brothers have forced her to labor for them. Her labor in vineyards is assumed. At 8:12, she is back in possession of her own vineyard and content, even to the point that she dismisses all that mighty King Solomon has. Her vineyard in this book functions as a metaphor for

---

[101]Chapter 5, pp. 179-80, 185-86.

[102]The verse is difficult because the גַּת lacks a כ, and ל is on בְּתוּלַת. The ל marks the direct object, namely, the "virgin daughter Judah." See P. Joüon, *A Grammar of Biblical Hebrew* (trans. and ed. T. Muraoka; Rome: Pontifical Biblical Institute, 1993) § 125k. The context argues for this reading as well, since it parallels with Yahweh's destruction of the young men in the beginning of the verse. See D. R. Hillers, *Lamentations* (Anchor Bible 7A; Garden City, N.Y.: Doubleday, 1972), 3; R. Davidson, *Jeremiah 2* (Daily Study Bible; Philadelphia: Westminster, 1985), 177; R. Gordis, *The Song of Songs and Lamentations* (2d ed.; New York: Ktav, 1974), 132.

her sexual power.   And, on that level, she is made to compare favorably to all that Solomon has, presumably his "thousand" wives and concubines (1 Kgs 11:3).   On both levels, the sexual and the agricultural, the value of a vineyard is apparent.   The vineyard is both valued above a king's wealth and controlled by a woman.   Her presence in the vineyard is, at any rate, a valued given for the author.

These biblical instances of a woman in a vineyard likely reflect agrarian custom even if the metaphoric potency obscures it. Israelite wives probably helped in the vineyards.   In modern Tuscany, Maria plays an active role in all stages of cultivation, from planting vines to the harvest.   In fact, her granddaughters also participate.   As they pick grapes at the harvest, "the girls trotted up and down behind them chatting and laughing to each other in high voices."[103]   The girls add to the labor resources of this Tuscany vintner, Adamo, and they add to his pleasure.   Their laughter, in fact, reminded Adamo and Maria of the past "when all the young people took part in the grape picking."[104]   Wives in Ain Beta and Karma, two West Bank villages studied by Alison Powell, were the primary laborers and decision makers for the fruit orchards next to their houses.   The men farmed the grain, while the women grew primarily grapes, olives, peaches, and apricots.[105]

The agrarian contribution of the wife is noted by classical authors, though in vague terms. In appointing an overseer for a farm of a city dweller, Columella recommends a woman companion "to keep him within bounds and yet in certain matters to be a help to him."[106] Varro too advises that the overseer have a mate, but he only mentions the steadying effect and childbearing usefulness of the woman.[107] In Columella and Varro, the woman's role on the farm as helper is tied to her other labor area, that of childbearing (cf. Gen 2-3).   At least in Varro's case, the wife's contribution to the farm is significant.   For it is to her that he wrote his treatise *On Farming*, so that she could run the farm after his death.[108]

---

[103] A. T. Calabresi, "*Vin Santo* and Wine in a Tuscan Farmhouse," in *Constructive Drinking: Perspectives on Drink from Anthropology* (ed. M. Douglas; Cambridge, Eng.: Cambridge University Press, 1987), 129.

[104] Calabresi, "*Vin Santo* and Wine," 129.

[105] A. Powell, *Food Resources and Food Systems in Two West Bank Villages* (Jerusalem: Arab Thought Forum, 1987), 39-40. The garden setting in Gen 2 is, then, a quite natural one for a wife's creation, if she was similarly so useful to the Israelite farmer in this area of agrarian life.

[106] Columella *De re rustica* 1.8.5. See also 1.8.19, where he suggests that women slaves be given exemption from farm work if they bear a lot of children: a mother of three sons was exempt from work; a mother of more than three gained her freedom.

[107] Varro *On Farming* 1.17.5. (To avoid confusion, I use the English title for Varro's work, rather than the Latin, *De re rustica*, since this is also the title for Columella's work on agriculture).

[108] Varro *On Farming* 1.1.3.

To conclude, owning and cultivating a vineyard in ancient Israel was a facet of the small farming through which a family made a living and gained its sustenance. Grain—bread—remained the mainstay of living as it does today in Middle Eastern rural communities, but grapes, along with olives, figs, and vegetables, complemented the work regimen and diet of the farmer. The man and his wife, their sons and daughters worked the land, and consumed the majority of its yield together. Peasant life is in fact typified by just this feature: that production and consumption of foodstuffs occur primarily in the same locus, the family farm. The family, as center for all this farming activity, constituted, then, the basic economic and social unit for ancient Israel.

The Israelite vintner was a small farmer who divided his labor, time, and land into several agricultural pursuits. Variations no doubt occurred in this portrait as some farmers would concentrate on, say, orchards or pastoralism, if mitigating circumstances allowed. Nabal, for example, might well represent a farmer who invested most of his energies in pastoralism (1 Sam 25). Naboth, for his part, devoted his land to vineyards, as this was the inheritance land in dispute (1 Kgs 21).[109] In the *Odyssey*, orchards dominated Laertes' farm, and consisted primarily of vines and olive, fig, and pear trees.[110] Land plots, wealth, and village interdependencies undoubtedly affected individual livelihoods in ancient Israel, but mixed farming remained the mainstay of the small farmer. The Israelite vintner, then, was the small farmer and his family. And the importance of the vineyard to the farming family is clearest in a biblical text about inheritance in 1 Kgs 21 and 2 Kgs 9:26, to which we now turn.

III. Patrilineal Inheritance

Land continued to be a family affair after a father's death through the means of inheritance. A man's farm passed through the generations of a family, and sanctions evident in the biblical record were in place to protect this tradition. The custom of family inheritance in ancient agrarian communities was a way to transmit economic livelihood from generation to generation. It offered some insurance for the succeeding generation against hardship, as the farm would already be productive and familiar when a couple's children took over. Starting a new farm required both land availability and an outlay of time and resources that poor (would-be) farmers would try

---

[109]See below, pp. 68–85.
[110]Homer *Odyssey* 23.226-27.

to avoid. Familial inheritance offered a fairly simple mechanism for land tenure. It kept families in food and fostered a social stability and endurance to the village in which generations of the same families coexisted.

Ancient Greece practiced a similar method of land tenure, by handing down the *kleros* (κλῆρος), or "ancestral portion," within generations of the family.[111]   The custom meant that land was considered inalienable, that is, that it did not leave the farmer or his family except under hardship. The Greek polis would apportion land and pasture for private exploitation by new citizens. These allotments "tended to remain in control of citizen families for generation after generation."[112]   Inheritance within the family from father to son insured the preservation of the *oikos*, the house, the base unit of the polis,[113] and so was guarded by legal sanction and tradition.

Biblical depictions show the transmission of land from generation to generation to be patrilineal, that is, through the sons. Inheritance entailed a patrilocal pattern of residence as well, since both land and the house were bestowed (Mic 2:2).[114]   Patrilocal residence made sense since the Israelite house, it will be recalled, was an agricultural facility, complete with stables and storage rooms, as much as it was living space for the family. Livelihood and family are so intertwined in peasant agriculture that it is not surprising that the terms for "house" and "family" are interchangeable in the modern rural communities of Egyptian Silwa Bahuri and the Greek village of Ambeli.[115] The biblical category *bêt ʾāb,* "father's house," reflects too a social unit based around the house structure owned by the father.[116]

The petition of Zelophehad's daughters, viz., that they inherit their father's estate, illustrates an exception to patrilineal inheritance, but this was because there were no sons to inherit (Num 27). Deuteronomy 21:15-17 marks a preference for the firstborn son for inheritance: among several sons, the eldest is to receive a double

---

[111]A. Burford, *Land and Labor in the Greek World* (Baltimore: Johns Hopkins University Press, 1993), 21; M. I. Finley, *The Ancient Economy* (London: Chatto & Windus, 1973), 105-110.

[112]Burford, *Land and Labor in the Greek World,* 21.

[113]Burford, *Land and Labor in the Greek World,* 34.

[114]Inheritance in the modern Middle East is traditionally figured through the father to the son. See R. Patai, "The Middle East as a Culture Area," in *Readings in Arab Middle Eastern Societies and Cultures* (ed. A. Lutfiyya and C. Churchill; The Hague: Mouton, 1970), 193. He describes Middle Eastern cultures as "patrilocal, patrilineal, patriarchal, and extended."

[115]In the Egyptian village of Silwa Bahari, the term is "bait": H. Ammar, "The Social Organization of the Community," in *Readings in Arab Middle Eastern Societies and Cultures,* 109. In Ambeli, the interchangeable term is σπίτι: du Boulay, *Portrait of a Greek Mountain Village,* 18.

[116]1 Sam 2:27, 30; 1 Sam 22:11.

portion. In both texts, however, inheritance from father to son is assumed to be customary.[117]

The Samaria ostraca reflect the practice of patrilineal inheritance in eighth-century BCE Israel.[118] Ostraca 1 (three examples), 24, 26 (partial), 30-31, 37-38, 41 (partial), 42, and 45 list personal names (of non-*l*-men) with patronymics. A personal name with patronymic identifies both a man and his father, thereby providing a two-generation roster of farmers in a given village. Patronymic naming is, of course, ubiquitous in the Bible, though its social function has often gone unexplained.[119] Patronymics were never simply names for the uninspired, too wearied by farm work to innovate.[120] They were means to carry patrilineal descent. It is possible, for example, that Pega Elisha from Poraim in ostracon 1 inherited his father's farm and name, Elisha. Both land and lineage were transmitted from father to son in ancient Israel.

In the modern Arab village of Baytin, transmission of authority from father to son is starker still. The son is named and he adds his father's name after his own, with or without an intervening *ibn*, "son." But the parents' names also change. Once a son is born, the parents become *Abu* ("father") x and *Umm* ("mother") x.[121] In a three-member household, then, family members would all be defined in terms of one another. Through such naming practices, the importance of the son to the family is clear. In fact, if a daughter is born first, the parents take on that name, but then drop it if a son comes after her.[122] The switch from daughter to son demonstrates the preference of the patrimonial principle in Baytin. A man's name endures both for memory and for land inheritance.

Family inheritance of the farm lent a social stability and endurance to a village, since the same families would work their land from generation to generation. Such persistence of land tenure is evident, as we saw in the discussion above, in the Samaria ostraca.[123] Many of the village and general clan locations mentioned in the ostraca are also mentioned in biblical texts (Num 27:1; Josh 17:2-3).[124]

---

[117]Cf. Gen 21:10; Gen 25:29-34. At Jacob's deathbed (Gen 49), twelve sons receive a blessing, while the daughter, Dinah (Gen 30:21; 34), goes unnamed.

[118]Borowski, *Agriculture of Iron Age Israel*, 25; Aharoni, *Land of the Bible*, 363; Stager, "The Archaeology of the Family," 24.

[119]See, however, R. R. Wilson, *Genealogy and History in the Biblical World* (New Haven, Conn.: Yale University Press, 1977).

[120]Though exhaustion no doubt plays its role in the formation of custom.

[121]A. Lutfiyya, "The Family," in *Readings in Arab Middle Eastern Societies and Cultures* (ed. A. Lutfiyya and C. Churchill; The Hague: Mouton, 1970), 506.

[122]Lutfiyya, "The Family," 506.

[123]See pp. 53-54.

[124]As Stager noticed, five of the territorial names in the ostraca were great-grandsons of Manasseh; two others, Noah and Hoglah, were, as it happens, daughters of Zelophehad: "The Archaeology of the Family," 24.

These names offer corroborating evidence of land tenure within clans in the eighth century BCE.[125] If this Priestly material from Numbers and the material in Joshua originate from any period other than the eighth century BCE, earlier or later, then the names reflect a persistence of families within that region. Clan settlement endured through the centuries, and, within those clans, families handed down their farms from father to son as the patronymic names make evident.

Land tenure through inheritance meant that simply growing up on a farm, helping his father, served as the son's apprenticeship for when the land became his own.[126] In principle, transmission of the land would go smoothly because of familiarity with the plots and chores. In practice, however, fractious sibling disputes and filial resentments could break out over the inheritance, and are evident in antiquity.[127] Patrilineal inheritance was a fairly simple, traditional means by which the father could give his son some stability for the future and gain for himself the comfort and satisfaction that his own life's work would continue after his death. Of course, lapses in wisdom could occur. In Aesop's fable of "The Hedge and the Vineyard," for example, the heir comes into his father's estate and rashly decides to get rid of the hedge, and the vines are destroyed.[128] He then exemplifies a "foolish young heir."

A son would figure prominently in the farmer's household, by providing labor as soon as he was of age to do so. In rural modern Egypt, for example, in Silwa Bahuri, the eldest son is favored by his parents precisely because he is the first to relieve his father of some of his tasks.[129] He begins field work around the age of fourteen. In the Greek village of Vasilika, boys also begin working around that age. At Tell Toqaan in Syria, boys are thirteen or fourteen when they begin helping their father, and the father can even afford an afternoon

---

[125] Stager, "The Archaeology of the Family," 24.

[126] In Western Iran, boys start helping their fathers in the fields at around seven years of age: Watson, *Archaeological Ethnography in Western Iran*, 209.

[127] Sibling dispute is the premise for Hesiod's *Works and Days*. His brother Perses took an unfair portion and then bribed the legal council to side with him. Hesiod takes the higher moral ground and produces a treatise on the wisdom of farm life. While he turned his agricultural misfortune to literary merit, such a route was not open to all cheated sons. Menecles and his brother fare better with their inheritance, but conflict nevertheless results long after their father dies: Hesiod *Works and Days* 3-65. Patrilineal conflict is evident in the Bible as well, e.g., Gen 25:30-34; 27; 35:22; 48:17-18; Judg 11:2. See also 2 Sam 16:21-22 and 1 Kgs 2:22, where sleeping with the king's women is a son's claim to his father's throne. In Luke 15:11-32, two sons get their inheritance while their father is alive. The elder works the farm and never disobeys his father's orders (v 29). He cannot help resenting his younger brother, the prodigal son, who returns having squandered his share of the inheritance.

[128] Aesop, *Aesop's Fables* (trans. T. James; London: John Murray, 1848), 146.

[129] Ammar, "The Social Organization of the Community," in *Readings in Arab Middle Eastern Societies and Cultures*, 120.

nap while the son plows.[130] In Karpofora, Greece, older children, beyond the age of fourteen, play a vital role in work and in wine production.[131] Karpofora children below that age are exempt from tasks save for errand running on the farm and in the village.[132]

If we can take these ethnographic parallels as suggestive, then an ancient Israelite male probably began substantial work in the vineyards and fields by around the age of fourteen, if not earlier. Sons are spied in various farming scenes in the Bible, e.g., Gen 34:5; 1 Sam 9:2; 16:11; 1 Kgs 19:19,[133] though their ages are not specified. These examples show sons, namely Saul, David, and Elisha, being elected for service away from the farm.[134] The only hint of age occurs in David's anointing, for he is almost bypassed by Samuel for being the youngest of Jesse's sons.

In the vineyards of a small farm, a son would help with pruning, harvesting, and wine production. These tasks would train him as a vintner for land he would one day inherit. He acquired skills, both general and idiosyncratic to his father's plots, to maximize one day his own family's economy. On Adamo and Maria's vineyard estate in Tuscany, sons, their wives, and their children help at the harvest. Primo, the aptly named eldest son, is intimately acquainted with his family's vineyard.[135] As death nears for Adamo, any last-minute tips can be transferred to Primo, so that his vineyard would thrive. In like manner, Elisha of Israelite Poraim (ostracon 1), could have taught his son, Pega Elisha, viticulture, farming, and (alas) something about taxes.

Because the grape harvest was often a festive occasion,[136] Israelite children may well have worked the vineyards before they were thirteen or fourteen. Children are recorded helping in modern vineyards during the grape harvest, in Palestine and Tuscany, for example.[137] They would also be prone to giggle and nibble at grapes,[138] so their presence in Israelite vineyards might have been

---

[130]L. Sweet, "A Day in a Peasant Household," in *Readings in Arab Middle Eastern Societies and Cultures* (ed. A. Lutfiyya and C. Churchill; The Hague: Mouton, 1970), 219.

[131]J. K. Campbell, "Karpofora: Reluctant Farmers on a Fertile Land," in *Regional Variation in Modern Greece and Cyprus: Toward a Perspective on the Ethnography of Greece* (ed. M. Dimen and E. Friedl; New York: New York Academy of Sciences, 1976), 211.

[132]Campbell, "Karpofora," 213.

[133]Cf. Luke 15:25.

[134]The first, Gen 34:5, shows Jacob's sons with the cattle in the field, unaware that Shechem has raped their sister.

[135]Calabresi, "*Vin Santo* and Wine," 123-32.

[136]Chapter 5, pp. 167-74.

[137]Turkowski, "Peasant Agriculture in the Judaean Hills," 27; Calabresi, "*Vin Santo* and Wine," 123, 129.

[138]Calabresi, "*Vin Santo* and Wine," 129.

delayed until an age when they offered a more sustained contribution
to the labor force. Perhaps not, though, for their levity would add to
the festivity of a family's grape harvest and wine production.[139]

After so many years of working his father's farm, as a child
and then a youth, a son would be well trained when his time came to
inherit. No glimpses of the son's presence in vineyards remain in
biblical tradition. Their importance in the patrimonial inheritance of
a vineyard is, however, evident in one biblical story, that of 1 Kgs 21.
The following section offers an analysis of how patrimonial
inheritance shapes the story of Naboth's vineyard.

a. Naboth's Vineyard

Patrimonial inheritance plays a central, deadly role in the story
of Naboth's vineyard (1 Kgs 21). A dispute erupts over this
Jezreelite[140] vineyard between a commoner and the crown, from which
none of the characters walks away unscathed. Indeed, the incident,
narrated by Dtr in a fairly virulent antimonarchic vein, is Melvillian
in its scope: Naboth dies early, right in the chapter; Ahab is sentenced
to die, earning a stay of execution only with his contrition; Jezebel is
later thrown through a second-story window by her own staff (2 Kgs
9:30-37); and their son, Joram, must die on the very ground that had
been Naboth's (2 Kgs 9:25-26). The Deuteronomist alone is left to
tell the tale and the exegetical task is to explore why he has done so in
this fashion. Throughout the narration of these fates "Naboth's
vineyard," כֶּרֶם נָבוֹת, is recalled as a chafing refrain. Naboth is named
fully nineteen times in this chapter, eight times after his death. Why
is this lone vintner given such a memory in Dtr? Why specifically is a
vineyard the occasion for dispute and then prophetic doom to King
Ahab?

Naboth's vineyard is, as Naboth himself defines it, "his
inheritance of my fathers," נַחֲלַת אֲבֹתַי. This becomes the sticking point
in negotiations that never do get very far. Ahab wants the vineyard
because it is next to his house, and Naboth cannot give it because it is
his inalienable inheritance. Its value as a vineyard, then, is not the
point of contention. Ahab does not want the plot of land because its

---

[139]Calabresi describes a touching seen of Adamo pressing grapes while his
granddaughter begs for a taste: Calabresi, "*Vin Santo* and Wine," 131.
[140]The Jezreel valley is one of the richest sections of the country, with a plain full of
rich alluvial soil, and hills which bring water. The name indicates its agricultural
esteem: "God will sow." See Baly, *Geography of the Bible*, 148; Orni and Efrat,
Geography of Israel, 85.

yield is starting to make a name for Naboth in Israelite wine circles. In fact, he intends to uproot the vines and make a vegetable garden:

וִיהִי־לִי לְגַן־יָרָק כִּי הוּא
קָרוֹב אֵצֶל בֵּיתִי

so that it will be my vegetable garden,
for it is near my house.

(1 Kgs 21:2)

Palace-side gardens are attested pleasures for ancient Near Eastern monarchs.[141] King Manasseh gets buried in one (2 Kgs 21:18); King Zedekiah has to flee right by his to leave a Jerusalem besieged and about to be destroyed (2 Kgs 25:4); King Hazael of Damascus has his destroyed by Shalmaneser III;[142] and Assurnasirpal II has exotic plants imported for his garden.[143] Palace gardens gave food for the royal table and lent prestige to the throne if stocked with plants from faraway places, as exotic plants signaled international trade.[144] Farther west, when Odysseus' travels take him to the Phaiakians, a large orchard abuts the palace of King Alkinoos. It is complete with fruit trees, grape vines, and a garden with rows of greens (*Odyssey* 7:112-132). Ahab may have been seeking through his home improvements to achieve a similar royal variety. The כִּי particle in v 2 introduces an explanatory clause, as an argument in support of Ahab's preceding stated wish for a garden.[145] Hence, for Ahab, location determines the land's function: since the plot is next to his house, it should be a garden.

Gardens sit next to homes, as well, in modern Middle Eastern villages, such as Karma in the West Bank, the Jebaliyah Bedouin settlements in the southern Sinai,[146] and in the Greek rural village of

---

[141]L. E. Stager, "Jerusalem and the Garden of Eden," *Eretz Israel* 26 (1999) (Frank Moore Cross volume), 183*-94*.

[142]D. D. Luckenbill, *Ancient Records of Assyria and Babylonia* (2 vols.; Chicago: University of Chicago Press, 1927) 1:672.

[143]Wiseman, "Mesopotamian Gardens," 137-44;, trans., "Babylonian and Assyrian Historical Texts," translated by A. L. Oppenheim (*ANET*, 280, 312).

[144]D. Stronach, "The Imagery of the Wine Bowl: Wine in Assyria in the Early First Millennium, B.C.," in *OAHW*, 175.

[145]Joüon, *A Grammar of Biblical Hebrew*, § 170da.

[146]Powell, *Food Resources and Food Systems in Two West Bank Villages*, 39-40; Perevolotsky, "Orchard Agriculture in the High Mountain Region of Southern Sinai," 331-58.

Ambeli.[147] In antiquity, the same practice occurred. Gardens added to the food regimen of farms, and, though they are not often mentioned in biblical texts, they were no doubt important facets of the Israelite farm.[148] For Herodotus, they were positive markers for people who were Greek-descended, namely the Geloni, who "cultivate the soil, eat grain, and keep gardens." [149] A tomb relief at Thebes, in Egypt, of an official named Ipui, shows a garden next to a house, with a man watering the plot.[150] Gardens were often the source of experimentation and more intensive cultivation, so they required frequent attention. In ancient Greece, the garden "was often situated close to the dwelling house, convenient to sources of fertilizers and attention; its water supply presumably being the household's source as well."[151]

From these parallels we can conclude that Ahab's desire for a garden next to his house is not what makes him a hated king. The desire is instead a fairly solid monarchic and agrarian impulse. Rather, the vineyard in question is caught between the competing interests of two men. Ahab wants the land for his own development; Naboth cannot give up an ancestral inheritance passed from previous generations to his father and down to him.

Ahab barters with Naboth for the vineyard, generously offering him either a superior vineyard, כֶּרֶם טוֹב מִמֶּנּוּ, or any amount of silver (v 2). Naboth's refusal is quick and absolute, almost cheeky, considering that he is speaking to his king:

חָלִילָה לִּי מֵיהוָֹה מִתִּתִּי אֶת־נַחֲלַת אֲבֹתַי לָךְ

> Yahweh forbid that I give you the
> inheritance of my fathers.
>
> (1 Kgs 21:3)

With this response, Naboth severs negotiations and redefines the desired commodity—his vineyard—as "the inheritance of my fathers." Ahab and Jezebel never do understand, or at least accept, this

---

[147]du Boulay, *Portrait of a Greek Mountain Village*, 31.

[148]Stager notes that houses and gardens would have lined terraces on the East Slope of Jerusalem: L. E. Stager, "The Archaeology of the East Slope of Jerusalem and the Terraces of the Kidron," *JNES* 41 (1982): 113; Hopkins, *Highlands of Canaan*, 243.

[149]Herodotus *Histories* 4.109.1.

[150]G. Steindorff and K. C. Steele, *When Egypt Ruled the East* (Chicago: University of Chicago Press, 1942), fig. 59.

[151]Burford, *Land and Labor in the Greek World*, 136.

conceptual shift. Throughout 1 Kgs 21, in vv 2, 6, 7, 15, 16, the
coveted land remains for them כֶּרֶם נָבוֹת.

b. חָלַל / חָלִילָה

חָלִילָה is an interjection often rendered "Far be it!" It expresses
a "negative wish or rejection."[152] With the preposition לְ, the formula
is rendered "Far be it from me" or even "God forbid me," and carries
the notion of "repulsion and aversion."[153] From the root חָלַל,
"profane," the formula connotes theological distaste as well.[154] For
Naboth, then, the proposal is preposterous, faintly blasphemous. He
adds further theological freight by mentioning Yahweh in the
clause.[155] His is a principled rebuff. It illustrates just how inalienable
he believed his land was.

The verbal root, חָלַל, "profane" or "pierce," from which this
interjection derives, in fact has a specific sense around vineyards in
Deut 20:6; 28:30; and Jer 31:5. Its coincidence with vineyards in
these three passages, and in Naboth's exclamation, therefore, merits
explanation. In the three passages, חָלַל denotes an action consequent
upon נָטַע כֶּרֶם in some form:

וּמִי־הָאִישׁ אֲשֶׁר־נָטַע כֶּרֶם וְלֹא חִלְּלוֹ יֵלֵךְ וְיָשֹׁב
לְבֵיתוֹ פֶּן־יָמוּת בַּמִּלְחָמָה וְאִישׁ אַחֵר יְחַלְּלֶנּוּ

Has anyone planted a vineyard,
but not profaned it? He should go

---

[152]Joüon, *A Grammar of Biblical Hebrew*, § 105f.
[153]Joüon, *A Grammar of Biblical Hebrew*, § 165k.
[154]F. Brown, S. R. Driver, and C. A. Briggs, *A Hebrew and English Lexicon of the Old Testament*, (Oxford: Clarendon 1907; hereafter BDB), 321; *ad profanum*! A negative oath, "Far be it from me!" B. K. Waltke and M. O'Connor, *An Introduction to Biblical Hebrew Syntax* (Winona Lake, Ind.: Eisenbrauns, 1990), 40.2.2c; *Gesenius' Hebrew Grammar* (ed. E. Kautzsch; trans. A. E. Cowley; 2d ed.; Oxford: Oxford University Press, 1910; hereafter GKC) § 105b: *min* + infinitive expresses an undesirable outcome. Here, the construction is *min* + Yahweh *min* + infinitive; cf. 1 Sam 24:7. For the interjection without the divine name, cf. 1 Sam 26:11; 1 Chr 11:19; Gen 44:7,17; Josh 24:16; 1 Sam 12:23; 2 Sam 20:20 (where Joab protests that he would swallow up the נַחֲלַת יְהוָה); Job 27:5. M. R. Lehmann suggests that לִי חָלִילָה is akin to Akkadian *elēlu/ullulu,* which he translates as "erase." Hence, he renders the formula "May I be erased if I do this": "Biblical Oaths," *ZAW* 81 (1969): 82. However, the Akkadian root meaning is "to make pure, clean" or "to make free," so his construction of the oath cannot work.
[155]Cf. 1 Sam 24:6; in 2 Sam 20:20, חָלִילָה occurs in response to a threat to another inheritance, נַחֲלַת יְהוָה.

and return to his house, lest he
die in the battle and another man
should profane it.

(Deut 20:6)

אִשָּׁה תְאָרֵשׂ וְאִישׁ אַחֵר יִשְׁגָּלֶנָּה בַּיִת תִּבְנֶה
וְלֹא־תֵשֵׁב בּוֹ כֶּרֶם תִּטַּע וְלֹא תְחַלְּלֶנּוּ

You will be betrothed to a woman,
but another man will sleep with her.[156]
You will build a house, but not live in it.
You will plant a vineyard, but not
profane it.

(Deut 28:30)

עוֹד תִּטְּעִי כְרָמִים בְּהָרֵי שֹׁמְרוֹן נָטְעוּ נֹטְעִים וְחִלֵּלוּ.

Again you shall plant vineyards
on the hills of Samaria;
the planters shall plant and shall profane.

(Jer 31:5)

חָלַל in these passages describes the activity of a farmer after he
has planted a vineyard. In Deut 20:6, the farmer has been mustered
for war right after he planted his vineyard and before he can
experience this חָלַל activity. In fact, the concern here is that the one
who planted the vineyard ought to be the one who would חָלַל it.
Many different tasks await the farmer after he plants his vineyard.
Transplanting, hoeing, pruning, press construction, harvest, treading,
guarding, and jarring the wine all follow the planting, so which task
or action is חָלַל meant to indicate in these passages?

In Deut 20:6 and 28:30 it is clear that חָלַל is some sort of
activity that is both important and anticipated after the planting of a
vineyard. In both texts, the adversative sense of וְלֹא sharpens the

___

[156]Reading with the *Qere* here, יִשְׁכָּבֶנָּה, "sleep with her," instead of the *Kethib*,
"ravish her." Cf. Isa 13:16; Zech 14:2.

threat that חָלַל might not follow נָטַע כֶּרֶם. In Deut 20:6, the fear is that a man (a survivor) other than the planter of the vineyard will חָלַל the vineyard, and in Deut 28:30, its loss is clearly a punishment for those who have planted vineyards. חָלַל itself is so vital to the individual farmer that it is grounds for exemption from military service, and its loss constitutes one of the blistering curses listed in Deut 28 for disobedience to Yahweh.

The consequences or rewards incumbent upon vineyard planting are somehow expressed in the verb חָלַל. In Deut 20:6, that reward is insured for the new vintner through military exemption. In 28:30, that same reward is denied the vintner and thus constitutes a curse. In both passages, then, חָלַל describes some anticipated or desired action, yet an English rendering of "profane" or "pierce" obscures this positive nuance. This sense is likewise obscured in the Jeremiah passage. Jeremiah 31:4 describes all the joys of restoration to the land after the exile. Out of his everlasting love (v 2) Yahweh will let the people return to the land where once again they will plant and "profane" their vineyards. Because the English rendering "profane" does obscure the positive sense in these passages, the NRSV translates all three passages with "enjoy its fruit" instead. This may well be the meaning, yet why חָלַל is associated with vineyards still remains an issue. Enjoying vineyard fruit is otherwise easily expressed. In 2 Kgs 19:29 (and Isa 37:30), for example, Isaiah gives Hezekiah the sign that

בַּשָּׁנָה הַשְּׁלִישִׁית זִרְעוּ וְקִצְרוּ וְנִטְעוּ כְרָמִים וְאִכְלוּ פִרְיָם

In the third year, they (the people) will sow,
harvest, and plant vineyards and eat fruit.
(2 Kgs 19:29)

And Isa 65:21 describes the joy of a transformed Jerusalem, where people will eat the fruit of their vineyards again:

וּבָנוּ בָתִּים וְיָשָׁבוּ וְנָטְעוּ כְרָמִים וְאָכְלוּ פִּרְיָם

They shall build houses and live (in them),
they shall plant vineyards, and eat (its) fruit.
(Isa 65:21)

Since אָכַל is used in these passages for the enjoyment of eating from one's own vineyard, the question remains, then, what meanings does חָלַל contribute in the passages cited above?

The contexts of the Deuteronomic passages suggest that חָלַל is an action consequent to having planted a vineyard.[157] The passages list three situations with undesired consequences. These are: betrothing a woman, but not lying with her; building a house, but not living in it; and planting a vineyard, but not "profaning" it. These situations bear at least two similarities. First, they entail initiation, the making use of something or someone for the first time. Newness is clear in the case of a wife with the verb אָרַשׂ, "betroth" (Deut 20:7; 28:30).[158] It is also clear in the case of house building, where those who have built a house have not yet lived in it (Deut 20:6; 28:30, and also Isa 65:21). Deuteronomy 20:5, in fact, sharpens the sense of newness with its use of חָנַךְ, "to dedicate" a house for the first time (see also 1 Kgs 8:63). In Deut 20:5-6, then, the Deuteronomist uses חָנַךְ and בָּנָה to describe the preparations of a new house. Here, and in Deut 28:30, the desired next step is "living" in that house, יָשַׁב. In the case of Deut 20:5, it is possible that the Deuteronomist notes the dedicating of a house to stress its newness. In like manner he may employ the rarer verb חָלַל in Deut 20:6 and Deut 28:30 in order to distinguish a vineyard's very first harvest from its subsequent annual yields. A vineyard, in other words, might be considered just as virginal as the new wife a farmer would enjoy. Enjoying its fruit for the first time is an anticipated pleasure and even human right, along with enjoying one's new house and wife. All three pleasures are grounds for military exemption and become targets of threat in punishment oracles.

Second, these three situations all required of the farmer effort and patience. Marriage negotiations and dowry arrangements typically had to be struck before a man and woman lived together.[159] Building a house took labor, planning, materials, and time, with the anticipation of getting to dwell in it. Planting a vineyard required devotion, time, and labor as well. With all three endeavors, then, labor and patience convert to anticipation, then enjoyment. The

---

[157]Jer 31:5 occurs in a list of different activities that resume in Jerusalem's restoration. Here, since restoration is greatly desired, חָלַל has to have a positive nuance.

[158]BDB, 77; See Exod 22:15 [ET 22:16], where the sense is most explicit. The case deals with a man who "opens a virgin," וְכִי־יְפַתֶּה אִישׁ בְּתוּלָה. Also, Deut 22:23, 28; Hos 2:21-22 [ET 2:19-20]. In 2 Sam 3:14, betrothal involves for David a grim bride-price of a hundred Philistine foreskins (1 Sam 18:27 has "two hundred").

[159]E.g., Gen 24; Gen 34, where Shechem lacks the customary patience and rapes Dinah; Exod 2:21; Judg 14:1-10; and 1 Sam 18:20-25, the aforementioned foreskin betrothal. For the biblical case *in extremis*, see Gen 28:1-5; 29:9-30 in which Jacob serves fourteen years for his Rachel.

Deuteronomic descriptions of these endeavors pair the labor with its anticipated fruit, and in the case of the vineyard, literally so. These descriptions suggest that houses, wives, and grape vines were valued and enjoyed, along with the possible daily irritations they brought. It is evident from the contexts that חָלַל describes a desired action, delayed through time and effort.

In Jer 31:5, the context differs from the two Deuteronomic texts. It is part of a list of joyful activities to be resumed in Jerusalem's restoration. These include music, dancing, profaning vineyards, and pilgrimages to Zion. Though the situation is different for this passage, patience and anticipation are nevertheless involved, though now on a national scale. Building and planting figure in all three passages, though in Jer 31:4 Yahweh is the builder. Jeremiah's call involves as much building and planting as it does tearing down and plucking up. Building, planting, and, for the Deuteronomic passages, betrothing, are investments in livelihood and prospects of stability.

חָלַל in the Piel as "profane" is overwhelmingly a negative act in the Bible, as it usually is directed towards Yahweh's name (e.g., Lev 18:21; Ezek 20:39), his sanctuary (e.g., Lev 21:12; Mal 2:11), and the Sabbath (e.g., Exod 31:14; Isa 56:2). It makes of the holy something common. The negative nuance is clear when the above three sacred spheres are subject to profanation, but what is conveyed when חָלַל is used in reference to a vineyard? S. R. Driver suggested that the term indicates that first fruits were reserved as sacred. And so, for him, the second harvest is the first one that a farmer could enjoy. The farmer "profanes" the vineyard precisely by initiating the crop for human, not religious, use (cf. Lev 19:23-25).[160] These texts may retain a hint of such an ancient practice, where the first fruits were considered Yahweh's, just as the firstborn of animals and humans were (Exod 23:16,19; Lev 23:9-14; Deut 26:2-11; see also Exod 13:1-16). However, the calendar texts depict an annual offering of first fruits, so that it seems that the first fruits of a yearly harvest were offered to Yahweh and not only the first fruits ever from a vineyard. I agree with Driver that חָלַל signals some sort of sacred meaning for harvesting a vineyard, but do not believe it can be pinpointed to a ritual of a first-time harvest offered to Yahweh. The stipulations regarding first fruits in the calendar texts (Exod 23:16, 19; Lev 23:9-14; Deut 26:2-11) describe just the annual offerings and nothing about the relatively rare occasion of a new vineyard. Exod 23:10-11 and Lev 25:2-7 offer stipulations regarding the sabbatical year, when the land was to remain fallow every seventh year, but they

---

[160]S. R. Driver, *A Critical and Exegetical Commentary on Deuteronomy* (3rd ed.; International Critical Commentary 3; Edinburgh: T. & T. Clark, 1902), 238.

do not include any other situations—such as the harvesting of a new vineyard—that were also not annual. Instead, I contend that the use of חלל embues the vineyard with a sacred value. It was a way to capture the esteem with which the vineyard was held, so that harvesting for the first time is conveyed to be a protected act. What is so protected in the Deuteronomic passages and in the Naboth story is who gets to enjoy the vineyard, namely, the one to whom Yahweh gave the land. The vintner enjoys his vineyard as a proxy to Yahweh, so that the loss of that right or the substitution of another is seen as offensive, as "profaning" the rightful owner.

In this reading, חלל captures the delayed enjoyment of the farmer to taste the second harvest from his new vineyard. The pleasure of a planted vineyard becomes both enjoying its fruit and the stability of agricultural years with two or more harvests.[161] Arguably, since it takes three to four years for new vines to develop, with the first harvest going to Yahweh, the military exemption in Deut 20:6 is as good as a college deferment was for the American draft.

Practically, though, this law, if it ever were enacted, could not have cut substantially into the Israelite militia. Most vineyards were inherited and then expanded by planting shoots off preexisting vines,[162] so that they remained ever old and renewed at the same time. A casuistic pacifist could parlay this military exemption into decades if he wanted to try, since some of his vines would always be new. A preferential deferment might be indicated in Deut 20:6, whereby those farmers just starting out were exempt from initial muster. Vineyards require ongoing care and attention in pruning and hoeing throughout the year, but so too do fruit trees in general. Hence, the preferential deferment for new vintners in Deut 20:6 cannot be based solely on the unique labor responsibilities they were facing. There must be additional reasons for why a new vineyard is cause enough to grant the farmer an exemption from battle.

The initial vineyards in these three passages are subject to profanation. No other crop is cause for military exemption or enjoyed through profaning. Grain, oil, and sheep are farm products subject to tithing along with wine in Deuteronomy, yet none takes verbs distinguishing between their cultic and common use. The other pleasures listed in the contexts of Deut 20:6 and 28:30, those of building a house and betrothing a woman, do not entail profaning as the act of their initiation. This alone is said of vineyards. The very act of "enjoying its fruit" requires an initiation, a breach, really, that for the Deuteronomist is unique to vineyards.

---

[161]Chapter 3, pp. 117-18.
[162]Chapter 3, pp. 100-101.

When Naboth refuses with the expression חָלִילָה לִי to give his vineyard, a Deuteronomistic trope of profaning vineyards is sounded again. "To profane" in these texts is not so much a trace of an early cultic practice of offering the very first yield of a vineyard to Yahweh, as Driver maintains. Instead, it is an active possibility in the present scene. Ahab is virtually asking to profane Naboth's vineyard, and Naboth is the only one to whom Yahweh has given that right. He, Naboth, in effect, "profanes" by right, by enjoying the esteemed pleasure granted by Yahweh of eating from one's own vineyard, while Ahab would "profane" in a grave sense by trying to purchase it. חָלַל, in other words, acts as a kind of fence or protection for the rights to enjoying a vineyard, so that any intruder or usurper virtually desecrates the right of a vintner to his vineyard yield.

חָלִילָה as an interjection of rejection is, then, for the Deuteronomist, an apt expression of Naboth's Israelite convictions on patrimonial inheritance. It is his way to convey that Ahab's request is blasphemous, because land—any land, but always a vineyard—is granted by Yahweh. The expression is rather effective too, as the king of Israel retreats home to bed, sullen. In recalling this trope of profaning vineyards, חָלִילָה also lends a theological freight not clear when חָלַל in Deut 20:6; 28:30; and Jer 31:5 is rendered into English as "enjoy its fruit."[163] In this Deuteronomistic tale, Ahab, of course, cannot own the כֶּרֶם נָבוֹת but neither is it really Naboth's. The כֶּרֶם is really, and always has been, Yahweh's, bestowed through the generations of Naboth's ancestors to him. The vineyard is essentially on loan from the deity. As such, it is exempt from exchange or sale. Custom and theology combine to thwart Ahab's designs.

Naboth cannot possibly give his vineyard: it is from his fathers and "from Yahweh," מֵיהוָה (v 3). It is "profane" to exchange or sell his vineyard. Indeed, in Deuteronomic theology (Deut 20:6; 28:30; and Jer 31:5), it is profane simply to enjoy a vineyard. And so, it is a protected and sanctioned profanation only for the rightful farmer of the vineyard. "Far be it" really for either man to enjoy the vineyard apart "from Yahweh."

The perspective that the land and all its fruits are Yahweh's is characteristic of Deuteronomy. All farm plots in Deuteronomic land theology are Yahweh's, and the vineyard especially so. The use of חָלַל / חָלִילָה in these four passages speaks worlds of the esteem with which vineyards were held in the Deuteronomistic theology. Of all the plots given on loan from Yahweh, the vineyard required special care, a tending that reached "profaning." On the pedestrian plane, such a theologically infused sentiment may well reflect the value

---

[163]NRSV.

placed on vineyards in Israelite society. Here in 1 Kgs 21 a vineyard,
namely Naboth's, makes the Deuteronomistic case study for customary
patrimony. It is, for the Deuteronomistic historian, a theologically
bolstered custom. Naboth's patrimony extended back to his ancestors
and his God.

c. Ahab's Crime

Ahab's request for Naboth's vineyard was an affront against
agrarian custom and Yahweh. In some ways the incident narrated in 1
Kgs 21 illustrates the Deuteronomistic skepticism over the Israelite
monarchy, whose *locus classicus* is 1 Sam 8:11-18. There the prophet
Samuel warns of a king:

> Your best fields, vineyards, and olive groves,
> he will take and give to his servants.
>
> (1 Sam 8:14)

Yet 1 Kgs 21 does not simply actualize this fear of royal confiscation
of land, for it begins with King Ahab bothering to barter for the
vineyard. Ahab is fairly polite with Naboth. Covetous kings of the
Bible are not always so gracious.[164] Ahab offers a better vineyard or
silver in an amount Naboth has only to name. He differs in this
respect from his father, Omri, who bought the Shemer estate (1 Kgs
16:24), and David, who purchased Araunah's threshing floor (2 Sam
24:18-24; 1 Chr 21:18-25). In those instances, both kings set their
price. Here, Ahab generously offers to pay whatever Naboth has in
mind. His openness to negotiation honors a characteristically Middle
Eastern custom of bartering, where the two participants arrive at
mutually agreed upon conditions for their transaction.[165] Next, he
rather conversationally tells Naboth of his plans for a palace-side
garden (v 2). When Naboth refuses, Ahab goes home sulking,
incapacitated with disappointment.[166]

---

[164]Gen 12:15; 20:2; 1 Sam 25:8 where David acts with entitlement even before he is
king; 2 Sam 11:4; cf. 2 Sam 24:24 and 1 Kgs 16:24 for contrasting portrayals of
kings purchasing lands.

[165]D. F. Eickelman, *The Middle East: An Anthropological Approach* (Englewood
Cliffs, N.J.: Prentice-Hall, 1981), 187.

[166]Bedridden, refusing to eat or look up, Ahab's behavior is one of the few biblical
portrayals of depression. Ahab's condition may be persistent; see v 27. Elijah may
suffer a bout of depression in his wilderness collapse in 1 Kgs 19:4-7.

While the Deuteronomistic historian despises all the Northern kings, the evil Ahab is, for him, in a league all his own:

> Indeed, there was no one like Ahab,
> who sold himself to do what was evil
> in the sight of Yahweh, urged on by his
> wife, Jezebel.
>
> (1 Kgs 21:25)

Still, this Deuteronomistic portrait of Ahab is of no fearsome monarch, brashly usurping custom by lopping off the best of farmers' fields, vineyards, and olive groves. That is, Ahab is not presented as the embodiment of all the fears expressed in 1 Sam 8:11-18. While this king has the intent so feared in Samuel's warning, in that he covets commoner land, he lacks the wherewithal to execute it. Dtr's portrait is, in the end, more damning than a fulfillment of all of Samuel's fears of monarchy, for it drips with ridicule. Ahab's paralysis contrasts pointedly with the decisiveness of both Jezebel and Naboth. Before both people, the king cannot wield the power that is his. The narrative showcases Ahab's own impotence for despotic rule with his wife's penchant for it. She is the one who secures Naboth's vineyard. Ahab becomes the king who *would* have taken a man's vineyard, had he his wife's strength. 1 Kgs 21 is no simple illustration of how well Samuel prophesied the tyranny of kings—queens maybe, but not Ahab.

We saw that Ahab did not by himself, that is, in vv 1-2, intend to usurp Naboth's vineyard. Still, he infringed on Israelite custom simply by approaching Naboth, since land was traditionally inalienable (Lev 25:23). Though Naboth rebuffs him rather handily, Ahab's act is not unconscionable. The Bible has instances of land sale, so it is not this negotiation itself that earns Ahab most hated monarch status in Dtr's account.

Land is traded or sold in the Bible in extreme cases, and traditions of redemption suggest that exchange outside of the family did occur in Israelite society. Ancient Greece shared the same principle of inalienability, yet nevertheless had legal sanctions for the trade and exchange of lands, should that become necessary.[167] Selling one's land was likely not met with enthusiasm, but was perhaps unavoidable under conditions of duress.

---

[167]Burford, *Land and Labor in the Greek World*, 29-55.

In the two narrative examples of selling land, Ruth 4:1-12 and Jer 32:6-15, the transactions involve distant relatives.[168] In this way, patrimony remained in the family. In Jer 32:6; Num 27:1-11; and Num 36:7 (the latter two, of Zelophehad), the father's brother(s), that is, uncles, figure into the inheritance first after sons. Hence, in Zelophehad's case, where there were no sons, the father's brothers inherit. His daughters then petition Moses to be included with their uncles in the inheritance. Jeremiah is asked to redeem the land of his cousin, that is, the son of his father's brother. In patrilineal societies, the father's brother plays an important role in inheritance and marriage. The daughter of a father's brother[169] is often the first choice for a son's marriage. Levirate marriage taps the husband's brother first and then extends to the nearest relative (Gen 38:3; Deut 25:5-6; Ruth 4:1-12). These marriage arrangements helped to keep the patrimony in the same family. We infer from these biblical examples that the father's brother was next in line if patrimonial inheritance could not follow through the son for some reason.

In Naboth's case, no family is mentioned in 1 Kgs 21, except for the fathers he so proudly proclaims to Ahab (v 3). Naboth is not under duress to sell his land. In fact, with no textual evidence to the contrary, we may infer from the detail that he was a Jezreelite, i.e., in fertile agricultural territory, so his vineyard was not in economic trouble. Perhaps Ahab violated custom by approaching the landowner first, instead of waiting for Naboth to come to him. It remains an irony that Ahab's name, אַחְאָב, means "father-brother" and that his initial intent had been to barter like an uncle, not usurp. In Jer 32:6, Hanamel, the son of Jeremiah's uncle, Shallum, came to ask that Jeremiah buy the family land.[170] Ahab wanted to buy the vineyard or exchange for another, better, vineyard. It is Jezebel who wants to usurp the plot as a king would. Nothing in Ahab's behavior up to now, though, differentiates him from an uncle or nearest relation. Like a "father-brother" rather than a monarch, Ahab barters with Naboth and honors Naboth's unwillingness to make a deal. As Dtr's foil for an evil king, Ahab's own name, if nothing else and if sheer coincidence, belies the king's obedience to the Israelite custom of land transaction through family. On the narrative plane, the coincidence undercuts Dtr's otherwise belittling portrait of a man who, in this scene, might have been expected to act more like a king. For Ahab

---

[168]For a possible third example in the case of Omri's purchase of Samaria (1 Kgs 16:24), see L. E. Stager, "Shemer's Estate," *BASOR* 277/278 (1990), 93-107.

[169]For a clear discussion of this custom, see Eickelman, *The Middle East: An Anthropological Approach*, 116, 129-31.

[170]In Ruth 4:1-12, the next-of-kin is not identified precisely, though he is in Elimelech's family and is the first with rights to redeem the family land. Once he declines, Boaz is the next eligible relation and opts to purchase the land and Ruth.

acts respectfully, deferring to Naboth and the custom he cites. Had Ahab had the despotic backbone, he could have seized Naboth's vineyard without ever having spoken to the little Jezreelite vintner.

By varying the theme of monarchic misuse of land in 1 Kgs 21, Dtr "kills two birds": he mocks the king bedridden by a commoner's rejection, and saddles the foreign, Baal-worshipping queen, Jezebel, with his contempt. Ahab's predilection for evil is, we recall, tied up with his choice of partner: "urged on by Jezebel, his wife." Jezebel is the engine force that secures the vineyard for the throne. What precisely was her role in relation to Naboth's vineyard?

Jezebel finds Ahab in bed, rebuffed by Naboth, and is not pleased with her husband's melancholy. Indeed, she cuts to the chase asking, rhetorically or genuinely, "Do you now govern Israel?" (v 7). She sees no problem in what has incapacitated Ahab, or at least sees the issue as a problem she can solve, and sets out to do so. As far as she knows, this problem hinges on a vineyard. Ahab has not recounted to her Naboth's reason for not selling, i.e., that it is more than a vineyard: it is his inheritance. Not beset by Ahab's own sense of gloomy defeat, Jezebel goes down in interpretive history as the hated, proactive woman. Dtr stresses the contrast between the coping strategies of the two royals. She wastes no time stewing over the dilemma, as Ahab does, and sets out to do what Naboth cannot, namely "give" her husband Naboth's vineyard (v 7).

Jezebel's solution to give the vineyard is ruthless. Her plan is to have Naboth killed for slander against the king and Yahweh, but what agrarian principles might underlie her strategy? Her crime in 1 Kgs 21 against Israelite custom is threefold: false accusation, murder, and usurping land from another. The first two are proscribed in the Ten Commandments, the third as well, as she covets her neighbor's vineyard (Exod 20:16, 13, 17; Deut 5:20, 17, 21, respectively). Often, her foreign status or downright meanness is offered as an explanation for this triple play of indignity, suffered first and foremost by Naboth, but also more generally by all Israelites.

Naboth's gentilic "the Jezreelite" (1 Kgs 21:1, 4, 6, 7, 15, 16) stands as homonymic reminder for what is really at stake in the incident: this royal couple violates Israelite custom. "Jezreel" does double duty in the narrative as historical breadbasket of the North and as homonym for national, "Israel," identity. All Jezreelites, as all Israelites, are undermined by such monarchy. Naboth stands in for them all, from the richest agricultural valley in the North. He is in some sense "agricultural man," more specifically located than was Adam before him. Since agriculture was Israel's lifeline, the fate of these characters does indeed matter. Naboth is sacrificed to the throne and Dtr, in essence, wants all Israelites to take note.

I have argued above that Israelite patrimony stayed within the family and that this was especially so for vineyard estates. The family in Israel, nuclear and by generations, was economically quite closely tied to their land. Subsistence, the survival of a family, was tied to its land inextricably as a general feature of peasant economies. Jezebel, the Phoenician, would seem to intuit that link immediately. She knows that Naboth and his vineyard are tied together, that they involve one another. She knows, further, that securing one means sacrificing the other.[171] Jezebel's solution, albeit ruthless, respects the Israelite custom of land ownership. She knows that she cannot just take the vineyard out from under Naboth, that to do so effectively kills Naboth, since his livelihood would be gone. She understands that she can have the vineyard only if Naboth is killed. She then uses Israelite law and custom to have Naboth framed for an Israelite offence—cursing God (Exod 20:7; Deut 5:11)[172]—and uses the requisite number of witnesses required by Israelite law to convict him (two or three, see Deut 17:6; 19:15).[173] Far from not knowing Israelite law and custom, Jezebel has worked them with remarkable efficiency. According to Israelite law, she is herself guilty of false accusation and murder. But these are her creative, brutal solutions to counter the Israelite tradition of land inalienability. She has not ignored that convention with despotic indifference at all; she has responded to it with deadly earnest.

d. The Blood of Sons

Naboth's vineyard, the "inheritance of my fathers," נַחֲלַת אֲבֹתָי, becomes the scene for Dtr's satire and condemnation of King Ahab. It is literally the scene as well, as Yahweh tells Elijah where Ahab can be found:

בְּכֶרֶם נָבוֹת אֲשֶׁר־יָרַד שָׁם לְרִשְׁתּוֹ

in Naboth's vineyard, which
he went down to possess.
(1 Kgs 21:18)

Dtr underscores the persistence of Israelite customary inheritance and further mocks Ahab by referring to the land

---

[171]Perhaps Ahab does too, and this explains his paralysis.
[172]He is accused of cursing the king too, so that he is framed for both political and religious offences.
[173]I am indebted to Peter Machinist for this insight.

throughout as Naboth's, not Ahab's. This he does in the narrative of
the incident (1 Kgs 21) and in Jehu's subsequent recollection (2 Kgs
9:21-26). Though a generation has gone by from the incident at the
time of Jehu, he still refers to the plot of land as Naboth's, three times
in six verses (2 Kgs 9:21, 25, 26). Even Jezebel, who could flout
Israelite customs, tells Ahab to go take possession not of *his*, but of
Naboth's vineyard, כֶּרֶם נָבוֹת (1 Kgs 21:15).

Both Jezebel and Yahweh explain Ahab's presence in the
vineyard with a phrase of movement there to go and take possession,
יָרַד שָׁם לְרִשְׁתּוֹ (vv 18, 15). Perhaps confiscation of land entailed some
ceremonial gesture requiring Ahab's presence before the כֶּרֶם נָבוֹת
could become a גַּן אַחְאָב. If so, the transfer of title never does go
through for poor Ahab. Naboth, Elijah, and Yahweh remain firm on
this by continuing to use Naboth's name with the vineyard.   A
generation later, the plot of land is still known as Naboth's (2 Kgs
9:21, 25, 26). This נַחֲלַת אֲבֹתַי was not for exchange or sale, nor can it
shake Naboth's memory, once confiscated.

In 2 Kgs 9:21, 26, the כֶּרֶם נָבוֹת is חֶלְקַת נָבוֹת, a different term
denoting land apportionment, stressing again Naboth's ownership. By
2 Kgs 9, it is no longer termed a "vineyard," but a "land plot." At
this point, perhaps the vines themselves of the חֶלְקַת נָבוֹת had been
stripped for garden preparation, or suffered neglect as they stood
intestate for a generation. At any rate, by 2 Kgs 9:21-26, there is no
reference to the land having been a vineyard, a heady economic waste
in its own right, though this is not what concerns Dtr.

Naboth's rightful, enduring inheritance is trumpeted by Dtr in
this later tale of 2 Kgs 9. Just as Naboth's vineyard came from his
father to him, Ahab's crime follows, and will finish off, his son
Joram.   The patrilineal inheritance of Naboth now dogs its
transgressor and his son. What should have gone to Naboth's sons,
namely the vineyard, yields neither grapes nor vegetables, nor even
ownership for Ahab and his son, Joram. It yields only trouble.

Jehu recalls Naboth's blood and that of his sons (2 Kgs 9:26) on
the ground of Naboth's plot. Blood imagery is used throughout the
Naboth story (1 Kgs 21:19; 22:35, 38; 2 Kgs 9:26, 38), and Ahab,
Jezebel, and Joram all die with their blood surrounding them. While
Ahab dies in battle, filling his chariot with blood (1 Kgs 22:35), and
Jezebel is thrown from a window, splattering her blood on the wall
and horses (2 Kgs 9:33), Joram's death on Naboth's plot (2 Kgs 9:25-
26) recalls not only his parents' crime, but also Abel's primeval
murder, where bloods seep into the earth and cry out to God (Gen
4:10).

Cain, a (less appreciated) farmer, had committed murder and
filled the land plot with blood. Here, Naboth, the vintner, is killed for

his land and his blood lingers, so that no one seems to cultivate it
again. The description of blood with soil in these two biblical scenes
is a graphic way to portray murder. It reflects as well the farmer's
inextricable tie to his land. In Cain's case, his farming days are over
as he is forced to wander the rest of his life. The blood-soaked
ground will no longer yield to him, and he is banished from it (Gen
4:11-12). In Naboth's case, Ahab turned a vineyard not into a garden
at all, but a blood plot, of the Naboth family and of Joram, his son.
Murdering Naboth wrecked his plot of land for use. All the royal
couple really confiscated, then, was their son's own burial spot.
Joram inherits his father's punishment and receives the disputed plot
only as his burial tarp, as he is thrown on its surface (2 Kgs 9:26).
Joram pays with his life for this thwarted, yet insistent "inheritance."
As verse 26 indicates, he was not the only son affected by this
incident.

Naboth's nuclear family is mentioned for the first and only time
later in Kings:

> For the blood of Naboth and for the blood of
> his sons that I saw yesterday, says the Lord,
> I swear I will repay you on this very plot of
> ground.
>
> (2 Kgs 9:26)

Naboth's sons are remembered by Jehu, along with Naboth. In 1 Kgs
21:19 Yahweh, through Elijah, speaks only of Naboth's blood. But in
2 Kgs 9:26, another generation is affected, and Yahweh's oracle
includes now Naboth and his sons. They presumably have come of
age and are without their patrimonial inheritance. They were,
according to Dtr, cut off along with Naboth, their blood recalled by
Jehu. Verse 26 may recall the literal death of Naboth's sons, by
starvation, or their figurative death, i.e., nonexistence at the vineyard,
by having to leave their home to seek livelihood elsewhere. They
became, in the latter case, *personae non gratae* in the village or
region, and perhaps sought a means of living through itinerant trade,
the military, or the cult.[174] In either case, they were cut out of their
agrarian community, and not remembered except by Yahweh.

What did Yahweh "see yesterday" (2 Kgs 9:26), the blood of
Naboth and of his sons, or Naboth's sons themselves? Did their
presence remind him again of the Naboth incident? That Dtr viewed
this erasure of Naboth as awful is clear by his insistence on Naboth's

---

[174] Stager, "Archaeology of the Family in Ancient Israel," 1-35.

name throughout the story, and his stress on Jezebel's fate. Her corpse is flung on the ground like Joram's, but the reason varies: "so that no one can say, this is Jezebel" (2 Kgs 9:37). In other words, her name vanishes with the corpse, as the names of Naboth's sons vanish with the stolen inheritance.

In Naboth's case, inheritance of a vineyard stringently went through generations from father to son. Fully three generations were affected by this royal confiscation: "my fathers" whom Naboth recalls in protest of Ahab's initial offer (1 Kgs 21:3); himself as he is killed (v 13); and his sons (2 Kgs 9:26). In effect, of course, many more generations were affected than these, for "fathers" means all the ancestors before Naboth, and, since his sons were killed, all the subsequent generations in his lineage remain unborn. Vineyards, for the Deuteronomist, clearly were the domain of the Israelite family. The narrative skill is evident in this story of a king and a vintner. Much theological and social critique is carried by this one man's vineyard. The value of the vineyard to ancient Israel is apparent in the story, and family ownership is assumed. What precisely was involved in cultivating a vineyard is the subject of the next chapter.

Chapter 3

The Cultivation of Grape Vines[1]

I. Starting a Vineyard

1 אָשִׁ֤ירָה נָּא֙ לִֽידִידִ֔י שִׁירַ֥ת דּוֹדִ֖י לְכַרְמ֑וֹ
כֶּ֛רֶם הָיָ֥ה לִֽידִידִ֖י בְּקֶ֥רֶן בֶּן־שָֽׁמֶן

2 וַֽיְעַזְּקֵ֣הוּ וַֽיְסַקְּלֵ֗הוּ וַיִּטָּעֵ֙הוּ֙ שֹׂרֵ֔ק
וַיִּ֤בֶן מִגְדָּל֙ בְּתוֹכ֔וֹ וְגַם־יֶ֖קֶב חָצֵ֣ב בּ֑וֹ
וַיְקַ֛ו לַעֲשׂ֥וֹת עֲנָבִ֖ים וַיַּ֥עַשׂ בְּאֻשִֽׁים

3 וְעַתָּ֛ה יוֹשֵׁ֥ב יְרוּשָׁלַ֖͏ִם וְאִ֣ישׁ יְהוּדָ֑ה
שִׁפְטוּ־נָ֕א בֵּינִ֖י וּבֵ֥ין כַּרְמִֽי

4 מַה־לַּעֲשׂ֥וֹת עוֹד֙ לְכַרְמִ֔י וְלֹ֥א עָשִׂ֖יתִי בּ֑וֹ
מַדּ֧וּעַ קִוֵּ֛יתִי לַעֲשׂ֥וֹת עֲנָבִ֖ים וַיַּ֥עַשׂ בְּאֻשִֽׁים

5 וְעַתָּה֙ אוֹדִֽיעָה־נָּ֣א אֶתְכֶ֔ם אֵ֛ת אֲשֶׁר־אֲנִ֥י עֹשֶׂ֖ה לְכַרְמִ֑י
הָסֵ֤ר מְשׂוּכָּתוֹ֙ וְהָיָ֣ה לְבָעֵ֔ר פָּרֹ֥ץ גְּדֵר֖וֹ וְהָיָ֥ה לְמִרְמָֽס

6 וַאֲשִׁיתֵ֣הוּ בָתָ֗ה לֹ֤א יִזָּמֵר֙ וְלֹ֣א יֵעָדֵ֔ר וְעָלָ֥ה שָׁמִ֖יר וָשָׁ֑יִת
וְעַ֣ל הֶעָבִ֗ים אֲצַוֶּ֔ה מֵהַמְטִ֥יר עָלָ֖יו מָטָֽר

7 כִּ֣י כֶ֜רֶם יְהוָ֤ה צְבָאוֹת֙ בֵּ֣ית יִשְׂרָאֵ֔ל
וְאִ֥ישׁ יְהוּדָ֖ה נְטַ֣ע שַׁעֲשׁוּעָ֑יו
וַיְקַ֤ו לְמִשְׁפָּט֙ וְהִנֵּ֣ה מִשְׂפָּ֔ח לִצְדָקָ֖ה וְהִנֵּ֥ה צְעָקָֽה

1. Let me sing of my beloved, a song of my beloved
Of his vineyard.
My beloved had a vineyard on a valued hill

---

[1]The field of ampelography, from the Greek *ampelos* "vine," is the science concerned with the identification, description, classification, and behavior of grape-bearing vines: de Blij, *Wine: A Geographic Appreciation*, 9.

2. He worked the soil and cleared it of stones.
He planted rare vines.
He built a tower in its midst and also
hewed a winepress in it.
He waited eagerly for it to yield grapes,
but it yielded fetid fruit.
3. And now dwellers of Jerusalem, people of Judah,
Judge, please, between me and my vineyard.
4. What more was there to do for my vineyard
that I did not do in it?
Why did I wait eagerly for it to yield grapes,
when it yielded fetid fruit?
5. Now, I will tell you what I will do to my vineyard.
I will remove its hedge and it will be consumed;
I will breach its wall, and it will be trampled down.
6. I will lay it waste; it will not be pruned or hoed,
briers and thorns will sprout up.
And to the clouds, I will command that they
rain no rain on it.
7. For the vineyard of the lord of hosts is the
house of Israel, and the people of Judah
are his pleasant planting;
he waited eagerly for judgment, but saw bloodshed;
for righteousness, but heard a cry!

(Isa 5:1-7)

This is Isaiah's Song of the Vineyard (Isa 5:1-7), the most detailed description of vineyard maintenance in the Bible. The poem is a sustained metaphor for the deity's care for the people, where Yahweh is portrayed as the meticulous, attentive vintner, the people are the vineyard, and their fruit is injustice (vv 2, 4). That this discrepancy is severe, costly, and even heartbreaking is clear from the prophet's repetition of the expectation and its reality in vv 2 and 4. The resultant produce after the list of manual labors in v 2 is not the expected עֲנָבִים, "grapes," at all, but only בְּאֻשִׁים, "fetid fruit" (vv 2, 4).[2] בְּאֻשִׁים is a *hapax legomenon*, but its root, בָּאַשׁ, is used elsewhere

---

[2]The Vulgate renders בְּאֻשִׁים *labruscae*, "wild vines": G. B. Gray and A. S. Peake, *A Critical and Exegetical Commentary on the Book of Isaiah* (International Critical Commentary 18; Edinburgh: T. & T. Clark, 1912), 86; O. Kaiser, *Isaiah 1-12: A Commentary* (Old Testament Library; Philadelphia: Westminster, 1972), 60. The precise classification of the uncultivated grape vine is, however, *Vitis vinifera silvestris*. Pliny notes a distinction between the *labrusca*, which the Greeks and Romans harvested before the grapes were ripened, and the *silvestris* which is more properly the wild vine, since it was not harvested or worked by humans. The *labrusca* of antiquity yielded a small, bitter grape and was uncommon in the Levant

in texts with the meaning to "have a bad smell" or "stink."[3]   So, the stricken Nile smells during the first plague (Exod 7:21), and the leftover manna in the wilderness (Exod 16:20) smells of rotting food.[4] In Isaiah's Song of the Vineyard, the people of Judah are the vineyard with its stinking fruit, and Yahweh is the disappointed vintner (v 7).[5]

---

and rare in Italy. It was viewed as wild because the plant itself was not cultivated, that is, planted, fertilized, or pruned, though its grapes were sometimes used by Greek and Roman vintners. Pliny terms the *labrusca* vine the "claret vine": cf. Pliny *Natural History* 12.130; 14.9, 98; 16.208; 23.19. This is not the same vine as the prominent North American species of the same name, *Vitis vinifera labrusca*. Ancient Roman vintners, such as Pliny, knew the difference between *Vitis vinifera* ("wine growing") and *labrusca* and would not likely have confused the two as Jerome has.   Neither would an Israelite vintner. Jerome's rendering is a mistranslation at Yahweh's expense, for it meant that Yahweh had planted the wrong vines. Much, if not all, of the prophetic sting is lost if Yahweh planted one kind of vines and then expected them to yield some other kind of grape. He becomes an ignorant or sadistic vintner, an unlikely subtheme to Isaiah's Yahwistic theology. These are not "sour grapes" either (e.g. M. Sweeney, *Isaiah 1-39* [The Forms of Old Testament Literature 16; Grand Rapids, Mich.: Eerdmans, 1996], 124) for similar reasons; then, Yahweh would be depicted as a vintner who harvested prematurely and then got angry at the bitter fruit. The grapes here had a chance to ripen.
[3]BDB, 92-93.
[4]בְּאֻשִׁים is used figuratively as well of people who have become offensive, such as when Absalom slept with his father's concubines (2 Sam 16:21), and Simeon and Levi reflected poorly on their father, Jacob (Gen 34:30). Ammonites and Israelites are described as odious to their enemies (1 Sam 13:4, 2 Sam 10:6). Presumably, offensive odor elicits the desire for distance.   In other words, odor engenders distance between the related parties; one wants distance from an unpleasant odor as one does from enemies or immoral sons. Isaiah could well be invoking this nuance in Isa 5:2, 4 also by suggesting that Yahweh wants nothing to do with the fetid fruit.
[5]During wine production itself, a strong pungency results from fermentation. Inside many of the vat rooms I visited in the Napa Valley of California in the fall of 1994, the odor was overwhelming and nauseating, because of the size of production and enclosed spaces. Even in antiquity, fumes from fermentation could overwhelm the laborers. A New Kingdom tomb painting from Beni Hasan shows two laborers being carried away: Unwin, *Wine and the Vine*, 72. Whether they were overcome by fumes or from tasting the wine they had just trod remains open to conjecture. In ancient Rome, presses were often indoors, and must, at times, have given off pungent odors during fermentation. The *villa rustica* two miles north of Pompeii in southern Italy, near Boscoreale, for example, had 83 vats for fermentation in one large room: A. Mau, *Pompeii: Its Life and Art* (Washington. D.C.: McGrath, 1973), 361. בְּאֻשִׁים of Isa 5:2, 4, though, cannot mean the fermentation odors of the wine-making process even as a prefigured hint for what the grapes will become, for several reasons. First, the fruit is still on the vine in Isaiah's Song. It is not crushed yet, so no pungent odor can occur. Second, most of the presses found from ancient Israel were outdoor presses. The cellars at Gibeon were enclosed, but they were found with closed amphorae in them to provide both storage and a second, less odiferous fermentation. See chapter 4, pp. 144, 158, 160, and chapter 5. pp. 187-90. Third, the disappointing fruit of Isa 5:2, 4 cannot even make it far enough to undergo fermentation: they are that worthless. The single form, בָּאְשָׁה in Job 31:40, refers to noxious weeds instead of barley. These two biblical instances of a noun form from בּאשׁ, in Isa 5:2, 4 and Job 31:40, indicate agricultural mishap, since the expected yield is replaced by stinking, unusable produce.

Isaiah's audience, from the eighth century on, likely would have understood the theological barb within the details of a cultivation gone awry. A vintner begins a vineyard in anticipation of ripe grape clusters for eating and turning into wine. Care extends beyond mere maintenance in the vineyard song, as the threefold use of יְדִיד, "beloved," in v 1 suggests. Isaiah is addressing Yahweh as his beloved and then singing about the latter's vineyard. In the Song of Songs, the beloved (דּוֹדִי)[6] is the object of desire, and a vineyard is the setting for consummation. While the term "beloved" is expected in a love song, it is no less apt for Isaiah's vineyard song. Love motivates Yahweh's actions. Through the details of vineyard maintenance, themselves symbols of divine attention to humans, Yahweh loves his vineyard as Isaiah loves his God. Vineyard care, in this sense, *is* a labor of love. Yahweh's destruction of his vineyard is total in vv 5-6; it is the expression of a passion disappointed. Isaiah's song, in addition to this poignant theological content, depicts both the details and the passion involved in viticulture. The Israelite farmers did not simply work the vineyards; they loved them to fruition. Yahweh, for Isaiah, can be no exception; in fact, he may be the rule.

Through an analysis of Isa 5:1-7, in combination with data from archaeology, classical sources on agriculture, and other biblical descriptions, I have, in this and the following chapter, reconstructed what the cultivation of vines likely entailed for the Israelite farmer. This first section of this chapter addresses the usefulness of Isa 5:1-7 as a source for the historical reconstruction of Israelite viticulture. The following sections discuss the main features involved in the cultivation of a vineyard, viz., agricultural terracing and soil work (Section II), vine planting (Section III), methods of training the vine (Section IV), and pruning (Section V). A final section (VI) focuses on the reasons for security, viz., military attack and intrusion by pests and unwanted guests, rather than the features erected for it, the walls and tower. The walls, since they were also a feature of terracing, are discussed in Section II, and the vineyard tower is addressed in detail in chapter 4. The sequence of activities, from selecting a hill slope, to working the soil, planting the vines, and constructing a tower and winepress, follows that of Isaiah's Song, but with considerable elaboration at points, particularly in the discussion of the type of vine used and in methods of training. The tower and winepress are the subject of the next chapter.

Since the Song of the Vineyard is, however, prophecy and not simply an ancient farmer's manual, some discussion of why Isaiah

---

[6]It is probable that דּוֹד is related to יָדַד as its nominal form: J. Sanmartin-Ascano, "דּוֹד" in *Theological Dictionary of the Old Testament* (ed. G. J. Botterweck and H. Ringgren; Grand Rapids, Mich.: Eerdmans, 1978) 3:143.

employed this image of a vintner for the deity is in order. Isaiah's credibility and power as a prophet are bound up in his attention to detail. Much of First Isaiah contains agricultural images as vehicles for theological and social critique and insight.[7] Accuracy in metaphoric skill, for a prophet, enhanced his or her credibility and potential effectiveness. If Isaiah had had only vague notions of farming, his persuasiveness as a prophet would have suffered considerably. I assume it did not, because of the preservation of his prophecies and those of his followers in the biblical canon.

The poem of Isa 5, as a judgment by God against the people, can best be understood as rhetoric. That is, part of the poem's goal was to persuade an audience of Isaiah's view. In vv 5-6, Yahweh announces what he will do to his vineyard, viz., destroy it. The verb forms primarily detail "future" actions, in contrast to the verbs of completed action in the description of Yahweh's construction of the vineyard (vv 2-4). Verses 5-7 announce impending action and function as a threat. The prophet's description of divine care and imminent (v 5: וְעַתָּה) destruction seeks to persuade, since presumably there is a delay between a threat and its execution. My contention is that Isaiah used viticultural detail for rhetorical intent, that he used an elemental feature of Israelite farm life to grab and hold his audience's attention. By making Yahweh a vintner, Isaiah aligns the farmers' own God with them. Isaiah could, then, elicit the empathy of vintner farmers who would identify with the agricultural waste Yahweh suffered. In v 7, he shatters that bond of empathy by revealing that the people in this Song are not the vintner at all; they are the vines of worthless fruit, the now horrified recipients of Yahweh's rage. Isaiah has taken the vineyard—something vital and dear in Israelite life, a part of farmers' livelihood and joy—and dangled it for theological challenge.

Tasks for starting a vineyard are listed here: the site is chosen—"a valued hill"; the soil is loosened and cleared of stones; the vines are planted, "rare" ones even; and a tower and winepress are constructed. After all that work, Yahweh plausibly asks in v 4: "What more was there to do for my vineyard that I have not done in it?" The Song, then, is not a description of a neglectful vintner, but of one who has been thorough. As such, it becomes a possible source for details about viticulture in ancient Israel. While classical antiquity produced

---

[7]For discussions of form and historical background of the Song, see Sweeney, *Isaiah 1-39*, 124; J. T. Willis, "The Genre of Isaiah 5:1-7," *JBL* 96 (1977): 337-62; G. Yee, "The Form Critical Study of Isaiah 5:1-7 as a Song and as a Juridical Parable," *CBQ* 43 (1981): 30-40; C. Seitz, *Isaiah 1-39* (Interpretation; Louisville: John Knox, 1993).

treatises describing viticultural practice,[8] no such works exist for ancient Israel during the Iron Age. Instead, our knowledge of Israelite viticulture is culled from biblical description and archaeological remains.

Isaiah's Song may offer a glimpse into viticulture not only for his own eighth century BCE, but also for much of the Iron Age in Israel. In a traditional society, such as ancient Israel's, where customs, religion, and agricultural wisdom are passed down from generation to generation, along with the land and the tools required to farm it, drastic innovations are unlikely. Basic vintner tasks, such as planting from shoots rather than seeds, pruning, and treading grapes by foot have persisted into the twentieth century of our era,[9] so it is reasonable to assume that these practices extended through the Iron Age. In addition, in Israel's subsistence economy, that is, where farmers grew enough to live on year by year, the threat of crop failure would rule out much experimentation in the fields. While an Israelite might have experimented on occasion with a jar of his wine by adding spices[10] or water,[11] he was less apt to tamper with the grape harvest itself.

The Israelites lived on an agricultural regimen that perhaps few of us can understand today.[12] It determined their lives and fundamentally shaped cultural attitudes about labor, family, and land. Richard Antoun, an ethnographer of the modern Arab village of Baytin (biblical Bethel), notes a social force in play that is applicable to ancient farmers as well. "Most farmers," he says, "are unwilling to invest either their time or money in any experiment unless it has been tried in the village and has produced good results."[13] The farmer's wariness is one of his survival skills. It is likely to have been in play for the farmers of Iron Age Israel as well. What worked for their fathers before them and brought in the best harvest likely remained

---

[8]*Enquiry into Plants*, Theophrastus (370-285 BCE); *Agriculture*, Cato (234-149 BCE); *On Farming*, Varro (116-27 BCE); *Georgics*, Virgil (70-19 BCE); *De re rustica*, Columella (mid-first century CE).

[9]Turkowski, "Peasant Agriculture in the Judaean Hills," 101.

[10]Prov 9:5: 23:30; Matt 15:23. See chapter 5, pp. 203-207.

[11]Isa 1:22. See chapter 5, pp. 203-207.

[12]While Braudel was working on his study of ancient France in the Regenstein Library at the University of Chicago, he reflected on the difference between modern life and that in antiquity. As he noted, he could write in shirtsleeves in that library, while outside, it was bitterly cold. He reminds historians of the ancient past beset by the tyranny of the seasons that we have now largely escaped. We take for granted, for example, that during frost or drought, "whatever the economic consequences, we are unlikely to starve." The ancient farmers likely did not often have that same leisure: *Identity of France*, 2:242.

[13]A. Lutfiyya, *Baytin: A Jordanian Village: A Study of Social Institutions and Social Change in a Folk Community* (The Hague: Mouton, 1966), 109-10.

the practice for generations.   Tradition in general, when backed against the threat of economic ruin, can become tenacious.

כֶּרֶם הָיָה לִידִידִי בְּקֶרֶן בֶּן־שָׁמֶן

My beloved had a vineyard on a valued hill
(Isa 5:1b)

The very first task in starting a vineyard was for the farmer to establish its location. Verse 1 of Isaiah's Vineyard Song provides that, and the vineyard is precisely where it should be—on a fertile hill. Vineyards, in antiquity and today, thrive on hill slopes, as we have seen.[14]   Roman writings on viticulture note the superiority of hill slope vineyards.[15]   Grapes, olives, and figs likely dominated the hill slopes leading up to Israelite villages and towns, both because they benefited from the drainage offered by the hill's slant and rocky soils, and because they were the most valued crops in terms of investment[16] and exchange.[17]

The value of hill slopes for the Israelites because of their agricultural yield in olives is at least hinted at with Isaiah's unique phrase of v 1, קֶרֶן בֶּן־שָׁמֶן literally, "a horn, a son of oil." A "horn of oil" is elsewhere used to anoint kings (e.g., 1 Sam 16:1, 13; 1 Kgs 1:39), and so here, both the importance of the hill as chosen by Yahweh, and the abundance of its oil are implied. In this phrase, the hill is basically a horn, filled with oil. In fact, the hill itself becomes the yield or "offspring" of oil, when in reality it would be the hill that yields the oil. This is a "very fertile hill" indeed, as the NRSV translates the phrase.   שׁמן is used often to denote oil, specifically olive oil (e.g., Exod 30:24; Deut 8:8; 1 Sam 10:1; Isa 41:19), and the

---

[14]See chapter 1, pp. 31-32; Baly, *Geography of the Bible*, 86; Moldenke and Moldenke, *Plants of the Bible*, 244; Amerine and Singleton, *Wine: An Introduction*, 46; Winkler et al., *General Viticulture*, 44; White, *Roman Farming*, 56; J. F. Ross, "Vine," *IDB* 4: 785. This viticultural wisdom may lie behind the Greek myth of Dionysus, the god of vines and wine, having been brought up by nymphs on slopes of a mountain, Unwin, *Wine and the Vine*, 87.

[15]Cato *Agriculture* 1.3; also Varro *On Farming* 1.6.5; Strabo *Geography* 16.2.9.

[16]Vines take 3 years of care before they bear fruit, while the olive tree takes 5-6 years, the fig 3-4 years. See Zohary and Hopf, *Domestication of Plants in the Old World*, 143, 137, 150, respectively.

[17]Cf. Amos 9:13 where the hills drip sweet wine, and Jer 31:5 where vineyards will again dot the hills of Samaria. Such hills enabled economic prosperity.

fat of both people (e.g., Judg 3:29; Ezek 34:16) and the land (e.g., Gen 27:28; Job 29:6; Ezek 16:13).

Here, and in two other vineyard texts, שֶׁמֶן is used to denote the fertility of land for viticulture (Isa 5:1; Gen 27:28; Num 13:20).[18] Its use in these contexts may be metonymic, invoking a general horticultural allusion while signifying the land's readiness for vine growing. It may also, of course, suggest the coincidence of olive orchards and vineyards near each other in land plots. Not only are orchards and vineyards often mentioned together in texts (e.g., Exod 23:11; Deut 6:11; Judg 15:5; 1 Sam 8:14; 2 Kgs 5:26), but also installations for the production of oil and wine are often found near each other in surveys and excavations.[19] To some extent, we expect this mixing of crops, since the same hill slopes conducive to the growing of vines were also desirable for tree crops.[20] Isaiah's description of a fertile hill is apt overall for horticultural growth and is here used specifically to describe a vineyard's location. Cultivating olive trees and grape vines on the slopes allowed the urban dwelling farmers to keep their investment close to their homes. This arrangement was also fortuitous because it saved the valley beds, typically the more fertile soils, for grain.

II. Agricultural Terracing

Because cultivation on hill slopes required special techniques to prevent soil erosion and water runoff, the Israelites practiced terrace farming, which made these slopes into a series of flat, narrow treads.[21] These terraces are evident for the Late Bronze environs of

---

[18]Cf. Deut 32:13; 1 Chr 4:40; Neh 9:25; Ezek 34:14;, for examples of the root שׁמן in descriptions of agriculturally rich lands.

[19]For examples, see Stager, "Shemer's Estate," 93-107; Y. Dagan, "Bet Shemesh and Nes Harim Maps, Survey," *Excavations and Surveys in Israel* 13 (1995): 94-95; Dar, *Landscape and Pattern*, 1:36-37, 148, 166-167, 180; G. Edelstein and M. Kislev, "Mevasseret Yerushalayim: The Ancient Settlement and Its Agricultural Terraces," *BA* 44 (1981): 54; G. Edelstein and S. Gibson, "Ancient Jerusalem's Rural Food Basket," *BAR* 8 (1982): 49; G. L. Kelm and A. Mazar, "Three Seasons of Excavations at Tel Batash—Biblical Timnah," *BASOR* 248 (1982): 32.

[20]The Jebaliyah Bedouin, for example, today intercultivate their grape vines with olive and other fruit trees: Perevolotsky, "Orchard Agriculture in the High Mountain Region of Southern Sinai," 331-58.

[21]C. H. J. de Geus, "The Importance of Archaeological Research into the Palestinian Agricultural Terraces with an Excursus on the Hebrew word *gbi*," *PEQ* 107 (1975): 67.

Jerusalem,[22] and Iron I at 'Ai.[23] Agricultural terraces thought to be from the eighth-seventh centuries BCE have been excavated seven km west of Jerusalem at Mevasseret Yerushalayim, and occur with installations characteristic of the olive oil and wine industries, such as cisterns, channels, and basins.[24] These may exemplify the kinds of terraces that would have lined Jerusalem slopes in Isaiah's time (eighth century BCE). They are found as well in the north around Shechem, also with industrial installations.[25] Agricultural terraces recently covered over 50 percent of the Judean hills around Jerusalem,[26] so their use in antiquity is probable for exploiting the hilly central highlands, as the Israelites did.[27]

Dating terraces is difficult, however, for several reasons. First, they are not sealed by debris layers above and below as are the strata of tells. Second, their installations and walls were built upon bedrock and so leave no foundation evidence. Third, ceramic finds are not expected for fields as they are for buildings. Where pottery shards are found lodged within the walls, they may have washed down from above in subsequent periods,[28] or been an element in soil refurbishing from the hilltop.[29] Nevertheless, since terraces are attested at the Iron I sites of 'Ai and Raddana,[30] at Mevasseret Yerushalayim for the Iron II period, and predominate today in the Judean highlands, we surmise that the Israelites continued to practice terrace agriculture throughout Iron II. Several biblical texts mention "terraces," שְׁדֵמוֹת, with vines (Deut 32:32; Isa 16:8; Hab 3:17).[31]

---

[22]L. E. Stager, "The Archaeology of the East Slope of Jerusalem and the Terraces of the Kidron," *Journal of Near Eastern Studies* 41 (1982): 111; Edelstein and Kislev, "Mevasseret Yerushalayim," 53-56.

[23]J. A. Callaway, "The 1966 'Ai (et-Tell) Excavations," *BASOR* 196 (1969): 16.

[24]Edelstein and Kislev, "Mevasseret Yerushalayim," 53-56.

[25]E. F. Campbell, "The Shechem Area Survey," *BASOR* 190 (1968): 19-41.

[26]Z. Ron, "Agricultural Terraces in the Judean Mountains," *IEJ* 16 (1966): 33-49, 111-22.

[27]de Geus, "The Importance of Archaeological Research," 67. Stager has discussed the significance of this technology for the Iron I settlements. Terracing the hills, making them agriculturally useful, he argues, helped to open up the highland frontier of Canaan. See his "The Archaeology of the East Slope of Jerusalem and the Terraces of the Kidron," 111-21, and "The Archaeology of the Family in Ancient Israel," 1-35; For a contrasting view of the importance of terraces for the highlands settlement, see I. Finkelstein, *The Archaeology of the Israelite Settlement* (Jerusalem: Israel Exploration Society, 1988), 202, 309.

[28]J. Bradford, "Fieldwork on Aerial Discoveries in Attica and Rhodes," *The Antiquaries Journal* 36 (1956): 172-80.

[29]T. Wilkinson, "The Structure and Dynamics of Dry-Farming States in Upper Mesopotamia," *Current Anthropology* 35 (1994): 483-520.

[30]Stager, "The Archaeology of the Family," 6-8; Callaway, "The 1966 'Ai (et-Tell) Excavations," 16.

[31]For a discussion of this term and the Ugaritic *šdmt* in *CTA* 23:11, see Stager, "The Archaeology of the East Slope of Jerusalem," 114-121.

The fourth terrace of Mevasseret Yerushalayaim, for example, is 5-7 m wide and showed evidence of "numerous plastered basins and channels—all findings characteristic of the oil and wine industries," according to their excavators.[32] Terrace walls, used as revetments for stopping water runoff, had to be built for hill slope agriculture. Terrace walls served to retain soil and yet allowed excess water to trickle down the slope to terraces on a lower level.[33] Isa 5:5 includes the detail of a vineyard "wall," גָּדֵר.[34] This, along with his description of the vineyard on a hill, indicates that Isaiah most likely had a terrace plot in mind. The גְּדֵרִים of mountain vineyards in Ps 80:13 have been broken down, enabling intruders to steal the fruit. Vineyard walls are also found in Num 22:24 (גָּדֵר), where Balaam's frightened ass is doing its best to walk between them. There, however, the walls occur alongside field paths near a road, so גְּדֵרִים need not be exclusively terrace walls. Two types of walls are, thus, suggested by the term גָּדֵר: one parallel to the terrace treads as the retaining terrace walls; and the other, at right angles from the first to help establish a path down the slope. Balaam's ass got caught in the latter kind, on a path; otherwise he would have trampled vines.

A thin hillside soil is best for vine growth, and this was known from early times.[35] Theophrastus advised a thin, hillside soil for vines because the richer, moister soils would yield not fruit, but only a lot of foliage and wood.[36] Greek and Roman farmers, like the Israelites, were cultivating their fruit trees and vines along the hill slopes of their lands. Yahweh in Isaiah's Song of the Vineyard is exercising sound agricultural judgment in his selection of locale for the vineyard. With this knowledge of hillside cultivation and terracing in mind, Amos's restoration scene also becomes clearer. The prophet notes a coming day when

> the mountains shall drip sweet wine,
> and all the hills shall flow with it.
> (Amos 9:13)

His image becomes a sensible agrarian wish, rather than a vision of cosmic disturbance. It recalls terraced vineyards, where wine

---

[32]Edelstein and Kislev, "Mevasseret Yerushalayim," 54.
[33]de Geus, "The Importance of Archaeological Research into the Palestinian Agricultural Terraces," 68.
[34]BDB, 165.
[35]White, *Roman Farming*, 224; Semple, *The Geography of the Mediterranean*, 390.
[36]Theophrastus *Enquiry into Plants* 2.4.2-3.

production was. Amos's wish, in essence, is that vineyards line every hill.

וַיְעַזְּקֵהוּ וַיְסַקְּלֵהוּ

He dug the soil and cleared it of stones.
(Isa 5:2a)

After the location of a vineyard has been selected, the next task is to prepare the ground of the vineyard by digging the soil and clearing it of stones. The root עזק of Isaiah 5:2a is a *hapax legomenon*, though it has cognates in Arabic and Aramaic for digging with an instrument.[37] עֲזֵקָה, the name of a Judean town in the Shephelah,[38] is from the same root. The Shephelah was a valued region of the Judean kingdom because of its roads through the west-east valleys, its fortified cities, and also its horticulture, specifically, olives and sycamore figs (1 Kgs 10:27; 1 Chr 27:28). The cognate meanings in Aramaic and Arabic are "to cleave or furrow the earth with an implement,"[39] and so, it seems best to translate וַיְעַזְּקֵהוּ in Isa 5:2a as "he tilled it (by hand)," that is, the vineyard soil.

The vintner used tools to work the soil. And for the vineyard, the primary tool was the hoe.[40] יֵעָדֵר, to "be hoed," is the activity mentioned later in Isa 5:6 for the soil work that will not get done in Yahweh's destruction of his own vineyard. Hoeing, then, would have been a regular facet of vineyard maintenance.[41] A hoe was used to loosen soil throughout the year and rid the vineyard of weeds. Virgil, believing "never is enough pains spent" on a vineyard, advises hoeing three or four times a year.[42] It controlled weeds, "thorns and thistles," שָׁמִיר וָשָׁיִת (v 6). The verb עָדַר is used only here and in a judgment in Isa 7:25, where hills will no longer be hoed. In this

[37]BDB, 740.
[38]Josh 10:10, 11; 15:35; 1 Sam 17:1; 2 Chr 11:9; Neh 11:3; Jer 34:7.
[39]BDB, 740; L. Koehler and W. Baumgartner, *The Hebrew and Aramaic Lexicon of the Old Testament* (Leiden: Brill, 1994-95; hereafter KB), 2: 810; E. D. Klein, *A Comprehensive Etymological Dictionary of the Hebrew Language for Readers of English* (New York: Macmillan, 1987), 468-69. See Gray and Peake, *A Critical and Exegetical Commentary on the Book of Isaiah*, 85; J. D. W. Watts, *Isaiah 1-33* (Word Biblical Commentary 24; Waco, Tex.: Word, 1985), 55.
[40]Turkowski, "Ancient Judaean Agriculture," 25; Hopkins, *Highlands of Canaan*, 228; Forbes, *Studies in Ancient Technology*, 3:109.
[41]Forbes mentions two or three times a year, *Studies in Ancient Technology*, 3:109; Borowski, *Agriculture in Iron Age Israel*, 107.
[42]*Georgics* 2.398; see also, Pliny *Natural History* 14.50.

judgment, all the mountain slopes that used to be filled with vineyards become only "thorns and thistles," שָׁמִיר וָשָׁיִת (Isa 7:23). Without hoeing, there is no vineyard, only thorns and thistles.

These two occurrences of עָדַר in the Bible are in contexts of vineyard maintenance. The hoe is a tool worked by hand, as opposed to the plow, which an animal could pull. Soil work by hand is characteristic of vineyard cultivation.[43] This was because the vineyards were terraced, and their limited space made hand cultivation easier. The same technology of terracing and cultivation by hand with a hoe seems to be evident in these two Isaiah passages. Even though the hoe is not explicitly mentioned until v 6 of Isaiah's Song, it was a primary tool in viticulture and so was likely used for the tilling mentioned in v 2a.

After having initially loosened the soil by tilling, Isaiah's vintner would next have to clear the soil of the larger stones. In terms of labor efficiency, vintners could use these stones to build walls for their terraces. Clearing the area of stones (וַיְסַקְּלֵהוּ) is infrequent in biblical agrarian scenes, yet it must have been a prerequisite to planting, especially in the hill country where there is a profusion of rocks and stones. The stones cleared from a terraced plot could have been used in construction of the vineyard walls.[44] Terrace construction, at least, seems to demonstrate Qoheleth's maxim of "a time to throw away stones, and a time to gather stones together" (Eccl 3:5).[45] The task of clearing stones occurs again in Isa 62:10 for the making of a road. Stone clearing for vine cultivation need not have been as thorough a process as it must have been for grains and legumes since some rocks aerate the soil and can strengthen the vine roots.

In fact, stones help to drain the soil, a prime consideration for vine growing.[46] Gravel soil is good for this. One of the finest vineyards in France today, Château Lafîte, is three-fourths gravel.[47] Virgil even advises the planting of 30 stones *into* the soil that is being

---

[43]Braudel, *Identity of France*, 2:254-56.

[44]Borowski, *Agriculture in Iron Age Israel*, 106.

[45]Some interpreters see a sexual nuance in the phrase, whereby the time to "throw away stones" signifies intercourse, and the time "to gather stones" is abstinence during the period in which the woman is ritually unclean: *Midrash Koheleth* (trans. A. Cohen; London: Soncino, 1951), 79-80; R. Gordis, *Koheleth: The Man and His World* (New York: Jewish Theological Seminary, 1951), 220; cf. M. Fox, *Qohelet and His Contradictions* (Sheffield, Eng.: Almond, 1989), 192.

[46]White, *Roman Farming*, 230; J. Renfrew, *Palaeoethnobotany: The Prehistoric Food Plants of the Near East and Europe* (New York: Columbia University Press, 1973), 131.

[47]K. D. White, "Farming and Animal Husbandry," in *Civilizations of the Ancient Mediterranean* (ed. M. Grant and R. Kitzinger; 3 vols.; New York: Scribner's, 1988), 1:227: Another French vineyard, Château Beaucaillou ("beautiful pebbles"), is even named for the stones in the soil: Cox, *From Vines to Wines*, 38.

readied for vines.[48] The verb סָקַל, then, in v 2a signifies some work with stones, but not their utter removal. In other biblical contexts, סָקַל denotes the stoning of a person (e.g., Exod 19:13; 1 Sam 30:6; 1 Kgs 21:13). This is stone work too, of course, though not of the agricultural sort. For the focus in these contexts is the target and reasons for punishment, rather than the labor involved in finding stones. In v 2a of Isaiah's Song, however, the focus is precisely that labor involved in removing some of the stones from the soil. As Cox explains, "one reason why steep hillsides are so good for grapes is that erosion has scoured the land to its poorest, stoniest constituents."[49] Thus the least productive soils for cereals become the most productive for the grape. None of the grain land need be taken up with vines. It is an irony or unexpected benefit that such a valuable crop would thrive in harsh soil. H. J. de Blij has noted just this feature for the wine lands around the Mediterranean: "It is one of the apparent anomalies of viticulture that certain of the world's greatest wines are produced from the grapes of vines that stand in soils that would, for other agricultural purposes, be described as 'poor' at best."[50]

III. Vines

The grape vine is made up of several parts: the roots, wood, leaves, and, of course, grape berries.[51] Branches grow out from the trunk, as shoots from bud beginnings, and tend to curl. The grape vine is a perennial plant, whose initial growth typically takes three to four years, but once mature it will fruit every year.[52] The cycle of a vine's growth is telescoped in the cupbearer's dream:

---

[48]Virgil *Georgics* 2.346 ff.; Theophrastus *Enquiry into Plants* 3.4.3.

[49]Cox, From Vines to Wines, 38.

[50]de Blij, *Wine: A Geographic Appreciation*, 4.

[51]Moldenke and Moldenke estimated that the grape vine of the Old World would have had a stem up to half a meter in diameter with bunches approaching 10-12 pounds in weight, and berries the size of small plums. Hence, for them, the story of the spies carrying a cluster large enough to require two men (Num 13:23) is not surprising: *Plants of the Bible*, 243. They do not explain what would still then be surprising though, viz., that two men were needed to lift a mere twelve pounds. Pruning grapes would not have required two men to carry grape clusters, so their presence in this scene is meant to convey the exaggerated abundance of the Promised Land. Only severe pruning, that is, of virtually all branches for several years, would produce a vine trunk so thick. The number of clusters on the vine depends on the pruning, but the average, balanced, pruned vine does yield eight to twelve pounds of grapes total. The berries, however, would have been the size they are today: Cox, *From Vines to Wines*, 32. The cluster in Num 13:23 is clearly extraordinarily large. It represents two things: the lavishness of the spies' deity; and the fact that the land is already inhabited, and the vines cultivated, since pruning is what increases fruit yield.

[52]Olmo, "Grapes," 294.

> There was a vine before me, and on the vine
> there were three branches. As soon as it budded,
> its blossoms came out and the clusters ripened
> into grapes.
>
> (Gen 40:9-10)

The cupbearer gives a fairly apt though terse description of vine growth. Blossoms develop on the branches, and these become the grape cluster, the אֶשְׁכּׂל. The three branches in the dream come to mean the three days that the cupbearer has remaining in prison, so the number is allegorical in Joseph's interpretation. Since a vine will sprout many branches if left to itself, the vine described in the cupbearer's dream has been pruned.

Isaiah's vintner has planted vine shoots, not seeds. Hence, Isaiah's use of נָטַע, "plant," rather than זָרַע, "sow," is accurate for starting a vineyard. Vine cultivation in antiquity, as today, was done from planting the cuttings and shoots, not pips. Classical writers knew that seedlings gave poor fruit.[53] Planting vine stock was established practice in Roman viticulture as described by Columella and Cato.[54] In fact, horticulture in general differs from grain and legume farming in just this way. With the exception of the almond tree, all vine and tree crops, including vines, olive, pomegranate, fig, sycamore fig, and date palm, are grown by planting shoots and branches, not seeds.[55]

The farmer undertakes horticulture by dismembering a part of an original plant in order to develop more plants. The cultivation of grapes, olives, figs, dates, and pomegranates required of the early farmers a fairly simple practice: that they plant cuttings of plants they had. Horticulture then was a kind of rudimentary cloning, mastered

---

[53]Theophrastus *Enquiry into Plants* 2.2.4: he notes that not only inferior berries result, but also sometimes none at all as the vine only makes it as far as flowering, forming no berries at all; Virgil *Georgics* 2.60: a vine grown from seed "bears sorry clusters, for the birds to pillage."

[54]Columella *De re rustica* 5.5; Cato *Agriculture* 47-48.

[55]Psalm 1:3 includes the image of a transplanted fruit tree to describe one who delights in the Torah.

in ancient Mediterranean countries like Israel.[56]   The economic efficiency in such cultivation is apparent. The farmer could increase his vineyard from within, by lopping off the shoots of his growing vines and planting them. In such a way, he could tend a balance of new, medium, and old vines each year.[57]   Jebaliyah Bedouin families in the Southern Sinai today, for example, retain a balance of new, medium, and old vines for the continual renewal of their vineyards.[58] Modern viticultural practice makes cuttings from the vine plant in the late fall or early winter, when the vine is dormant.[59]   Virgil recommended that the best time for planting is in the spring,[60] or in the autumn before the first cold.[61] Timing of planting the vines, then, seems to allow some room for individual preference.

Vines grown from seeds are vastly inferior in fruit and bulk to vines grown from shoots.[62] The domestication of the vine altered the plant in two ways. First, it shifted the manner of propagation from sexual to clonal. The wild vine, *Vitis vinifera silvestris*, had sexual reproduction by seed growth.[63]   Second, the domesticated grape vine was changed from a dioecious plant (separate male and female individuals) into a hermaphroditic crop, one that was able to pollinate

---

[56]Zohary and Hopf, *Domestication of Plants in the Old World*, 135: "Five fruits—grapes, figs, pomegranates, olives, and dates—were 'preadapted' for domestication since they lent themselves easily to vegetative manipulation. This trait was probably decisive for their success in becoming the first domesticated fruit trees in the Near East." These, not coincidentally, are precisely the five fruits mentioned in Deut 8:8: "a land of wheat and barley, of vines and fig trees and pomegranates, a land of olive trees and honey" (the last, made from dates). All seven crops result from cultivation in this way. Deut 8:7-10 acts, in a sense, as a farmer's inventory or agrarian hymn to the achievements on Yahweh-given land.

[57]After prohibition and again in 1994, after severe crop damage from *Phylloxera vitifoliae*, Robert Mondavi and other vintners in Napa Valley, California replanted their vineyards completely from the few vines still growing on their estates. Fully one-third of Mondavi's estate in the harvest of 1994 was not available, since it was still of young vines. His example is representative of the extent of damage, then renewal, of the central Napa Valley during 1994.

[58]Perevolotsky, "Orchard Agriculture in the High Mountain Region of Southern Sinai," 349. A vine on average yields fruit for twenty years. The first four years of its growth are typically concentrated in the plant, with no fruit.

[59]Winkler et al., *General Viticulture*, 200: the size of cuttings used for nursery plantings varies, but on average in California, they are .30-.50 m. For the sake of comparison and simplicity, I have converted measures, where possible, into the metric system.

[60]New vines in the Napa Valley of California are planted in the spring: Lyons, *Vine to Wine*, 7.

[61]*Georgics* 2.319-20.

[62]Winkler et al., *General Viticulture*, 197: seed propagation is impractical, for it results in vines that are poor in vigor, productivity, and quality of fruit.

[63]Zohary and Hopf, *Domestication of Plants in the Old World*, 135; Unwin, *Wine and the Vine*, 33.

itself.[64] The fruit then would develop by itself without the need of
male pollinators. As wild plants, grape vines were cross-pollinated,
"widely heterozygous," and economically useless because their berries
were small, varied, and inconsistent in number.[65]   Domestication
regulated the produce because it entailed the selection of examples
with good fruit for clonal reproduction.[66]

Other biblical texts besides Isaiah's Song also reflect this
propagation of vines through shoots rather than seeds with the use of
the term נָטַע, rather than זָרַע, in vineyard settings (e.g., Gen 9:20;
Deut 20:6; Josh 24:13; 2 Kgs 19:29; Pss 80:9; 107:3; Jer 31:5; Amos
5:11).[67]   Two texts, however, do mention "seed," זֶרַע, for vine
planting and so merit attention: Jer 2:21 and Ezek 17:5-8.

וְאָנֹכִי נְטַעְתִּיךְ שֹׂרֵק כֻּלֹּה זֶרַע אֱמֶת
וְאֵיךְ נֶהְפַּכְתְּ לִי ⁸ˢᵒᵘⁱרֵי הַגֶּפֶן נָכְרִיָּה

I planted you a rare vine,
all of it, of true stock
How were you changed from me
(and) become a putrid and foreign vine?
(Jer 2:21)

Jeremiah uses the same verb (נָטַע) and vine term (שֹׂרֵק)[69] as are
found in Isa 5:2. It is his use of the phrase "true seed," זֶרַע אֱמֶת, that

---

[64]D. Zohary, "Domestication of the Grapevine Vitis Vinifera L. in the Near East," in
OAHW, 26. For a concise treatment of the plant's transformation under
domestication, see H. P. Olmo, "The Origin and Domestication of the Vinifera
Grape," in *OAHW*, 31-43.

[65]Zohary and Hopf, *Domestication of Plants*, 135; Winkler et al., *General
Viticulture*, 197.

[66]Winkler et al., *General Viticulture*, 197. Cuttings yield plants exactly like the
parents, so the latters' beneficial characteristics, such as abundant berry growth, get
selected.

[67]Semple argued that ancient farmers maintained a radical distinction between crops
that were sown and those that were planted. The latter, she noted, were cultivated on
the hills, the former on valley beds: *The Geography of the Mediterranean*, 389.

[68]The meaning of the MT is uncertain. סוּרֵי is from סוּר, "to turn," so the term is
perhaps a participle with the sense of "those turning (away)": see W. Holladay,
Jeremiah I, 1-25 (Hermeneia; Philadelphia: Fortress, 1986), 53; J. Bright redivides
the consonants, attaching the ה to the end of the verb, and argues for the meaning,
"foul-smelling thing," based on a Syriac root, *sry*, "become putrid": *Jeremiah*
(Anchor Bible 21; Garden City, N.Y.: Doubleday, 1965), 11. His translation
remains speculative, though plausible, in that Jer 2:21 is a variation of Isa 5:1-7, and
so might be expected to have a variant of בְּאֻשִׁים, "fetid fruit."

[69]See below.

is problematic. Has his vintner—again it is the deity—"planted" a
"vine" or broadcasted grape pips as seed? זֶרַע אֱמֶת is parallel with שֹׂרֵק
and so would seem to illuminate what the vintner has planted. If
Yahweh has planted from seeds, then, the harvests will not be
bountiful, and so Yahweh's disappointment here is rendered naive.
Such cannot be Jeremiah's theological goal. The clue to the meaning
of זֶרַע אֱמֶת lies in the precise expression of Yahweh's disappointment.

The expectations of the divine vintners in Isa 5:1-7 and Jer 2:21
are different. Isaiah's vintner waits for the fruit, namely, the grapes.
Jeremiah's vintner waits for the plant, the vine's growth, and neither
vintner gets what he expected. In the Jeremiah verse, there is no
mention of fruit; it is the plant itself with which the vintner is
concerned. In fact, it is not even what he planted: Yahweh's שֹׂרֵק has
become a גֶּפֶן. It is possible that a vintner could plant from seed and
then discover a plant he had not expected. Such would explain this
verse, but it would go against viticultural practice and thereby cast
aspersions on the deity's skill. Also, the first part of the verse, with
its use of נָטַע, is accurate on planting practice, so it is clear that
Jeremiah knows the viticultural tasks. Hence, it is unlikely that he
would use נָטַע with זֶרַע and unknowingly confuse two agricultural
practices, namely "planting" a shoot and "sowing" seed. Instead, his
use of the unexpected term, זֶרַע, must have another agenda behind it.

Since Jeremiah knows that viticulture involves the planting of
vine shoots, he uses זֶרַע to describe something extraordinary, not
quotidian, in this specific planting. He captures divine foreknowledge
and previous care by having Yahweh know from where that vine
shoot came. While farmers might know the fairly recent background
of the vine shoots they were planting—be it through trade or from
their own vineyards—they would not be able to testify to the
historical origins of the plant itself. But Yahweh, as creator of the
world, would have just that kind of botanical knowledge. I suggest,
then, that זֶרַע is not grape pip,[70] but "stock,"[71] thereby also signaling
the pure or "true" origins in Yahweh's rare שֹׂרֵק. Jeremiah's use of
זֶרַע, in other words, is not agricultural, but theological. This vintner
is planting a vine whose entire history he has known from its botanical
origins. זֶרַע serves Jeremiah's pronatally infused imagery for
Yahweh. Present at Jeremiah's conception in his mother's womb
(1:5), the same Yahweh is present in this vine's "birth," that is, its
initial seeding, before domestication.

Ezekiel 17:5-8 is the other description of vine cultivation from
seed. This passage describes the vine cultivation of the great eagle:

---

[70]Which might possibly have been named by the Hebrew פֶּרֶט (Lev 19:10; Amos
6:5), though this meaning is dubious.
[71]This is the translation of the NRSV as well.

Fruit of the Vine

5 וַיִּקַּח מִזֶּרַע הָאָרֶץ וַיִּתְּנֵהוּ בִּשְׂדֵה־זָרַע
קָּח עַל־מַיִם רַבִּים צַפְצָפָה שָׂמוֹ"
6 וַיִּצְמַח וַיְהִי לְגֶפֶן סֹרַחַת שִׁפְלַת קוֹמָה
לִפְנוֹת דָּלִיּוֹתָיו אֵלָיו וְשָׁרָשָׁיו תַּחְתָּיו יִהְיוּ
וַתְּהִי לְגֶפֶן וַתַּעַשׂ בַּדִּים וַתְּשַׁלַּח פֹּארוֹת
7 וַיְהִי נֶשֶׁר־אֶחָד גָּדוֹל גְּדוֹל כְּנָפַיִם וְרַב־נוֹצָה
וְהִנֵּה הַגֶּפֶן הַזֹּאת כָּפְנָה שָׁרָשֶׁיהָ עָלָיו
וְדָלִיּוֹתָיו שִׁלְחָה־לּוֹ לְהַשְׁקוֹת אוֹתָהּ מֵעֲרֻגוֹת מַטָּעָהּ
8 אֶל־שָׂדֶה טּוֹב אֶל־מַיִם רַבִּים הִיא שְׁתוּלָה
לַעֲשׂוֹת עָנָף וְלָשֵׂאת פֶּרִי לִהְיוֹת לְגֶפֶן אַדָּרֶת

5. He took some of the seed of the land and put it in a field of
seed. He took (it) by abundant waters, a willow twig, he set it
(there).
6. It sprouted and became a vine. Spreading out low,
its branches turned toward him, its roots remained where
it stood. So it became a vine; it put forth branches, put forth
foliage.
7. There was another great eagle, with great wings and much
plumage. Behold, this vine stretched out its roots toward him;
it shot out its branches toward him, so that he might water it.
8. From the bed where it was planted, it was transplanted
to good soil by abundant waters, so that it might produce
branches and bear fruit and become a noble vine.

(Ezek 17:5-8)

From the seed mentioned in v 5, a vine grows, spreading out
low on the ground, with firm roots, and much foliage. Ezekiel's
allegory of the eagles casts Nebuchadrezzar as the great eagle of vv 3-
6. His earlier horticultural endeavor in v 4 was to lop off the top of a
cedar tree (Judah's royal house) and set it in a city (Babylon). In v 5,
the great eagle resorts to seed planting to produce a vine (v 6). Since
this technique was economically useless, Ezekiel might well be
mocking Nebuchadrezzar. In the next verse (7) the vine (Zedekiah)
grows, with roots and branches mentioned, but no fruit. The great
eagle has planted, then, but not with skill. He has managed to produce

---

[72]W. Zimmerli argues, and I concur, that this unusual form is likely the result of a
vertical dittography from וַיִּקַּח in v 5a: W. Zimmerli, *Ezekiel I, 1-24* (Hermeneia;
Philadelphia: Fortress, 1979), 355.

a plant, as seed planting will do, but not what is desired from a vine, namely, its fruit.[73]

In v 8, this vine is "transplanted," שְׁתוּלָה. Ezekiel's description here of a replanting may represent the only biblical example of a vine nursery, where a young shoot or even smaller bud was planted first, before being transplanted as a vine.[74] Transplanting was a vital part of viticultural practice described in Roman sources,[75] and is evident in two other biblical texts, Ezek 19:10 and Ps 80:9.[76] Planting vine shoots was usually done in a section of the vineyard or in nursery beds first so that the plants could develop strong root systems. After that, the shoots would be transplanted into the area in which they would grow.[77] Ezekiel's second eagle (Psammetichus II), then, plants the vine correctly; he transplants it and can expect fruit (v 8).[78]

In the allegory, the vine planted by Nebuchadrezzar is Zedekiah. It is a full and noble vine, גֶּפֶן אַדָּרֶת (v 8). Zedekiah, after all, was Judah's king. But, it does not prosper under siege conditions. It is the conditions of planting, then, that finally condemn the plant, rather than its method of cultivation. Zedekiah was Judah's last king, under vassalage to Babylon.   His "transplanted" loyalty to Psammetichus II initiated his rebellion against Nebuchadrezzar. Zedekiah's rebellion, during a two-year siege, marked the end of Judah. His punishment before Nebuchadrezzar was to watch his sons slaughtered, and then be blinded himself (2 Kgs 25:3-7; Jer 39:1-7, 52:5-11). This plant, then, while it once had given the "fruit" of potential successors, would yield no more.

Ezekiel's allegory of the end of Judah, thus, is a powerful one. Discussion of it here, however, has been restricted to the descriptions of vine planting. Ezekiel gives, in essence, a depiction of the incorrect and correct ways to cultivate a vine. The Israelite vintner, of course, would have practiced the correct method, which was planting vine shoots and then transplanting them once the roots were established.

---

[73]Zohary and Hopf, *Domestication of Plants*, 135; Winkler et al., *General Viticulture*, 197.

[74]Psalm 1:3 reflects a fruit tree nursery when it likens a person who delights in the Torah to a "transplanted tree."

[75]Cato *Agriculture* 32; 40; 43; 46-49; Columella *De re rustica* 5.9.3-4; see White, *Roman Farming*, 225-26.

[76]Psalm 80:9 extends this viticultural datum metaphorically to depict Israel as a transplanted vine brought out of Egypt to Canaan. The vine's method of cultivation, then, being transplanted from nursery beds, lends itself quite effectively to this theology of Yahweh's care and deliverance of the people.

[77]White, *Roman Farming*, 225; Semple, *The Geography of the Mediterranean*, 390-391.

[78]A vine first grown from seed would never produce consistent fruit yields. Transplanting would do nothing to help that fact. The second eagle has then the right practice, but the wrong plant.

Isaiah's vintner had cultivated properly by planting a vine (5:2), though the Song of the Vineyard does not mention transplanting. Like Jeremiah's vintner in 2:21, the vine used was a שׂרֵק, rather than the more common גֶּפֶן. Since this term for vine is rare, occurring three times in the Hebrew Bible in contrast to the fifty-five uses of גֶּפֶן, it requires discussion. The three instances of שׂרֵק are Isa 5:2, Jer 2:21, and, as a feminine form, Gen 49:11.[79] In the two prophetic contexts for שׂרֵק, Yahweh is the vintner; in the third, Judah receives the blessing that he will be

אֹסְרִי לַגֶּפֶן עירה [עִירוֹ] וְלַשֹּׂרֵקָה בְּנִי אֲתֹנוֹ

> binding his foal to the vine
> and his donkey's colt to the rare vine.
> (Gen 49:11)

שׂרֵק is nearly always translated "choice vine"[80] in these three passages, without much ensuing commentary or question.[81] This sense, though, is not at all apparent from the root שׂרק, whose Mishnaic Hebrew and Arabic meanings are "light red" and "rise and shine (of sun)," respectively.[82] The infrequency of שׂרֵק in the Bible, the fact that Yahweh is the vintner in two out of three contexts, and that Judah as the favored son benefits in the third—probably determined its translation as "choice." שׂרֵק came to mean "choice vine" to biblical scholars, then, not from etymology, but by the company it kept. A variant, שׂרוּקֶיהָ (Isa 16:8), might also hint at the elite status of this vine, in that lords, rather than mere men, attack it: בַּעֲלֵי גוֹיִם הָלְמוּ שְׂרוּקֶּיהָ, "lords of nations struck her vines." The Moabite terraces (שְׁדֵמוֹת)[83] seem to have both vines, גֶּפֶן and שְׂרוּקֶיהָ.

In his analysis of biblical agricultural terms, Oded Borowski concludes: "There is a strong possibility that שׂרק was not just the term for "choice vine," but was the name of a variety producing red

---

[79]A variant, שְׂרוּקֶיהָ occurs in Isa 16:8. See below.
[80]RSV, NRSV, JB, NAB, JPS: Isa 5:2 "choice vine," Jer 2:21 "noble vine"; NEB: Isa 5:2 "red vines," Jer 2:21 "choice red vine."
[81]Bright, *Jeremiah*, 98; Borowski, *Agriculture in Iron Age Israel*, 104.
[82]BDB, 977; KB, 2:1269; M. Jastrow, *A Dictionary of the Targumim, The Talmud Babli and Yerushalmi, and the Midrashic Literature* (New York: Putnam, 1903), 1030, 1634.
[83]For a discussion of terraces, see above, pp. 94-96.

grapes."[84] His explanation, however, is deficient since all grapes and vine leaves of *Vitis vinifera* turn red in autumn.

The grape vine is a perennial, deciduous plant whose fruit and leaves turn from green to red in its yearly growth cycle. The chlorophyll that gives grapes their green coloring fades and other colors, viz., black and red, appear.[85] The main red pigment (or anthocyanin) of *Vitis vinifera,* called oenidin,[86] and the tannins on the grape's skin account for its color during ripening. Indeed, the onset of the harvest is determined, then and now, on the basis of the grapes turning a dark red. The dominant genes of the grape are for black and red colors, while "white-fruited grapes are considered to be recessive for both genes."[87] White grapes were possible in antiquity, but are not widely attested until the Hellenistic period,[88] when maritime trade throughout the Mediterranean increased the varieties and dispersal of *Vitis vinifera* and led to viticultural experimentation and hybridization, well attested by Greek and Roman authors.[89] The fall color change earned a technical term in viticultural practice, *véraison,* and is eagerly awaited each year by all vintners.[90] The leaves turn color slightly later, usually after the harvest, and then are shed for winter's dormancy. Hence, that a vine produced red grapes in the Iron Age does not distinguish it at all.

The grapes of the שֹׂרֵקָה vine mentioned in Gen 49:11 certainly were red, as Judah would wash his garments in the "blood of grapes" (v 12: דַּם־עֲנָבִים). The robes of the grape treaders in Isa 63 are splattered "red": חָמוּץ (v 1) and אָדֹם (v 2). Elsewhere blood is used of grapes (Deut 32:14; Ezek 19:10). By extension, blood becomes itself the source of intoxication in prophetic judgment (Isa 49:26; Jer 46:10; Ezek 39:17-19). It is the red color of wine that makes this metaphor and its prophetic literalization forcibly apt.[91]

---

[84]Notice that in his rendering the vine is, then, named for the fruit it yields, rather than for its general qualities as a plant: Borowski, *Agriculture in Iron Age Israel,* 104. Rainey's translation is even further removed from the plant itself. For him, שֹׂרֵק is "red wine": "Wine From the Royal Vineyards," *BASOR* 245 (1982): 59; cf. Ross, "Vine," 784-86: "A superior vine producing grapes with a rich, dark hue."

[85]Unwin, *Wine and the Vine,* 36.

[86]Unwin, *Wine and the Vine,* 36; Winkler et al., *General Viticulture,* 159.

[87]Einsett and Barritt, "Fruit Colors," 87-89.

[88]Athenaeus *Deipnosophistae* 1.33d; Virgil *Georgics* 2.91-92; Pliny *Natural History* 14.4.20-43.

[89]Theophrastus *Enquiry into Plants* 2.3.2; 2.5.7; Cato *Agriculture* 24; 51; Columella *De re rustica* 3.2; 12.44.

[90]See chapter 5, pp. 167-71.

[91]At the Last Supper, depicted in the New Testament Gospels, Jesus equated the table wine with his blood. Had the wine been colorless, the symbolism of blood and his incipient sacrifice would have been lost. He had not asked his disciples to *imagine* that the wine was his blood, but rather to drink it as such (Matt 26:27-29; Mark 14:22-25; Luke 22:14, 21-23).

The Bible does not always mention the color of wine, but when it does, it is red. The extrabiblical evidence for wine in the ancient Near East and Mediterranean suggests that it was primarily, if not exclusively, red. Egyptian tomb paintings depicting wine production and pouring show a dark liquid in their colored scenes.[92] A seventh-century BCE ostracon from Ashkelon mentions a "red wine," *yn 'dm.*[93] In the *Iliad*, Hektor will not drink wine during battle, when he is a "bespattered man, and bloody," because his association of the two liquids is too close.[94] The vineyard on Achilles' shield has grape clusters "hung dark purple."[95] Later, the confluence of wine and blood is evident in Dionysian rituals of intoxication and animal dismemberment.[96] In general, the symbolism of wine and blood in the sacrifices of ancient Israel, Greece, and in Christianity[97] is generated by the shared color of these liquids. It would be much more difficult to account for such persistent symbolism if white wine had been as common as red wine in antiquity.

The chemical process of fermentation itself indicates that preclassical wine production from grapes almost certainly must have been red wine.[98] Wine gets its color from the skins bleeding in the juice during treading and fermentation. Since white-skinned grapes are not attested until the Hellenistic period, the color from the skins would turn the wine red. White wine can still be made from red grapes, by separating the skins from the juice after a first press and fermentation begins. This is easy to do in modern wine making, because vintners simply add yeast to the freshly pressed, light-colored liquid. Ancient Israelites had knowledge of leavening in bread, so clearly they understood that yeast was an active agent, which changed the substances of which it was a part (e.g., Gen 19:3; Exod 12:15, 19-20; 12:34, 39; 13:3, 7; 23:18; Lev 7:13; Deut 16:13; Judg 6:21; 1 Sam 28:24; Hos 7:4; Amos 4:5). However, there is no indication that they knew how to separate the leavening agent, the yeast, and apply it to other products, such as wine.

The primary reason for assuming that wine in ancient Israel was red is simply chemical. The yeast for fermentation is on the

---

[92]Tomb of Khaemwese, British Museum K89048, reprinted in Unwin, *Wine and the Vine*, 70; tomb of Nahkt, reprinted in Darby et al., *Food: The Gift of Osiris*, 2:561, fig. 14.1; Johnson, *Vintage*, 31.

[93]Stager, "The Fury of Babylon," 66. See chapter 5, pp. 201-2.

[94]Homer *Iliad* 6.267. There were practical reasons too; he did not want to lose his nerve for battle: 6.264-67.

[95]Homer *Iliad* 18.531.

[96]Euripides, *Bacchae* (trans. E. R. Dodds; Oxford: Clarendon, 1960).

[97]The Austrian Catholic Church uses only white wine today, perhaps because the equation made between wine and blood in transubstantiation is considered too graphic.

[98]See chapter 5, pp. 187-90.

grape skins. Had Israelite vintners separated juice from skin before fermentation, they would have produced a juice light in color and nonalcoholic. Nowhere is a term for juice used in the Hebrew Bible.[99] The only example of grape juice occurs in Gen 40:11: Pharaoh's cupbearer dreamt that he squeezed grapes directly into Pharaoh's cup. Lighter, unfermented juice would have been the result, by virtue of its short contact time with the skins.[100]

The textual evidence for שֹׂרֵק denoting a color is slim, simply because the term is rare. If color is involved, we can rule out white wine and see the limitations inherent in Borowski's suggestion that it simply entails red. We are left discerning where on the spectrum between red and white the shade might be. It is reasonable to conclude that some red is present because of the "blood" mentioned in Gen 49:11 and the chemical necessities involved in fermentation. If the שֹׂרֵק vine did get its name from a lighter shade of red than that expected from the גֶּפֶן we have to assume an original meaning that is now not easily discernible because of the few occurrences of שֹׂרֵק. After these three biblical occurrences, it is lost until its emergence in one late text, Sir 50:7, and in the Arabic term for "sunrise."[101]

A possible hint of this meaning as a light shade of red may lie in Zech 1:8, where horses are listed presumably by color. Among the red and white horses are those of שְׂרֻקִּים, connoting perhaps a color shade in between red and white or, at least, distinct from them. The translation, "sorrel," has been proposed here.[102] If so, then it is also possible, though unlikely, that the Israelites harvested the grapes from some vines early, before the skins were fully red. Such שֹׂרֵק could have been valued, "choice," for its rare color.

There is another solution to the background of שֹׂרֵק, and this may be the most promising: שֹׂרֵק may denote, not a color of the leaves, grapes, or resulting wine, but rather the location of its primary cultivation, the Valley of Sorek. Both biblical and archaeological materials reflect evidence of horticulture in the region. Ekron (Tel Miqne) is the site of the largest olive oil production center ever found

---

[99]The prominence of red wine seems to be true for the New Testament period as well. Even Jesus' miracle of turning water into wine in John 2:9 does not rest on a visual deception. That is, the jars of water are turned into wine, but the surprise is over how good the wine tastes. No one seems to have realized that the jars were originally filled with water. The steward's surprise is due to the *taste* of the wine—that it is the best and has been saved for last—and not that it looked like water.

[100]Oddly, in this prison scene, the one who does not use leavening, the cupbearer, is restored by Pharaoh.

[101]BDB, 977; Holladay suggests that since שֹׂרֵק is associated with the brightness of the sun, it evidently refers to a particularly luscious variety of red grape: *Jeremiah I, 1-25*, 98.

[102]BDB, 977; KB, 2:1269.

in the Mediterranean basin.   So far, 115 oil presses have been
excavated.[103]   Timnah (Tel Batash), also located in the Valley of
Sorek, had oil and (perhaps) winepresses.[104]   In Judg 14:5-6, Samson
kills a lion in the Timnite vineyards.[105] The Valley of Sorek offered
sloping hills near a water source, and shared in the primarily Eocene
limestone of the Shephelah. This limestone then became the *terra
rossa* soil so prevalent in Israel.   As such, it made an excellent
cultivation area for tree and vine crops.

The overall incidence of horticultural activity reflected in the
archaeology and textual materials suggests that the שֹׂרֵק vine might
have been named for the location of its origin or cultivation, more
than for its color shade.   Vines grown there, namely in the best
agricultural land, would be esteemed by both farmer and prophet
alike. The three occurrences of שֹׂרֵק as vine—again, Isa 5:2; Jer 2:21;
and Gen 49:11—are all in Judean contexts. If the Valley of Sorek was
known for its horticultural success, then שֹׂרֵק vines would achieve a
status understood as "choice" to the Judeans. That these vines had
high potential for yield is clear in all three texts—in the bitter
disappointment of the vintner in Isa 5:2, 4 and Jer 2:21, and in Judah's
abundance of wine in Gen 49:11.

IV. Methods of Training the Vine

Viticultural practice today in America favors planting vines
about 1.83 m apart in rows 3 m wide.[106]   This allows each vine
adequate spacing for root and branch growth. The plants do best with
this spacing, but crowding does not substantially affect grape
quality.[107]   Hence, French and other European vineyards tend to be

---

[103]Personal communication with the excavator, S. Gitin. For discussion of the site,
see T. Dothan and S. Gitin, "Tel Miqne," *The New Encyclopedia of Archaeological
Excavations in the Holy Land* (4 vols.; ed. E. Stern; Jerusalem: Israel Exploration
Society and Carta, 1993; henceforth *NEAEHL*), 1058.
[104]Kelm and Mazar, "Three Seasons of Excavations at Tel Batash—Biblical
Timnah," 34-35.
[105]Samson's feat is recalled in the narrative because it showed Yahweh's spirit within
him and provided material for a difficult riddle. Often overlooked is his sheer
agricultural instinct to save the valued vineyards of Timnah from possible damage by
the lion. Anger later overrides that instinct when he is barred from visiting his wife,
and then burns Philistine vineyards, olive groves, and grain (Judg 15:5). His actions
demonstrate that he knows the value of vineyards and orchards and what effect their
devastation is likely to have on the Philistines.
[106]Cox, *From Vines to Wines*, 33; Lyons, *Vine to Wine*, 7.
[107]Cox, *From Vines to Wines*, 34; Ross, "Vine," 785. Many growers in the Napa
Valley of California space their vines 2.44 m apart in rows 2.44 m apart. Variations

much more crowded than California vineyards, in order to "squeeze every last drop of wine from the acreage."[108]   Where a typical vine acre in the Napa Valley of California will carry 440 to 600 vines, Cox notes, the same acre in the Champagne region of France will have 3000.[109]

Vine branches and leaves need room to extend so that sunlight is equally distributed along the plant and has access to berries for their ripening. The roots also need room to extend as much 1.83 m deep and .91 m wide.[110]   Spacing vines less than .91 m apart would force the plants to compete for the soil's available water and nutrients. Irrigation and nutrient replenishment can offset that hazard, but adequate spacing would have been a more efficient practice.

Ancient Greek and Roman agricultural manuals record spacing requirements for vineyards. Theophrastus recommends on average 2.74 m between each fruit-bearing tree, though he does not single out the vine.[111]   Columella advises that vines be planted two paces apart, roughly a meter.[112]   Varro suggested that there be enough room for a yoke of oxen to plow between the rows of vines.[113]

The use of a plow for spacing may also be indicated in Isaiah's description of a vineyard unit in 5:10: צִמְדֵּי־כֶרֶם. This phrase is usually translated as "acre," with צֶמֶד, "pair," referring to the amount of land a pair of animals can plow in a day.[114]   צֶמֶד could have been the general term for field measurement which Isaiah applied to the vineyard. The entire vineyard, unlike a grain field, does not get plowed anew each year as the plants remain standing. If plowing is meant in this passage, then it has to be between rows wide enough to fit a pair of animals and the yoke connecting them.[115]

Isaiah 5:10 mentions a vineyard of ten such "acres."   A vineyard of ten acres would be approximately forty dunams. This is

---

in spacing have to do with maximizing water reserves in the soil. Vines in hotter regions would need more space optimally.
[108]Cox, *From Vines to Wines*, 34.
[109]Cox, *From Vines to Wines*, 34.
[110]Cox, *From Vines to Wines*, 41.
[111]Theophrastus *Enquiry into Plants* 2.5.6.
[112]Columella *De re rustica* 5.3.2.
[113]Varro *On Farming* 1.8.5; White estimates this to be about 1-3 meters: *Roman Farming*, 237.
[114]The English term, "acre," from the Greek ἀγρός, developed, too, from the amount of land a pair of draft animals could plow in a day. Later English standardizations render an acre to be 4,047 m². See *Webster's Third New International Dictionary* (3d ed.; ed. P. Gove; Springfield, Mass.: Merriam, 1965), 19.
[115]Plowing is not mentioned for Yahweh's vineyard in vv 1-7. This vineyard of v 10 would be large, approximately forty dunams, much larger than the two-dunam vineyard we estimate for the average vintner farmer of ancient Israel (p. 112 below).

quite large for just the vineyard in comparison to the total farm sizes
of modern rural communities in the Middle East and Mediterranean.
In Baytin (biblical Bethel), for example, the farms average twenty-
five dunams.[116] In Greece, farm sizes at Vasilika and Messenia range
from twenty to thirty-five dunams.[117] Baruch Rosen estimates a sixty-
dunam farm for a family of five in the Iron I period.[118] Isaiah's
vineyard, then, is larger than the entire farm plots estimated for a
family farm, and so is not representative of Israelite viticulture in
general. His exaggeration is not at all surprising in the passage, for he
is contrasting how little such a large vineyard would nevertheless
yield, only a bath of wine (5:10), which was roughly twenty-two
liters.[119]

       From his survey of farms in Western Samaria, S. Dar estimated
that the Israelite vineyard was anywhere from 1.5-10 dunams.[120]
Assessing the size of an Israelite vineyard is quite difficult.
Ethnographic parallels of modern, rural vintners can provide at least
an approximation.   The vintner farmers of Vasilika, on average,
cultivate a vineyard of two dunams.[121] This size of a vineyard can be
harvested by a family in a day, and it provides them with enough wine
for home consumption for the year, though the ethnographer does not
indicate what that amount is. M. Finley mentions that for a small
Greek community founded in the Adriatic island of Curzola in the
third or second century BCE, peasant settlers were each allotted an
unspecified amount of arable land and the equivalent of three dunams
of vineyard to sustain their families.[122] Since this is the vineyard size
given to meet the family needs of Greek settlers, and also falls within
Dar's estimate for the ancient Samaria vineyards, two dunams can
stand as a reasonable estimate for a small-farmer vineyard. Wine
yields from such a vineyard are discussed in chapter 6.

       Virgil's advice for planting vines is more general. He suggests
a close planting if the vines are on a level plane, and more room for
vines on hills. Though he does not specify dimensions, he is firm that
the vines be evenly spread apart.[123] Biblical literature nowhere
provides the actual layout of a vineyard. Deuteronomy 22:9 only

---

[116]Antoun, *Arab Village*, 12.
[117]Friedl, *Vasilika*, 30; Aschenbrenner, "A Contemporary Community," 178;
Campbell, "Karpofora," 212.
[118]He cites twenty families for 120 hectares. Each family then would have six
hectares, or sixty dunams: Rosen, "Subsistence Economy in Iron Age I," 347.
[119]See chapter 6, p. 218.
[120]He arrived at this estimate by averaging the number of farmsteads of an area with
the number of presses discovered: Dar, *Landscape and Pattern*, 1:6.
[121]Friedl, *Vasilika: A Village in Modern Greece*, 19. Each vineyard averages 1-2
stremmata. Four stremmata equal one acre.
[122]Finley, *Ancient Economy*, 99.
[123]Virgil *Georgics* 2.273-76.

mentions that a vineyard ought not to have a second crop planted within it. An instance in the Talmud deviates from this biblical prohibition, by allowing a second crop in the case where the vineyard rows are sixteen cubits wide. [124] Since a cubit is roughly half a meter long, the rabbis were talking about an eight-meter row between vine rows. Under those spacious conditions, they allow a farmer to plant a second crop in the rows. This instance contrasts with the restriction against the intercultivation of a vineyard in Deut 22:9, and includes an exceptional amount of space between rows. With terracing and dividing his plot for grain and fruit crops, the farmer with so much space between his vine rows would have been the rare exception. We can assume that he would plant in such a way as to maximize his land and crops. He would presumably give the vines the room they needed for beneficial growth, while making do with whatever land limitations he had.

The methods of vine training varied considerably in antiquity. Vines could be sprawled out on the ground, propped on individual stakes, or trained up trees or poles, a technique known as trellising. All of these methods are attested in ancient materials from Egypt, Mesopotamia, Greece, and Rome. On Egyptian tomb paintings and an Assyrian palace relief, growing vines are depicted as trellised.[125] In the tomb paintings of Kwamwese at Karnak and of Sennufer at Deir el Medineh, vines are trellised on poles, and their intertwining branches form an overhead canopy. The Assyrian relief shows Ashurbanipal and his queen seated under just such a canopy of vines, sipping, presumably, wine.[126]

The earliest Greek description of vines occurs in Book 18 of the *Iliad*. There, on Achilles's divinely crafted metal shield, depicted vines are propped on silver stakes. Highly trellised grapes tease Aesop's (mid-sixth century BCE) fox to frustration, causing him to mutter that they were sour anyway.[127] Roman viticulture also often involved some form of trellising. Columella advised that trellised vines would do best if they were planted in deep holes.[128] This advice,

---

[124]The Talmud notes the unusual case where vineyard rows are sixteen cubits wide: *b. Kil.*4.9. A cubit is roughly the length of a forearm from elbow to fingertips: M. Powell, "Weights and Measures," *ABD*, 5:899.
[125]J. B. Pritchard, *The Ancient Near East in Pictures Relating to the Old Testament* (3d ed.; Princeton: Princeton University Press, 1969), photo 451, p. 155. New Kingdom tomb scenes depict vines on poles or trellised: Lesko, *King Tut's Wine Cellar*, 15; Unwin, *Wine and the Vine*, 70; Darby et al., *Food: The Gift of Osiris*, 2:561, fig. 14.1; Johnson, *Vintage*, 31.
[126]P. Albenda, "Grapevines in Ashurbanipal's Garden," *BASOR* 215 (1974): 5-17.
[127]"The Fox and the Grapes," in *Aesop's Fables* (trans. J. Keller and L. C. Keating; Lexington, Ky.: University Press of Kentucky, 1993), 108.
[128]Columella *De re rustica* 4.1, 4-6; 5.5.2.

if followed, meant that the Roman vintner ought to have decided on a method of training before he planted his vines.

Roman vintners would train vines to grow up trees and even devote entire orchards to this type of intercultivation. This particular form of tree-lined vineyard was termed an *arbustum*. For Cato, a vineyard on its own was superior to an arbustum. In his list of best farms, a vineyard is the first, with *arbustum* only a distant eighth.[129] Varro describes both staked and trellised vines and the best tree type for each: oak or juniper wood for stakes; maple or fig trees for *arbusta*.[130] The Roman preference for trellised vines may be due in part to the moister soil of Italy during the summers. Italy does get rain in the summers, so its soil is moister than that of Israel. The summers are the ripening time for berries and if the fruit is left on a moist soil, it is subject to rot.

We discern the prevalent Israelite method of vine training from biblical descriptions of vines. We are fortunate in this instance to have as well one iconographic representation of Judahite vines in the Nineveh relief of the capture of Lachish in 701 BCE. The rarity of iconography for ancient Israel, in contrast to Egyptian and Mesopotamian cultures, makes this relief an invaluable resource for the daily features of Israelite life. Judahite captives are depicted with their clothing, hairstyles, animals, and chariots. Vines and olive trees dot the landscape as the captives descend outside their city walls. The vines are freestanding and low to the ground.[131]

The vine planted in Ezek 17:6, as we saw above, spreads out unrestrained and low to the ground:

$$\text{גֶּפֶן סֹרַחַת שִׁפְלַת קוֹמָה}$$

a vine, spreading out, low in height
(Ezek 17:6)

This vine is apparently freestanding. No mention is made of stakes or trees in the divine vine plantings of Isa 5:1-7 or Jer 2:21-22, though prophetic silence is certainly no proof of their absence. In Jotham's fable (Judg 9:8-15) trees approach first the olive tree, then the fig tree, then the vine, and finally the bramble in their quest for a king to rule over them. The sequence is by descending order of height.

---

[129]Cato *Agriculture* 1.7.
[130]Varro *On Farming* 1.8.3-4; Columella *De re rustica* 5.6.5.
[131]Bleibtreu, *Die Flora der neuassyrischen Reliefs*, 131-39; Russell, *Sennacherib's Palace Without Rival at Nineveh*, photos 203-7.

Olive trees average 4-6 m, while the fig tree is typically somewhat smaller, averaging anywhere between 3-6 m. The vine, as a shrub plant, is much shorter. The trunk of a freestanding vine grows to a little over half a meter.[132] The precise botanical identification of "bramble" (אָטָד) is uncertain, though it is widely held to be some form of thorn shrub. In any case, its stature was the least of the four candidates.

The vine's place in this sequence allows for a height somewhere between the olive tree and bramble. Within these dimensions, the vine could be freestanding, staked, or trellised. A freestanding vine is, however, most likely in Jotham's fable, since the needed support of a stake or tree would vastly undermine the vine's leadership potential. Further, nothing about the vine here indicates that it was a poor choice for king. In fact, it is already suitably employed cheering gods and men with its wine (v 13). Rather, it is only when the trees ask mere bramble to lead that their desperation becomes apparent, and the fable as a whole thereby ridicules Abimelech's reign.

The benefit to growing vines on the ground for ancient Israel would have been twofold. First, the foliage would shade the soil from the sun, thereby lessening the evaporation of its moisture. Israelite soil went without any rainfall in the summer months, when berry development occurred. The preservation of soil moisture would be vital. With the ground so dry in the summer, berry rot would not be the danger that it was for ancient Roman viticulture. Second, freestanding vines dispense with the additional labor and timber needs necessary for *arbusta* or staking each vine.[133] Varro apparently assumed freestanding vines for Palestine when he described the kind of vineyard in which "the ground serves as a bed for the grapes, as in many parts of Asia."[134]

About the only real threat to such a system of cultivation is nevertheless a serious one, that of pests. Foxes[135] are twice noted in the Bible for their destructive propensity in vineyards: in Judg 15:4-5, albeit with Samson's help; and in Cant 2:15, where they come out when the vines blossom. Varro was even aware of their delight in the freestanding vines of Asia: "Foxes often share the harvest with man in such vineyards."[136] Worms curse the vineyards in Deut 28:39 and destroy the entire crop. The lion that Samson met in a vineyard

---

[132]Cox, *From Vines to Wines*, 48.

[133]Timber was largely imported from Syria and Phoenicia, for, by the Iron II period, the forests were denuded: Stager, "Agriculture," 13; Borowski, *Agriculture in Iron Age Israel*, 5; Semple, *The Geography of the Mediterranean*, 399; Ross assumes forked sticks for propping the vines, but cites no evidence: "Vine," 785.

[134]Varro *On Farming* 1.8.4-6.

[135]Forbes, *Studies in Ancient Technology*, 3:109; Semple, *The Geography of the Mediterranean*, 398.

[136]Varro *On Farming* 1.8.5-6.

undoubtedly would have caused some damage had he been allowed to live (Judg 14:5-6). Still, the risk of pests to freestanding vines need not have offset the ease of cultivation that they offered. Given their feasibility in the dry Israelite summers, as well as their attestation in the Lachish siege relief, freestanding vines were probably the most common in Israelite vineyards. Prophetic passages, particularly Ezek 17:6, but also Isa 5:1-7 and Jer 2:21-22, support this hypothesis, but are themselves not conclusive proof of it. Other vine-training methods are hinted at as well in biblical literature.

In several biblical texts, the vine's stature, more so than its flourishing, is stressed for metaphoric effect. Literary attention to vine height may well indicate knowledge, if not use, of trellising. In Ps 80:10, the vine that Yahweh transplanted from Egypt (i.e., his people) looms overhead to shade the Palestinian mountains. Ezekiel, whose vine in 17:6 was low to the ground, later likens Israel to a vine that "towers"(19:11: וַתִּגְבַּה קוֹמָתוֹ) and is known for its height (19:11: וַיֵּרָא בְגָבְהוֹ).

One image in the Song of Songs may also reflect vines that are trellised. In 7:8, the woman's height is likened to a palm tree, which the lover wants to climb. Her breasts are lower, indeed chest high, and are to her lover "clusters of the vine" (v 9: אֶשְׁכְּלוֹת הַגֶּפֶן).[137] The imagery of course evokes more than just stature. Throughout the song, horticulture is in the unabashed service of erotica. The ripenings of fruit and love are paired and awaited. Still, in this passage, the image is of a woman, upright as a tree, with her lover standing close enough to smell the fruity scent of her breath (v 9: וְרֵיחַ אַפֵּךְ כַּתַּפּוּחִים). At that angle, vine clusters are her breasts, and so are too high up to be freestanding.

Trellising may be implied in another biblical image, used in 1 Kgs 5:5 [ET 4:25]; Mic 4:4; and Zech 3:10. These verses express periods of prosperity wherein:

אִישׁ תַּחַת גַּפְנוֹ וְתַחַת תְּאֵנָתוֹ

every man under his vine
and under his fig tree
(1 Kgs 5:5)

---

[137]Varro's recommended limit for Roman trellising is the height of a human, in his case, a man: *On Farming* 1.8.5; cf. Pliny *Natural History* 14.15.

will enjoy peace and hospitality in the shade. For the 1 Kgs passage, that period is during Solomon's rule, one lauded by the Deuteronomistic historian for its peace and economic expansion. In Micah, such prosperity is an eschatological hope of the "last days" (4:1: אַחֲרִית הַיָּמִים). Zechariah also depicts a future day of peace, and he adds social gatherings in his vision:

אִישׁ לְרֵעֵהוּ אֶל־תַּחַת
גֶּפֶן וְאֶל־תַּחַת תְּאֵנָה

every man and his neighbor
under vine and fig tree.
(Zech 3:10)

This biblical vision of prosperity is an apt one, for any man so situated under vine and fig tree was privy to shade and valued crops. The image of an Israelite sitting under vines and fig trees connotes relief from the hot Israelite sun, rest from work, and enjoyment of their fruit.[138]   It may also indicate a practice of trellising vines, or that such a practice was part of an idealized prosperity, one that allowed every man to cultivate vines and figs and to trellis them. This is particularly the case with Zech 3:10, since there is room for one's neighbor under the vine as well.

At this point, it is worth pausing for a moment to discuss why the vine and fig tree are paired in the three passages just discussed (1 Kgs 5:5; Mic 4:4; and Zech 3:10), as well as elsewhere (e.g., Deut 8:8; 2 Kgs 18:31; Neh 13:15; Ps 105:33; Isa 36:16; Jer 5:17; 8:13; Hos 2:14 [ET 2:12]; Joel 1:7, 12; 2:22; Hag 2:19; cf. Matt 7:16; Luke 6:44). Stager's analysis of ancient horticulture is instructive for understanding these biblical traditions of the vine. He argues that there are three prerequisites for horticulture: permanent fields, residential stability, and tranquility.[139] These prerequisites allow the farmer to begin the long-term work of horticulture, with its heavy investment in time, capital, and labor, with some chance of success.[140] The existence of horticulture in ancient Israel, then, suggests that it

---

[138]Hesiod describes a similar scene as the reward due for the hard-working life on the farm: *Works and Days* 590-93.
[139]Stager, "Firstfruits," 177.
[140]Stager, "Firstfruits," 181.

had progressed beyond mere subsistence in order to be able to provide this level of stability.[141]

Permanent fields, residential stability, and tranquility were, of course, esteemed in their own right and, as prerequisites, were particularly associated with horticulture. Hence, in the biblical texts, the mention of fruit growth signaled these social conditions and became a trope for them. I suggest that these prerequisite conditions were carried in the symbolic freight of vines, olives, and fig trees throughout biblical traditions. The promise of fruit growth, then, always implied these beneficial social conditions. So, for example, for the prophet Hosea, Yahweh will lure Israel back to him by giving her vineyards, and this is her hope (Hos 2:17 [ET 2:15]). The pairing of the vine and fig tree, in particular, became a kind of shorthand for the social and economic stability horticulture evinced. So, for example, when the Rabshakeh tempts the people of Judah with a peace and relocation that include vines and fig trees (2 Kgs 18:31-32), these plants signal at once their treasured fruits as well as the regional security to produce them.

The opposite is true as well, namely, that the destruction of vines and fig trees signals an end to prosperity and peace (Ps 105:33; Jer 5:17; 8:13; Hos 2:14 [ET 2:12]; Joel 1:7). Military campaigns in Syria-Palestine often included vine and orchard destruction.[142] Israelite prophets like Isaiah, for example, could effectively, even menacingly, employ viticulture in metaphor, precisely because of what it symbolized in daily farm life. Relatively minor crops, such as onions or lentils, carry nowhere near the same symbolic freight, and so are scarce in biblical texts. These biblical texts on vines, then, reflect viticultural practices and the value of the vine in ancient Israel.

Biblical descriptions of vines, as we saw above, often occur with other fruit trees, but this in no way indicates a practice of intercultivation. Thus, evidence of trellising by itself does not imply intercultivation; there would have to be an added detail, such as vine branches growing up a tree or the tree's support of the vine. Orchards and vineyards too are paired, and they seemed to stand as discrete, not mingled, agricultural plots (Exod 23:11; Deut 6:11; Josh 24:13; Judg 9:8-15; 15:5; 1 Sam 8:14; Neh 9:25; Cant 6:11; 7:12). These pairings may simply offer an agricultural taxonomy of the land's products and have nothing to do with methods of cultivation. Vines in fields are mentioned only twice, and the intercultivation of grain with vines, as we saw above, is prohibited in Deut 22:9.[143]

---

[141] Badler, McGovern, and Michel, "Drink and Be Merry! Infrared Spectroscopy and Ancient Near Eastern Wine," 32.

[142] See below, pp. 123-26.

[143] See p. 113.

Because olive trees, fig trees, and vines thrive under similar geographic conditions, a chalky, rocky soil with low moisture on sloping hills, they may well have been grown near one another in ancient Israel.   They would, then, quite naturally be mentioned together in biblical land descriptions. Still, vines are paired with trees in these instances because they were both cultivated, not because they were trained to grow together.   In sum, the biblical coincidence of vines with trees or other crops cannot be used as evidence of either their intercultivation or training in *arbusta*.

V. Pruning

Pruning is an important facet of training grape vines.   It is mentioned only in Isaiah's Song, as one of the activities that will cease in Yahweh's destruction of his vineyard (5:6).   Its importance in the Song, then, is glimpsed by the withdrawal of this divine care. Pruning enhances the fruit growth over wood or leaf growth.   Cutting back on the wood growth of the branches diverts the plant nutrients into the fruit.[144]   Pliny recommended that most of the vine be pruned at the end of the first and second years to encourage the roots and branches, for a strong plant.[145]   The Bible nowhere mentions a time for pruning.   In the Napa Valley of California today, pruning is done in March and also in December through February.[146]   The vintner prunes when a bud looks plump, but before it breaks. It is the buds formed on the wood of the previous season's growth that will yield fruit.[147]   Columella and Virgil mention two pruning times in the year, once in the spring and once in the fall after the grape harvest.[148]

Pruning requires some management decisions for the next year's harvest.   That is, if the vintner decides to prune lightly to reap an abundant crop one year, the fruitfulness of his buds for the following year diminishes.   "An unpruned vine will have from 10 to 100 times the buds necessary for a good crop of quality grapes."[149] The vintner, then, must exercise a certain amount of discipline for the

---

[144]Semple, *The Geography of the Mediterranean*, 392; Olmo, "Grapes," 294; Weinhold, *Vivat Bacchus*, 83; Columella provides detailed descriptions of ancient pruning techniques: *De re rustica* 4.7-11, 23-24.

[145]Pliny *Natural History* 17.35. Columella also recommends pruning the new vine down to a small rod: *De re rustica* 4.9.1.

[146]Cox, *From Vines to Wines*, 75.

[147]Cox, *From Vines to Wines*, 75.

[148]Columella *De re rustica* 4.10.1; 4.23.1; *On Trees* 10.1; Virgil *Georgics* 2.406-7, 410.

[149]Cox, *From Vines to Wines*, 77; Pliny stresses the general importance of pruning for abundant fruit, rather than tall, thin wood: *Natural History* 17.35.173-81.

sake of longer-term goals. Pruning is briefly described in the New Testament, where again the vintner is divine: "Every branch that bears fruit he prunes to make it bear more fruit" (John 15:2). Conservative pruning enhances stability in production from year to year and so makes the most sense as a long-term strategy for the ancient vintner. While a one-time big harvest would have delighted the first-time vintner, he would probably not repeat the mistake in subsequent years. In addition, village or family custom would be sure to educate the new vintner against not pruning his vines.

After a year of growth, a small vine trunk will have on average three buds. Three buds will produce three canes by the next year. Pruning today typically begins in the vine's second year. The buds on the trunk after the first year are left alone. By the second year lateral shoots grow, and these are pruned off.[150] Pruning off the side shoots allows concentration of growth on the main cane. Also, during the first two years any flower clusters are pruned.[151] The reason, again, is to concentrate the growth into a hardy vine trunk, before fruiting is allowed to begin.

Biblical attestations of pruning are infrequent and vague. Leviticus 25:3 lists pruning a vineyard along with the yearly sowing of grain. Both activities are to cease in the seventh sabbatical year (v 4). Isaiah 5:6 lists Yahweh's cessation of pruning as part of the vineyard's destruction. In addition to these passages, the root זמר occurs in line 6 of the Gezer calendar: "his two months of pruning."[152] A pruning tool, מַזְמֵרָה, occurs in four passages.[153] Yahweh uses the tool in Isaiah's metaphor of the divine destruction of nations:

כִּי־לִפְנֵי קָצִיר כְּתָם־פֶּרַח וּבֹסֶר גֹּמֵל יִהְיֶה נִצָּה
וְכָרַת הַזַּלְזַלִּים בַּמַּזְמֵרוֹת וְאֶת־הַנְּטִישׁוֹת הֵסִיר הֵתַז

For before the harvest, when the blossom is
over and the flower becomes a ripening grape,
he will cut off the shoots with pruning hooks,
and the spreading branches he will hew away.
                                   (Isa 18:5)

---

[150]Cox, *From Vines to Wines*, 67.
[151]Cox, *From Vines to Wines*, 72.
[152]See chapter 1, pp. 37-38.
[153]See chapter 5, pp. 171-74, for a discussion of the use of this tool during the harvest.

The other three passages are related and also involve a military context.

וְכִתְּתוּ חַרְבוֹתָם לְאִתִּים וַחֲנִיתוֹתֵיהֶם לְמַזְמֵרוֹת
לֹא־יִשָּׂא גוֹי אֶל־גּוֹי חֶרֶב וְלֹא־יִלְמְדוּ עוֹד מִלְחָמָה

They will beat their swords into plow points,
their spears into pruning hooks;
nation will not lift sword against nation,
nor will they learn war any more.

(Isa 2:4)

Mic 4:3 contains the same tradition with only minor grammatical differences.[154] Joel 4:10 [ET 3:10], however, reverses the imagery, so that pruning hooks are beaten into spears in his call to arms against nations, and so he dispenses with the second half of the verse:

כֹּתּוּ אִתֵּיכֶם לַחֲרָבוֹת
וּמַזְמְרֹתֵיכֶם לִרְמָחִים
הַחַלָּשׁ יֹאמַר גִּבּוֹר אָנִי

Beat your plowshares into swords,
and your pruning hooks into spears;
let the weak one say, "I am a warrior!"
(Joel 4:10 [ET 3:10])

The tool itself was a most likely a blade attached to a handle, though no examples from ancient Israel have thus far been identified in the archaeological record.[155] Examples have been found, however, from Roman period Egypt.[156] These examples demonstrate that the pruning

---

[154]וְיִשָּׂא occurs in the plural, יִשְׂאוּ, and לָמַד has a paragogic nun, יִלְמְדוּן. This nun often occurs on third and second person plural yiqtol forms and often in pause, as it is here: Joüon, A Grammar of Biblical Hebrew, § 44e.

[155]King states that it was "simply a blade attached to a handle," though he offers no evidence for how he arrived at this description: Amos, Hosea, Micah, 113. A broken bronze sickle, found in an Iron II house from Gibeon, may have been a pruning hook: J. B. Pritchard, Winery, Defenses, and Soundings at Gibeon (Philadelphia: University Museum, 1964), 130-31.

[156]W. M. F. Petrie, Tool and Weapons (London: University College Press, 1917), 47. For a helpful display of seven such tools, see photo 59, nos. 30-36.

hook was a metal tool, with a sharp concave edge, and perhaps had a wooden handle attached:

Fig.1. Pruning Hook[157]

Since metals were still not inexpensive in the Iron II period, tools would likely have been forged into weapons and then back again after the battle ended.[158]   Virgil contains a similar tradition in his description of war: "pruning knives are forged into stiff swords."[159] From these texts and examples, we infer that the Israelites had a metal tool for their pruning, but not much about how rigorous and frequent their pruning practice was.

In sum, pruning was one of the essential practices of vine tending. Its importance is suggested in Isaiah's Song of the Vineyard, where negligence toward pruning is a means to destroy Yahweh's vineyard (5:6). Pruning was a vital aspect of vine tending because it enabled the vintner to produce consistent fruit yields from year to year.   Other essential practices in cultivating the vine were the planting and transplanting of vines from shoots, rather than pips, and some training of the vine as it grew. The bulk of the biblical evidence suggests that the majority of vineyards probably consisted of freestanding vines. These practices were the essential ones for vine cultivation. Isaiah's Song has mentioned some of them, and other biblical texts have helped to illuminate vine tending in ancient Israel. Protecting a vineyard was another essential practice, and to it we now turn.

---

[157]Redrawn from Petrie, *Tools and Weapons*, photo 59, no. 34.
[158]King, *Amos, Hosea, Micah*, 113.
[159]Virgil *Georgics* 1.508.

VI. Security of the Vineyard

Two features are mentioned in Isaiah's Song as a security system: the walls, discussed above in section II,[160] and the tower, מִגְדָּל, discussed in chapter 4. In 5:2b Yahweh builds a tower to guard the vineyard and, in v 5 the walls surrounding the vineyard are dismantled during his disappointed rage.

Vineyard walls in Iron II would serve primarily as a deterrent to animals and hostile intruders, while the towers afforded a vantage point over the vineyards and for scanning the surrounding area. Small and large animals, such as foxes or oxen, would be discouraged by these installations, as would people. In Aesop's fable "The Hedge and The Vineyard," a vintner tears down the walls and all his vines are destroyed by intruders, both human and animal. The moral of the fable is simply that "it was quite as important to protect his vineyard as to possess it."[161] Without hedges, intruders could not resist a vineyard.

Deuteronomy 23:25 [ET 23:24] allows the possibility of entering a neighbor's vineyard. One may even eat as many grapes as one wants during the walk. However, carrying away grapes is prohibited. In Lev 19:10, the grapes unpicked during harvest are left in the vineyard for the poor and foreigners to collect.[162] These texts suggest that some intrusion into a farmer's vineyard was tolerated, even sanctioned on humanitarian grounds. Taking more than gleanings, however, was likely met with stricter punishments. In Athenian law of the sixth century BCE, for example, the destruction of a vineyard was punished as severely as temple robbery, treason and murder, that is, by death.[163]

Perhaps the biggest threat to a vineyard in the ancient Near East lay in its being vandalized by an attacking force. Fields, vineyards, and orchards were often the first things to be destroyed in an attack. Siege warfare of fortified towns would start with this devastating ploy. This ploy made sense, as it cut off the food supply and quickly weakened the morale of besieged inhabitants. The victims of this tactic would lose not only their immediate food supply, but also their investment in future crops. Since Israelite farmers were dependent on

---

[160]See above, p. 96.
[161]"The Hedge and the Vineyard," in *Aesop's Fables* (trans. T. James; London: John Murray, 1848), 146.
[162]See chapter 5, pp. 174-77.
[163]Weinhold, *Vivat Bacchus*, 66. He cites as well a German medieval law, where both murderers and vine destroyers were to have their right hands cut off.

their fields and vineyards for their livelihood and very survival, siege destruction must have been a painfully effective inducement to surrender.

Several biblical stories detail this threat to fields and vineyards. In the premonarchic period of the judges, the Israelites are depicted as sorely harassed by Midianite invaders, who destroy their fields as soon as they plant their crops. The Israelites are forced to hide in caves and watch as their crops are destroyed (Judg 6:3-6). Gideon takes to threshing wheat inside a winepress,[164] so that he can salvage some of his crop from Midianite incursion.

Samson, in a fit of rage, performs the odd revenge of tying foxtails together and loosing the small animals on Philistine wheat fields (Judg 15:3-5).[165] Such vandalism was memorable not only for its shocking oddity, but for the crop damage it caused. Wheat was destroyed, and so too were the vineyards and olive orchards (v 5). As noted earlier,[166] this act would have done irreparable harm to an enemy's economy. It would have effectively knocked out that year's crop, leaving years before vineyards and olive orchards would produce again. In essence, Samson's act, though bizarre, constituted an act of war.

In 1 Sam 25, David and his men have been protecting Nabal's property. We are not told what crops Nabal grew, but his stock and provisions were formidable. When David turns from protector to attacker, Abigail, Nabal's wife, quickly offers David two containers[167] of wine, five roasted sheep, two bushels of grain, 100 bunches of raisins, and 200 fig cakes to stave off David's avenging wrath. David's horticultural take alone—in wine, raisins, and fig cakes—demonstrates Nabal's wealth. There is no doubt that hyperbole is employed to highlight Abigail's willingness to appease David. Still, the lavish bribe indicates that Nabal's estate was prosperous and in need of security, even if offered in response to David's strong-arm tactics.

Jerusalem's vineyards, according to the Deuteronomistic historian, were threatened during the 701 BCE Assyrian campaign under Sennacherib. The danger to Hezekiah's reign is recorded in 2 Kgs 18 and Isa 36. The military threat to the city is voiced by the Assyrian official, the Rabshakeh, who explicitly states that vines, fig trees, and wells will be spared if the Jerusalem inhabitants surrender to the Assyrian army (2 Kgs 18:31/Isa 36:16). This particular

---

[164]See chapter 4, pp. 141-42, 162.
[165]Foxes even without fire were known as destructive pests of vineyards. Cf. Varro *On Farming* 1.8.5-6; "The Fox and the Grapes," in *Aesop's Fables* (trans. J. Keller and L. C. Keating; Lexington, Ky.: University Press of Kentucky, 1993), 108.
[166]Pp. 110n105, 115, and chapter 4, p. 141.
[167]See chapter 6, p. 217.

military crisis is averted, and Jerusalem does not fall until 114 years later, at the hands of the Babylonians.   Then, Nabuzaradan, the Babylonian army captain of that conquest, is remembered by the Deuteronomistic historian specifically for sparing the vineyards and orchards and leaving poor Judahites to work them (2 Kgs 24:12). His action in the capture and destruction of a city is remarkable precisely because it was at variance with customary ancient Near Eastern siege tactics.

Deuteronomy 20:19 stipulates that fruit trees are not to be felled during siege campaigns, by reminding the Israelites that trees are not the enemy.   The assertion of such a law indicates that fruit trees were often targets in ancient warfare.   Marvin Powell termed vineyard destruction a virtual favorite pastime of first-millennium armies.[168] The Israelites violate the custom described in Deut 20:19 when they fight against Moab; they attack Moab and fell every good tree (2 Kgs 3:25). And in Num 21:22 the Israelite camp reassures King Sihon that it wants only to pass through his land, not touch the vineyards.[169]   Apparently, armies and the threat of vineyard destruction were frequent fears in antiquity.

Egyptian Asiatic campaigns often mention destruction of vineyards or gardens.   Pepi I (2375-2350 BCE) destroys both vineyards and its walls when he

> crushes the land of the sand dwellers.
> After it had thrown down its enclosures,
> the army returned in safety; after it had
> cut down its figs and its vines.
>   (25)[170]

Thutmose III (1490-1436 BCE), the pharaoh who campaigned in Syria-Palestine far more than any other, also made a habit of destroying the vineyards. During his fifth campaign, year 29, he

> destroyed the town of Ardata with its grain.
> All its pleasant trees were cut down.

---

[168]Powell, "Wine and the Vine in Ancient Mesopotamia," 119.

[169]During Emperor Probus' rule in Rome (276-282 CE), the ruler tired of battles and landscape destruction, so he encouraged the legions to build up areas of the Empire instead. He had the soldiers set up a vineyard in Sirmium (modern Serbia), and they killed him for it: Weinhold, *Vivat Bacchus*, 37.

[170]"Asiatic Campaigns under Pepi I," translated by J. A. Wilson (*ANET*, 228). See also, "The Campaign of Weni," *Ancient Egyptian Literature* (ed. M. Lichtheim; Berkeley: University of California Press, 1973), 1:20.

> Now [his majesty] found the entire
> [land of] Djahi, with their orchards
> filled with their fruit.[171]

In this detailed account Thutmose III is described as cutting down the vineyards and trees and making liberal use of the products. The next year, year 30, in his sixth campaign, Thutmose destroys Kadesh by "felling its trees, cutting down its grain."[172] Thutmose III was clearly aware of the effect his tactics had on his victims. In his campaign against Mittanni, preserved in the Barkal Stela, he boasted: "I took away the very sources of life (for) I cut down their grain and felled their groves and all their pleasant trees."[173] He seems to have never wavered in this tactic, as his final campaign, year 42, at Tunip, shows. For that campaign the annals of Karnak describe his arrival, the destruction of the town, cutting of its grain, and felling its trees.[174]

The Assyrians seem to have employed this same siege tactic. Shalmaneser III (858-824 BCE), in the fight against the Aramean coalition of his eighteenth year detailed in the Black Obelisk, stated:

> I followed him [=Hazael] and besieged him
> in Damascus, his royal residence. (There)
> I cut down his gardens (outside of the
> city, and departed).[175]

When Sargon's army attacked Urartu, it cut down vineyards.[176] And again, Assyrians threatened the vineyards and olive orchards when they laid siege to Jerusalem (2 Kgs 18:31). The destruction of fields and vineyards was evidently a menacing tactic in ancient Near Eastern warfare.

---

[171]"The Asiatic Campaigns of Thut-mose III," translated by J. A. Wilson (*ANET*, 239).
[172]Wilson, "The Asiatic Campaigns of Thut-mose III," 239.
[173]Wilson, "The Asiatic Campaigns of Thut-mose III," 240.
[174]Wilson, "The Asiatic Campaigns of Thut-mose III," 241.
[175]"Babylonian and Assyrian Historical Texts," translated by A. L. Oppenheim (*ANET*, 280).
[176]Luckenbill, *Ancient Records*, 2:76-90.

Chapter 4

Installations of the Israelite Vineyard

וַיִּבֶן מִגְדָּל בְּתוֹכוֹ וְגַם־יֶקֶב חָצֵב בּוֹ

He built a tower in its midst
and hewed a winepress in it.
(Isa 5:2b)

Isaiah's vintner next constructs two installations in his vineyard, a tower and a winepress. These complete the preparations of a vineyard, and he has only to await the grape harvest. He had begun, as we described in chapter 3, with a selection of the location for the vineyard. Then he worked the soil and planted the vines. Now come the vineyard tower and winepress. Their presence and function in the Israelite grape harvest are the subject of the present chapter.

A press was essential since most of a grape harvest would go to the production of wine. Some of a vintner's harvest, no doubt, would be for unpressed foodstuffs in the form of grapes and raisins. These fruits were valued as energy sources[1] and were storable.[2] The bulk of

---

[1] Hence raisins and raisin cakes are often listed as provisions for travel in biblical stories (1 Sam 25:18; 30:12; 2 Sam 6:19; 16:1; 1 Chr 12:40; 1 Chr 16:3; Cant 2:5).

[2] Large quantities of charred grape seeds were found in Stratum XIV (tenth century) at Tel Michal: Z. Herzog, "A Complex of Iron Age Winepresses (Strata XIV-XIII)," *Excavations at Tel Michal, Israel*, 74. Many grape pips were found at Shiloh, Stratum V (Iron I): Kislev, "Food Remains," 356. The pips from EB Arad were probably raisins rather than grapes since, as Hopf notes, they had blisters equally spread over their surface, an indication that they were probably dried before being charred. Fresh berries would have been spotted irregularly with larger blisters or splits: Hopf, "Plant Remains, Strata V-I," 73-74. Raisins, rather than grapes, were also evident from the pips from EB Lachish: Helbaek, "Plant Economy in Ancient Lachish," 309-17.

the grape harvest, however, went to the production of wine. Grape wine was also storable, a valued commodity, and, of course, enjoyable as an intoxicant. Hence, the purpose of a winepress is apparent. The presence and function of a tower in Israelite vineyards is, however, less evident in the biblical and archaeological sources. Isaiah's vintner might well be exercising his divine prerogative here. There are, in addition, biblical texts that refer, not to a tower per se, but to a "booth," סֻכָּה in vineyard contexts. The first section of this chapter is an analysis of the textual and archaeological evidence for vineyard towers in Iron Age Israel. A discussion of the vineyard booth is also included, since this structure may have functioned in ways similar to the tower. The second section is devoted to the winepress, with a discussion of the archaeology and biblical terms that denote this installation.

I. The Vineyard Tower

a. Agricultural Towers of the Bible

The tower, like the winepress, could be built after the vines had been planted. There was no need to build either of these installations before planting, and so the vintner had up to four years to do so before his first grape harvest. The sequence of Isaiah's description of the tasks in vv 1-2, then, is apt. Isaiah's tower was *within* the vineyard, בְּתוֹכוֹ, and not on its walls (v 5), or, say, between fields and vineyards. We may assume from its location, then, that its function primarily served that vineyard.

The question raised from Isa 5:2b is this: is the tower a requirement for the Israelite vineyard as was the winepress, or does its construction represent the extraordinary measures of Isaiah's divine vintner? If tower and winepress were the standard installations of Israelite viticulture, then their traces may be left in the archaeological record. The verb חָצֵב, "hew," used of the winepress here, most often has stone material as its object, implied or explicit,[3] so the winepress at least ought to be evident in the material culture of ancient Israel. It is, in fact, but the tower is less so. Let us turn first to a discussion of the tower to discern whether Isaiah's description is at all representative of Israelite vineyards in the Iron Age.

---

[3]E.g., Exod 20:25; Deut 6:11; 7:5; 12:3; 2 Chr 20:25; Ezra 5:8; Neh 9:25; Prov 9:1; Isa 51:1; Lam 3:9; Amos 5:11.

Yahweh built his מִגְדָּל, "tower," within his vineyard. No further description is given of the tower's makeup or its function. מִגְדָּל is a *maqtal* noun formation[4] from the verb גָּדַל, "grow" or "be great."[5] It denotes a structure of some height, literally a grown or great thing, or that which protects a great thing, such as a city or vineyard.[6] Most frequently מִגְדָּל in the Bible signifies a defensive structure for cities, where it serves as a guard post and battle station if need be (e.g., Judg 8:9, 17; 9:51; 2 Kgs 9:17; 17:9; 18:8; 2 Chr 14:7; 26:9; 32:5). The tower was part of the city's fortifications, along with walls, gate entryways, and perhaps earthen glacis, which increase the slope for approach. Archaeological examples of Iron Age towers for city fortification include Hazor, Tell Beit Mirsim, Timnah, Tell en-Nasbeh, and Jerusalem.[7] In addition, towers are depicted iconographically for two cities of the southern Levant: Ashkelon in the New Egyptian Karnak relief,[8] and Lachish in the Nineveh relief of Sennacherib.[9] Towers, then, were a facet of Iron Age city fortifications in Israel. Their use as agricultural installations, however, is far less evident.

This Isaiah passage, in fact, is unique in the Hebrew Bible for its inclusion of a tower within a vineyard. The New Testament contains two examples of vineyard towers—also built as winepresses are dug—in Mark 12:1 and Matt 21:33, but these writings are eight centuries later than First Isaiah. The question for historical reconstruction is whether the vineyard tower in the literary material was instrumental or rare to Israelite viticulture in the Iron Age.

Only one other biblical example exists of an agricultural tower. In 2 Chr 26:9-10, King Uzziah builds two kinds of towers, one for the gates of Jerusalem (v 9) and another out in the wilderness, בַּמִּדְבָּר (v

---

[4]Called a "substantive of location," with *a* shifting to *i* according to Barth's law of dissimilation: Waltke and O'Connor, *An Introduction to Biblical Hebrew Syntax*, 90.

[5]BDB, 152.

[6]The biblical tower most known for its imposing height is that built in Gen 11 as the tower of Babel.

[7]Y. Yadin, "Hazor," in *NEAEHL*, 2: 601; W. F. Albright, "The Excavation of Tell Beit Mirsim" *Annual of the American Schools of Oriental Research;* vol. 3; nos. 21-22; (New Haven, Conn.: American Schools of Oriental Research, 1943), 40-41; Kelm and Mazar, "Three Seasons of Excavations at Tel Batash—Biblical Timnah," 20; Zorn, "Tell en-Nasbeh," in *NEAEHL*, 3:1099; Y. Shiloh, "Jerusalem: The Period of the Monarchy (Strata 14-10)," in *NEAEHL*, 2:706. For a discussion of towers as a feature of Iron Age period fortifications in ancient Israel, see Z. Herzog, "Fortifications: Bronze and Iron Ages," in *The Oxford Encyclopedia of Archaeology in the Near East* (ed. E. M. Meyers; 5 vols.; New York: Oxford University Press, 1997), 2:325.

[8]L. Stager, "Merenptah, Israel and Sea Peoples: New Light on an Old Relief," *Eretz-Israel* 18 (1985) (Avigad volume), 57*, fig. 2.

[9]Russell, *Sennacherib's Palace Without Rival at Nineveh*, 205-9.

10).[10] The latter group of towers indicates an agricultural function, inclusive of viticulture, both by its context and by its location:

וַיִּבֶן מִגְדָּלִים בַּמִּדְבָּר וַיַּחְצֹב
בֹּרוֹת רַבִּים כִּי מִקְנֶה־רַּב הָיָה לוֹ
וּבַשְּׁפֵלָה וּבַמִּישׁוֹר[11] אִכָּרִים
וְכֹרְמִים בֶּהָרִים וּבַכַּרְמֶל
כִּי־אֹהֵב אֲדָמָה הָיָה

He built towers in the wilderness and hewed
out many cisterns, for he had large herds;
and in the Shephelah and in the plain, farmers;
and vintners in the hills and in the fertile lands,
for he was a lover of the soil.

(2 Chr 26:10)

Two כִּי clauses provide the rationale for Uzziah's building campaign. Towers and cisterns are built *because* he had large herds, and farmers and vintners out in the hills.[12] The second כִּי clause, that he loved the soil, could as well explain his tower and cistern construction, or it may explain the entire agricultural enterprise of towers, cisterns, herds, farmers, and vintners. In either case, the context is decidedly agricultural. In fact, the passage represents the mixed economy characteristic of Israel, with its herds, farmers, and vintners. Uzziah's towers here are in service to agricultural development away from Jerusalem. Viticulture was a feature of this agricultural development as the presence of vintners "in the hills and in the fertile lands" indicates. Hill slopes were used for viticultural cultivation, while the plains were left to grain.

The language of the description for construction of these agricultural towers is similar to that in Isa 5:2b. Both subjects "built", וַיִּבֶן, a "tower(s)," (יִם) מִגְדָּל, and "hewed," חָצֵב, an installation—a cistern for Uzziah and a winepress in Isaiah. The language in both passages thus depicts the construction activities for agricultural

---

[10]For a discussion of the possible locations of Uzziah's construction, see Rainey, "Wine From the Royal Vineyards," 57-62.

[11]The *athnach* should probably be moved to the וּבַכַּרְמֶל. In this way, the motive for all the activities listed is stressed, viz., Uzziah loved the soil. See S. Japhet, *I and II Chronicles* (Old Testament Library; Louisville, Ky.: Westminster/John Knox, 1993), 881.

[12]J. N. Graham, "'Vinedressers and Plowmen': 2 Kings 25:12 and Jeremiah 52:16," *BA* 47 (1984): 55-58.

pursuits. In the case of Isaiah's vintner, that construction went exclusively to a vineyard. In King Uzziah's case, it served a mixed economy, inclusive of grape-growing, herds, and more generally, farming.

Second Chronicles 26:10 occurs within a passage primarily concerned with military buildup during Uzziah's reign (vv 6-15). Some scholars, therefore, read v 10 as evidence of garrison stations.[13] The suggestion is possible even though no military hint exists in the verse, for soldiers could have farmed as part of their military service. The verse would then reflect Uzziah's combining of military and agricultural activities. However, all our other evidence for army provisions at present, both biblical and extrabiblical (1 Sam 17:17; 1 Sam 25:18; 2 Sam 17:27-29; the Arad letters; and *lmlk* jars), indicates that foodstuffs were sent to garrisons ready-made, rather than produced on site. This 2 Chronicles passage, then, may provide a glimpse into a new type of provision delivery system during Uzziah's reign: either "farmers and vintners" gave from their own yields in exchange for the continued royal support of their endeavors, or some garrisons farmed while they defended the land. Another possibility is that soldiers were given grants of land in exchange for their military service. This custom would parallel the Old Babylonian (and Kassite) *ilku* service, whereby soldiers were given land grants and then expected to farm and perform military duty for the state.[14] A similar type of military service is hinted at in 1 Sam 8:12 as sons are conscripted to plow the king's fields and make his weapons. While the precise custom in the Chronicles passage remains unknown, royal support of farming is clearly indicated by the verse.

The Chronicler describes Uzziah's domestic policy, one uniquely divided between defense (vv 6-9, 11-15) and agricultural expansion (v 10). The Dtr account of Uzziah's (Azariah's) reign in 2 Kgs 14:21-15:4 is silent on his domestic policy, save for the recapture of Elath, a southern port city. The motive and source for the Chronicler's rendition of Uzziah's reign, one inclusive of agricultural enterprise, cannot be determined. Second Chronicles 26:10, though, does contrast sharply with other Dtr materials concerning the monarchy and agriculture.

Neither Uzziah nor any other king is otherwise noted for his agricultural development in Dtr. In fact, 1 Sam 8:10-18 (of the antimonarchic source) cautions against kings because they would seize vineyards, fields, and orchards and redistribute them to their servants. Kings here and in the story of Naboth's vineyard (1 Kgs 21)[15] are

---

[13] Aharoni, *The Land of the Bible*, 314.
[14] J. Oates, *Babylon* (2d ed.; New York: Thames & Hudson, 1986), 72-73.
[15] See chapter 2, pp. 68-85.

deemed outright threats to the agrarian livelihood of the Israelite citizenry. According to the Chronicler, part of what was noteworthy about Uzziah was his support of agriculture. Uzziah is the only person in the Bible described with the phrase, "he was a lover of the soil" (v 10), and so his agricultural devotion is clearly being emphasized.[16] Chronicles was most likely written in the Persian period (538-333 BCE),[17] so we have a literary continuity of vineyard towers from Isa 5:2 to 2 Chr 26:10 to the New Testament examples in Mark 12:1 and Matt 21:33. It is time now to assess the archaeological evidence for agricultural towers to establish their prevalence or rarity in the Iron Age.

b. Archaeology of the Field Tower

Information about ancient cultivated areas is largely available through survey work, and these areas have not received the same degree of archaeological attention as have city and village sites.[18] Fields, orchards, and vineyards often occurred at a distance from dwelling sites and they leave few architectural and ceramic traces for the archaeologist. Within the last thirty years, survey study of Israel has increased. Adam Zertal has surveyed Manasseh,[19] Israel Finkelstein and Shimon Dar have surveyed the broader outline of Samaria,[20] Ze'ev Herzog has investigated the north coastal range,[21] and Moshe Kochavi and recently Avi Ofer have surveyed Judah.[22] These survey studies, along with site reports, detail few towers in Iron Age fields. One stands atop a mound at Giloh, south of Jerusalem, dated to Iron II.[23] Another tower stands at Deir Baghl in the Judean hills,

---

[16]Japhet, *I and II Chronicles*, 881.

[17]Japhet, *I and II Chronicles*, 28.

[18]R. Frankel, *The History of the Processing of Wine and Oil in Galilee in the Period of the Bible, the Mishna and the Talmud* (Hebrew with English summary) (Tel Aviv: Tel Aviv University Press, 1984), ii.

[19]A. Zertal, *The Israelite Settlement in the Hill Country of Manasseh* (Hebrew) (Haifa: Haifa University Press, 1988).

[20]Finkelstein, *The Archaeology of the Israelite Settlement*; Dar, *Landscape and Pattern*.

[21]Herzog, et al., *Excavations at Tel Michal, Israel*.

[22]M. Kochavi, ed., *Yehudah, Shomron ve-Golan: Seker arke'ologi bi-shenat tav-shin-kaf-het* (Jerusalem: Hotsaat ha-Agudah le seker arkhiologi shel Yisrael al yede Karta Yerushalayim, 1972); A. Ofer, "'All the Hill Country of Judah': From a Settlement Fringe to a Prosperous Monarchy," in *From Nomadism to Monarchy: Archaeological and Historical Aspects of Early Israel* (ed. I. Finkelstein and N. Na'aman; Jerusalem: Israel Exploration Society, 1994), 93-121.

[23]Mazar, "Giloh: An Early Israelite Settlement," 5.

attached to a square building.[24] There are none at Timnah, Ekron, or Lachish, sites whose horticultural activity is otherwise apparent.

Zvi Ron has conducted an extensive study of field towers in modern Israel. He discovered over 10,000 stone huts in fields today in the West Bank. His results are informative: 89 percent of the stone huts he studied were in vineyards and olive orchards.[25] Ron argues that these stone huts are an expression of the traditional agriculture of the rocky, hilly area of Judea and Samaria since earliest times.[26] In addition, he discovered that "winepresses are frequently found hewn in the vicinity of the stone huts."[27] These field huts, it would seem, became at some point features specifically of grape and olive growing in Palestine and Israel. Ron reasons that as vineyards and orchards were planted at some distance from the village, the need to protect them increased. Hence, a primary function of the field tower for these valued crops was as a guard post.[28]

A tower also provided shade during the day for those guarding the plots and acted as a deterrent to potential intruders who saw it from a distance. Ron tested the temperatures of the huts and found that on average, the huts were eight to thirteen degrees cooler than the outside temperature in the early afternoon heat.[29] The laborers at the grape harvest likely also enjoyed the shade and cooler temperatures provided by vineyard towers.[30] They may have elected to spend the night in the fields as an added security measure,[31] and then these same towers could have provided shelter from the cooler night temperatures. Isaiah terms one of the field huts a מְלוּנָה, a lodging place (1:8), so these towers might also have been used for temporary lodging.

Because towers offered cooler temperatures than the field, they could have played a role in storing jars of wine after pressing. Dar has suggested as well that part of the wine's fermentation in jars could have taken place in these cooler towers.[32] Some of the towers he investigated in western Samaria had, for example, Hellenistic amphora and juglet shards inside them.[33]

---

[24]A. Mazar, "Iron Age Fortresses in the Judaean Hills," *PEQ* 114 (1982): 105.

[25]Z. Ron, "Stone Huts as an Expression of Terrace Agriculture in the Judean and Samarian Hills," (Ph.D. diss., Tel Aviv University, 1977), viii.

[26]Though he does not specify historical periods: Ron, "Stone Huts," ii.

[27]Ron, "Stone Huts," ix.

[28]Ron, "Stone Huts," vi.

[29]Ron, "Stone Huts," xi-xii.

[30]S. Applebaum, S. Dar, and Z. Safrai suggest that these towers may also have been used for storing tools or dried fruit: "The Towers of Samaria," *PEQ* 110 (1978): 97.

[31]Ron, "Stone Huts," ix. See below, pp. 136-42.

[32]Dar, *Landscape and Pattern*, 1:110, 157-58.

[33]Dar, *Landscape and Pattern*, 1:108.

Field tower remains occur primarily in the north. Dar surveyed western Samaria from 1974 to 1981 and found no field towers dating from the Iron Age. In all, however, Dar found ancient remains of 1,200 agricultural towers dotting the landscape from Gilboa in the north to Nahal Beit ʿArif, just northeast of Lod.[34] It may well be that towers from the Iron Age were dismantled in subsequent periods for their stones, and so are not evident in the archaeological records. This is, in fact, a good possibility, since scavenging for stones was much easier than quarrying new stone. Because of their sheer numbers in Dar's survey, it is evident that field towers were prominent features of the region of Samaria in antiquity, and so we infer their presence for the Iron Age as well.[35] The quantity of ancient field towers in Samaria was noted even back in 1881 by Kitchener and Conder.[36] Their survey of Judah, by contrast, found no field towers.[37]

Most of the field towers investigated by Dar are from the Hellenistic and Byzantine periods. Nevertheless, his investigation of towers is significant for our understanding of Iron Age viticulture in two respects. First, it demonstrates that agricultural towers did exist in ancient fields and vineyards. Because of the limited arable land of ancient Israel, fields tended to be used throughout historical periods and were not abandoned or covered by subsequent occupation to the same extent as were strata on tells. We can reasonably infer that some field installations, such as towers or winepresses, would be reused or renovated through historical periods and so leave little or no evidence of their origins. Second, the proximity of wine and olive presses and cisterns to many of the 1,200 towers further indicates that towers were agricultural, more specifically, horticultural, in function.[38] The Bene Hassan Tower 80 in western Samaria, for example, has a nearby winepress, cup-mark, threshing floor, and oil presses.[39] Dar even found vine remains in two of the towers, that of 49 Qarawat Bene Hassan, south of Tul Karem, and nearby Tower 69 near Shufah.[40]

---

[34]Dar, *Landscape and Pattern*, 1:92; 2, fig. 64. For a detailed study of three such towers, see also Applebaum et al., "The Towers of Samaria," 91-100.

[35]Applebaum et al., "The Towers of Samaria," 91. Fifty-eight towers have also been found in Iron Age sites in the Negev with evidence of animal husbandry and grain farming, but no horticultural activity. These, then, served an economy specific to this drier region. See Haman, "The Iron Age II Sites of the Western Negev Highlands," 36-61.

[36]C. R. Conder, and H. H. Kitchener, *The Survey of Western Palestine* (vol. 1; London: Palestine Exploration Fund, 1881).

[37]Applebaum et al., "The Towers of Samaria," 95.

[38]Applebaum et al., "The Towers of Samaria," 94, 97. Ron adds that sometimes they are found thatched with growing vines: "Stone Huts," i.

[39]Dar, *Landscape and Pattern*, 1:110-11.

[40]Dar, *Landscape and Pattern*, 1:111; 2, figs. 107, 130.

Dar investigated 170 towers around Qarawat Bene Hassan and 30 more further south in the Khirbet Sur sector near 'Azzun.[41] He discovered around many of them pottery from periods extending from Iron II through Byzantine. The predominance of pottery for these towers was from the third to second centuries BCE.[42] He surmises from this that the towers themselves were Hellenistic in design, but that they occur in agricultural plots that were in use since Iron II.[43] He, like Ron, views the tower as a particular feature of the vineyard: "its invention was connected with the rapid development of viticulture and wine production in early Iron II."[44] I concur and suggest that Isa 5:2b and 2 Chr 26:10 represent the literary traces of these stone structures in the vineyard.

While the Iron Age towers are no longer standing, the Iron II pottery found near later field towers suggests that they had been. The stones of a tower would have been prime targets for removal and reuse in antiquity, and so, in this instance, the absence of evidence need not imply evidence of absence. When land plots underwent grape cultivation in successive periods, as they undoubtedly would have, these makeshift towers would have been dismantled to make room for the hewn stone structures so characteristic of the Hellenistic period. A field stone tower would be replaced by its Hellenistic or Byzantine counterpart without leaving any archaeological evidence such as a foundation. During the Hellenistic period, improved technology and available labor produced quarried stone towers in association with agricultural installations, and these then survived in significant numbers in the archaeological record.

It is unlikely that Isa 5:2b reflects an agricultural tower otherwise rare in eighth-century BCE Israel. Isaiah's description of the construction retains a certain verisimilitude that suggests that it was not a rare undertaking for the farmers of Isaiah's audience. For, if a tower were to be built, the Israelite vintner would most likely have used the fieldstones collected from his vineyard, rather than quarried for them. In Isa 5:2a, it will be recalled, Yahweh does just that; he clears the vineyard soil of stones. In v 2b, these stones, it may be inferred, then get used for tower construction.

Much of the material for building a vineyard tower would have been at hand from having cleared the soil of stones, and the tower's usefulness for security,[45] storage, and cooling is apparent. Those stones collected from the vineyard soil would otherwise have to be hauled away. The latter activity is nowhere described in the Bible and

---

[41]Dar, *Landscape and Pattern*, 1:108-9; 2, figs. 83, 130.

[42]Dar, *Landscape and Pattern*, 1:108.

[43]Dar, *Landscape and Pattern*, 1:109.

[44]Dar, *Landscape and Pattern*, 1:114.

[45]Broshi, "Wine in Ancient Palestine," 23; Ron, "Stone Huts," iii.

would be an unnecessary effort if a structure could be built from them. The field tower in some sense, then, was not a sophisticated installation, but could have developed from the preparation of a vineyard. Isaiah 5:2, then, offers evidence for vineyard towers in the eighth century BCE. Second Chronicles 26:10 may also, as it comments on Uzziah's reign.[46]

If Uzziah did build agricultural towers during his reign, Isaiah, the Jerusalem prophet, would have been aware of the king's activities. By describing the construction of a vineyard tower, Isaiah could have drawn on farmer (i.e., commoner) custom and monarchic policy. If so, the image would have been particularly apt for the Jerusalem audience. It was also a subtle theological comment on the power of the king: a reminder that Uzziah does not do anything, viz., tower building, that the deity cannot. The potential impact of such prophecy would then be to remind the people that their God loved them just as their king Uzziah "loved the soil." In this reading, towers instance the agricultural devotion of king and Yahweh.

c. Biblical *Sukkah*

While the tower thus predominated in the Hellenistic and Byzantine vineyards of ancient Israel, Isa 5:2 and 2 Chr 26:10 suggest the existence of vineyard towers in the Iron Age as well. Other biblical texts indicate that another structure present in the vineyard might have functioned in ways similar to the tower. The vineyard booth, סֻכָּה, *sukkah*, may have been a precursor or alternative to the stone tower in Israelite viticulture. These booths, perhaps made of branches or animal skins, could have performed many of the functions that were (later) the province of stone towers.[47] Since the booths were probably made of perishable materials, they do not leave archaeological traces. Ron discovered some modern-day field towers whose roofs were thatched with grape vines.[48]

Booths were most likely temporary or, better, seasonal, structures erected for the harvest. At such time, they too could have been used as both guard posts and cooling areas for the overheated laborers. For Isaiah, the booth offered shade by day and refuge from storms (Isa 4:6). Once the grapes were pressed and the juice put into containers, the jars could sit in the booth until the next trip to the

---

[46]He is also mentioned as the first king under whom Isaiah prophesied (Isa 1:1).

[47]A stone structure would typically have been cooler than one made from branches or animal skins.

[48]"Stone Huts," i.

farmhouse. The booth would have been flimsy architecturally, especially in comparison to a stone tower, but nevertheless functional.

Isaiah himself employs the image of a booth in a vineyard when he compares it to daughter Zion in 1:8. A second makeshift field structure likewise guards the cucumber field:

> And daughter Zion is left like a booth (סֻכָּה) in a vineyard,
> like a shelter (מְלוּנָה) in a cucumber field, like a besieged city.
> (Isa 1:8)

So Isaiah draws upon two traditions of vineyard structures: a booth here, and a tower in 5:2. In 1:8, it is the booth's very vulnerability that qualifies it for metaphor. Daughter Zion (Jerusalem), Isaiah suggests, is as precarious as a vineyard *sukkah*. She is a besieged city, whose fortifications are not going to hold. Zion herself, i.e., the eighth century BCE *fortified* city, is closer in image to a מִגְדָּל, a stone structure, yet Isaiah chooses סֻכָּה instead precisely for its vulnerability, its risk of being utterly crushed or blown away. He demonstrates at once his metaphoric dexterity and viticultural knowledge. Obliquely, he has also informed us of the limitations inherent in a vineyard booth and suggested the superiority of a tower by Yahweh's choice in 5:2b. The stone tower offered improvement over a booth in terms of permanence and as a cooling agent. It might have carried some prestige value as well, demonstrating the commitment of the vintner to his cultivation. This prestige is glimpsed in Yahweh's choice of מִגְדָּל over סֻכָּה, שֹׂרֵק over גֶּפֶן, and hill slope for his vineyard. Isaiah's divine vintner at all points was using the best options within Israelite viticulture.

Additional evidence of the vineyard booth is culled from the biblical calendars marking the fall harvest festival: four biblical agricultural calendars—Exod 23:14-17; Exod 34:18-23; Deut 16:1-17; and Lev 23—and one extrabiblical agricultural calendar, that of Gezer unearthed in 1908. Exod 23:14-17 and 34:18-23 are widely considered to be the earliest biblical calendars,[49] dating anytime after the settlement to a period well before Deuteronomy of the seventh century BCE.[50] They are followed, next, by the calendar in Deut 16

---

[49]G. W. MacRae, "The Meaning and Evolution of the Feast of Tabernacles," *CBQ* 22 (1960): 252.
[50]B. S. Childs, *The Book of Exodus: A Critical, Theological Commentary* (Old Testament Library; Philadelphia: Westminster, 1974), 483; J. Wellhausen, *Prolegomena to the History of Ancient Israel* (New York: Meridian, 1957), 85-86; R. de Vaux, *Ancient Israel: Its Life and Institutions* (New York: McGraw-Hill, 1961), 470.

and, then, that in Lev 23. The dating of this latter calendar is uncertain, since the Priestly source (P) and the Holiness Code (Lev 17-26), of which Lev 23 is a part, reflect traditions that can be either preexilic or postexilic.[51] However, since the latest Priestly redaction does occur in the postexilic period, I place Lev 23 as the latest of the biblical calendars, while recognizing that this conclusion is tentative. Apart from the issue of chronology, the Priestly calendar provides additional details of the fall harvest festival not contained in the other calendars. The Gezer calendar itself, it is recalled, is dated to the tenth century BCE[52] and may represent the earliest of Israelite calendars.

Barley, we recall, was harvested in the spring and marked with the festival of unleavened bread. The wheat harvest followed in late spring/early summer with its festival of weeks. Figs ripen in the summer and must be picked as they ripen to avoid rot. Their harvest, then, occurred throughout the summer, before the fall harvest.[53] Grapes and olives, then, were the dominant crops of the fall harvest.

The term for the fall harvest in the earliest calendars of Exod 23:16, Exod 34:22, and Gezer, line 1, is אסף/אָסִיף,[54] "ingathering." The festival for this harvest, according to Exod 23:16 and 34:22, occurs as the last of three in the year, as noted earlier. Grapes and olives are not named in these passages, but are the vital "first fruits of your labor" (Exod 23:16) for the fall harvest. Ingathering is the activity of the harvest, the gathering in of fruit crops. The festival in Exod 23:16 and 34:22, therefore, is known by the term of its central activity, the actual *gathering* in of the fruit. Zechariah, a postexilic prophet, deftly expands this gathering of fruits to include the survivors of all nations in his eschatological vision of a final fall harvest (Zech 14:16).

Deuteronomy 16:13-16 marks the fall harvest in different terms. Here, it is known by the phrase חַג הַסֻּכֹּת, "festival of booths" (v 13), yet it occurs בְּאָסְפְּךָ, "in your gathering," (v 13). In the Deuteronomic calendar, then, the fall festival is known by its field structure, rather than by the process of gathering in the fruit. Grapes remain a primary crop, as this calendar makes explicit by naming the other installation of a vineyard, the winepress:

$$\text{בְּאָסְפְּךָ מִגָּרְנְךָ וּמִיִּקְבֶךָ}$$

---

[51]J. Milgrom, "Priestly ("P") Source," in *ABD* 5:454-61.
[52]Albright, "The Gezer Calendar," 16-26; Cross and Freedman, *Early Hebrew Orthography*, 45.
[53]Hopkins, *The Highlands of Canaan*, 228. Figs are likely the primary "summer fruit" mentioned in line 7 of the Gezer calendar. See chapter 1, pp. 35, 38-39.
[54]אסף occurs in the Gezer calendar.

> when you gather from your threshing
> floor and your winepress.
>
> (Deut 16:13)

And, the coincidence of booths with the grape harvest indicates their presence in Israelite vineyards. Deuteronomy 16:13, thus, mentions two features to a vineyard, the booth and winepress. Isaiah 5:2b notes two installations for his vineyard: the tower and winepress. The installations represented in these texts are functional equivalents for an Israelite vineyard.

The tradition of the Israelite grape harvest involving a booth continues or is likewise evident in the Levitical calendar. Particularly, the calendar in Lev 23:33-36, 39-42 describes the form of worship during the festival of booths. Here too the festival of booths occurs at the end of the year, in the fall. Two differences about the booths are evident in this material. First, it enjoins the Israelite citizens to live in the booths during the fall festival. While Deut 16 prescribes keeping the festival of booths for seven days, there is no mention of having to live in them. Leviticus 23, however, commands that the booths be inhabited for that period of time (v 42). The booths not only provide the name of the harvest; they also become a dwelling place during it. Second, the Priestly version provides a historicized rationale for this tenancy:

> so that all your generations may know that I
> made the people of Israel live in booths (סֻכּוֹת)
> when I brought them out of Egypt.
>
> (Lev 23:43)

Inhabiting booths during the fall harvest is, then, understood in religious terms as a trust and remembrance of Yahweh's guidance out of Egypt. This is still the dominant memory reenacted in Judaism today, during the holiday of Sukkot, and it undercuts the agricultural significance of the booth.

These booths, then, as Lev 23:42 suggests, might have been dwelt in during the grape harvest.[55] The Jebaliyah Bedouin of the southern Sinai who practice horticulture live in huts near the

---

[55]Hopkins, *The Highlands of Canaan*, 229; Wellhausen, *Prolegomena*, 85; Baly, *The Geography of the Bible*, 85.

vineyards for the summer months, from June through November.[56] Yet, since they are Bedouin, their transhumance is somewhat expected. In the modern rural Greek village of Karpofora in Messenia, however, small family vintners built second houses near their vines to dwell in for the summer months and until the grape harvest.[57] We have no evidence that Israelites were constructing second houses in their fields to live in for four or five months out of the year. Amos 3:15 mentions winter and summer houses, but this seems to be directed at the wealthy, i.e., those whose houses have carved ivory in them. Most of the tower remains from Hellenistic times encircle a space too small to have housed a family, and they do not contain significant pottery evidence of dwelling. But the Priestly legislation that farmers dwell in booths for the grape harvest festival may indicate that at least part of the family stayed out in the vineyard during this time. That is, the cultic legislation may have arisen out of an agricultural practice, even as the Levitical calendar offers an historical motive for the practice. Ron notes both an agricultural and a social benefit to such a practice for the traditional farmer of antiquity:

> The use of a stone hut as a temporary dwelling
> for the whole family during the fruit picking
> season maximized the use of family labour resources
> including children and aged without interfering with
> the normal course of family life together and with
> even the additional recreational pleasure of living
> in ... nature.[58]

The Levitical calendar is the least explicit of the four about a harvest context. There is no mention of ingathering or of winepresses as there is in Exod 23, Exod 34, and Deut 16. Instead, the focus of this calendar remains on the worship for the festival. Verse 40 contains the only agricultural detail of the festival:

> On the first day you shall take the fruit
> of the splendid tree, branches of palm trees,
> boughs of leafy trees and willows of the brook;
> and you shall rejoice.
>
> (Lev 23:40)

---

[56]Perevolotsky, "Orchard Agriculture in the High Mountain Region of Southern Sinai," 343.
[57]Aschenbrenner, "A Contemporary Community," 60.
[58]Ron, "Stone Huts," ix.

These branches may have been used to build the booths. Verse 42 contains the instruction to live in booths, so the sequence may imply that these tree materials went to the construction of booths. But, they are only mentioned in v 40 as accoutrements for rejoicing, somewhat like party favors. In addition, fruit is mentioned, and this could serve no purpose in the actual construction of a hut, but would obviously symbolize the fertility hoped for in the harvest. To conclude, then, we cannot know what vineyard booths were made from, only that their presence in vineyards is reflected in the persistent tradition of the festival of booths.

The functions of the vineyard tower and booth are hinted at in the biblical material. Deuteronomy 16:13 and Lev 23:42 likely reflect a tradition of living in booths during the grape harvest. A booth offers, as we have seen, protection from sun and rain in Isa 4:6. Four other texts demonstrate the cost of not guarding one's field or vineyard. In Job 27:18 a booth is inhabited by a guard who falls asleep. As a result, his wealth will not "be gathered," according to the MT.[59] The guard's failure and its consequence illustrate the important agricultural function of, in this case, the booth.

Samson attacked the Philistine vineyards, orchards, and wheat fields (Judg 15:3-5). Such vandalism resulted in the loss of harvests, as it does in Job 27:18 above. As noted earlier,[60] Samson's act would have done irreparable harm to an enemy's economy. It effectively knocked out that year's crops and subsequent harvests, since it would be years before vineyards and olive orchards would produce again. A security system was unlikely to have stopped the rampaging Samson, but it would have been a deterrent to vandals in general. The vintners outside Shiloh lose their women to the abducting Benjaminites in the vineyards (Judg 21:19). No guard is there to thwart this marriage campaign.[61]

Finally, Gideon's plight in Judg 6:11 may or may not have been eased by the presence of a tower or booth. Gideon took refuge in a winepress against Midianite incursion.[62] Either he had no vineyard structure, or it proved paltry against the force of these invading bands. Alternately, his winepress was inside a building or within the city walls, as Oded Borowski argues.[63] I suggest instead that an outdoor winepress could still be indicated by this passage. The press is

---

[59]The LXX and Syriac reflect a different root, that of יסף rather than אסף, so that the wealth would be no more. The lack of an alert guard in a booth, in both senses, is clearly detrimental.

[60]Chapter 3, pp. 110n105, 115, 124.

[61]See chapter 5, pp. 185-86.

[62]Borowski, *Agriculture in Iron Israel*, 111.

[63]See pp. 142-57 below for a discussion of Israelite winepresses.

located somewhere near Ophrah's oak tree, where the angel is sitting
(vv 11-12). In my reading, Gideon can hide from the Midianites
because the winepress is not near the threshing floor. The grain and
wine installations were at some distance from one another, as were the
fields and vineyards. The Midianites, apparently, had raided at the
time of threshing, viz., the summer, which was months before the
grape harvest. Gideon hid, in other words, where he would not have
been expected to be. A winepress basin would partially obscure him,
as would the vines.

II. The Winepress

Isaiah's winepress was "in" the vineyard, בּוֹ, and the verb for its
construction is חָצֵב, "hew." While these are the only details provided
for the winepress in Isaiah's Song, archaeology corroborates that the
description, while terse, is accurate for Iron Age winepresses, viz.,
that they are found in field areas and are hewn from rock. The
characteristic location of the Israelite winepress also helps to explain
why more grape pips have not been found in excavation. They would
have been out by the presses, in fields where archaeologists have not
tended to concentrate.[64]

The construction of the winepress differed from that of the
vineyard tower. The farmer built, וַיִּבֶן, the tower with stones. This
entailed arranging the stones and piling them up into a tower. The
winepress, by contrast, was hewn or carved directly from the bedrock
of the surface. In the Hellenistic and Byzantine periods stones were
used to build the walls of winepresses and were then plastered, but not
in the Iron Age.[65] A fragment from Qumran (4Q500 1) reflects the
later practice of building a winepress with stones:

---

[64]Hopf, "Plant Remains, Strata V-I," 74.

[65]Exceptions are a press at Jebel Carson in Samaria that had stone walls around it,
and the Ashkelon presses, which had stone walls and plastering. For these two
presses, see Dar, *Landscape and Pattern*, 1:41, 148; and Stager, "The Fury of
Babylon," 64. Some of the Beth-Shemesh presses, too, had one or two courses of
stone built up around them and were plastered, but the site report does not mention
how many of these were so constructed: E. Grant and G. E. Wright, *Ain Shems
Excavations (Palestine)* Part 5 (Text) (Haverford: Haverford College, 1939), 76. For
other later examples see I. Roll and E. Ayalon, "Two Large Winepresses in the Red
Soil Regions of Israel," *PEQ* 113 (1981): 113: Here small stones were carefully
covered with mortar and then plaster. Also, two Stratum I Hellenistic presses at Tell
en-Nasbeh were partially masonry built: J. R. Zorn, "Tell en-Nasbeh: A Re-
evaluation of the Architecture and Stratigraphy of the Early Bronze Age, Iron Age
and Later Periods" (4 vols.; Ann Arbor, Mich.: UMI, 1993) (Ph. D. diss.,
University of California, Berkeley), 1:234.

[יקב תירושכה [ב]נוי באבני]

Your winepress [bu]ilt in stones[
(line 3)[66]

In the Iron Age a farmer would carve an area of the bedrock to create a flat surface surrounded by short walls. The flattened surface was for treading and the walls kept the grape juice within the press. This construction technique was successful largely because of the nature of the bedrock involved. Cenomanian limestone and Senonian chalk are the dominant bedrock materials of the central highlands.[67] These are both malleable to tooling and impermeable, provided that there are no fractures present. Hence, many installations such as winepresses, cisterns, mortars, cupmarks, and oil presses were carved right out of this bedrock in ancient Israel. The bedrock of the coastal plains differed, consisting mainly of Kurkar sandstone,[68] which was permeable. Hence, winepresses there had to be plastered before they were used. Plastering presses is more a mark of regional adaptation to bedrock than of technological innovation.[69] The practice of carving winepresses enabled the farmer to make use of the bedrock outcroppings already likely present in his vineyards. It also helps to explain the second detail of Isaiah's press—that it was in the vineyard.

Today, winepresses are inside the buildings of wineries and often at some distance from the vine area. In ancient Israel, however, grape pressing was most often done outside in the field along with other agricultural chores, such as olive oil pressing and threshing grain. Locating the press within the vineyard, in fact, as Isaiah's vintner has done, meant that harvest and wine production could occur at the same time and that grape transport was minimal. Grape skins are delicate when ripe, so locating the press near the picking minimized fruit damage.[70] A concomitant harvest and wine production also maximized the labor force and enhanced social

---

[66]G. J. Brooke, "4Q500 1 and the Use of Scripture in the Parable of the Vineyard," *Dead Sea Discoveries* 2 (1995): 268-69. The text is dated to the first half of the first century BCE.

[67]Orni and Efrat, *Geography of Israel*, 51; Stager, "The Archaeology of the Family," 10.

[68]M. Avimelech, "Geological History of the Yarkon Valley and its Influence on Ancient Settlement," *IEJ* 1 (1950-51): 77. Orni and Efrat, *Geography of Israel*, 39.

[69]W. F. Albright, *The Archaeology of Palestine* (Baltimore: Penguin, 1960), 113; G. W. Ahlström, "Wine Presses and Cup-Marks of the Jenin-Megiddo Survey" *BASOR* 231 (1978): 23; Stager, "Archaeology of the Family," 10; Dar, *Landscape and Pattern*, 1:150; Mazar, *Archaeology of the Land of the Bible*, 345.

[70]Zorn, "Tell en-Nasbeh: A Re-evaluation of the Architecture and Stratigraphy," 1:230.

cohesion through hard work and celebration. It also provided an added security measure, as the presence of workers thwarted intruders.

Surveys throughout the land of Israel have uncovered hundreds of ancient winepresses. Dar's survey of Samaria yielded 300 field winepresses and G. Ahlström's survey of the Jenin-Megiddo area netted another 117 presses, separate from any building remains.[71] These presses range from Iron II to the Byzantine period in date. Individual sites contain multiple winepresses as well. Beth-Shemesh has nine winepresses,[72] with another twelve wine and oil presses discovered in the region south of the site.[73] The site of Gibeon has ten outdoor winepresses.[74]   Because Gibeon also has considerable additional evidence for wine production in the form of rock-cut cellars for storage and epigraphic notations on jars, it will be discussed at some length below in section III. Only a few winepresses have been found in the remains of Iron Age houses. The winepress of Tell Qasile, Stratum IX$_2$ (tenth century BCE)[75] and several at Beth-Shemesh[76] are located within dwelling structures.   A large Iron II winery within a building has been discovered at the Philistine site of Ashkelon.[77]   Its vats and basins were lined with stones and high-quality plaster.[78]

The sheer number of ancient field presses found by survey suggests that ancient Israel practiced outdoor grape pressing. The ratio of outdoor presses to those found within buildings can serve as a reliable indicator of a practice even while areas remain to be excavated, since, if anything, it is the field installations that are underrepresented in the archaeological literature.   As archaeology continues to incorporate the survey work of fields and terraces along with site excavation, our picture of agricultural practices should sharpen. Outdoor pressing enabled the Israelites to make use both of daylight and of the fresh air to offset the strong fumes of fermentation, and, of course, it spared them the labor of constructing

---

[71]Ahlström, "Wine Presses and Cup-Marks," 19; Dar, *Landscape and Pattern*, 1:147.
[72]Grant and Wright, *Ain Shems Excavations (Palestine)* Part 5 (Text), 75-77; Part 4 (Pottery) (1938), plates 18-20.
[73]Dagan, "Bet Shemesh and Nes Harim Maps, Survey," 94-95.
[74]Pritchard, *Winery, Defenses, and Soundings at Gibeon*, 10.
[75]B. Maisler [Mazar], "The Excavations at Tell Qâsile: Preliminary Report," 130. Another three collection vats, these outdoors, have been found at Tell Qasile dating to the Iron Age: E. Ayalon, "Tel Qâsile," *Excavations and Surveys in Israel* 13 (1995): 51.
[76]Grant and Wright, *Ains Shems Excavations*, 4, plate 18: 2-4, three winepresses; plate 20: 2, a large rectangular press; E. Grant, *Rumeileh being Ain Shems Excavations (Palestine)* (Part 3) (Haverford: Haverford College, 1934), 64, 65, 79.
[77]Stager, "The Fury of Babylon," 56-69, 77; photo on p. 67.
[78]Stager, "The Fury of Babylon," 64.

a building. Some winepresses of ancient Greece were found within building structures.[79] Roman wine making seems to have entailed indoor pressing,[80] though, according to Cato[81] and Columella,[82] these press-buildings were ideally built out in the field areas. In this viticultural custom, ancient Israel differed from classical antiquity and typically pressed its grapes outdoors.

Modern vineyards now have machines that first remove stems and leaves from machine-harvested clusters, but there is still no delay in crushing grapes in the press. Today, the pneumatic, or Wilmes, press is used extensively. It is a stainless steel drum with a long rubber bag inside, which is blown up and then presses the grape mash against the metal walls. In this way, juice is extracted without breaking grape pips—the inner part consisting of oils, seeds, and tannins.[83] The inner bag or bladder of the machine performs a gentle and firm juice extraction from pip and skins. Pips are avoided since the oils, tannins, and seeds lend bitterness to the juice. Ancient Egypt used a cloth press that somewhat resembled the bladder press. The tomb painting of Baot, from Beni Hasan (1890 BCE), depicts a bag press.[84] This bag was twisted to express juice, rather than used to push mash against another surface. Though the modern Wilmes press is a technological advance for viticulture, it does not differ from the dominant press method of antiquity—treading—in its principle of providing a gentle, firm press.

Grape pressing before mechanization was done by foot. This was an effective method because it too was gentle enough not to break the pips. Human feet and their tendency for slippage allowed the pips to stay intact while they expressed the juice. Treading, in fact, was a method that did not really need much improvement, save in hygiene, and so was widespread from the Early Bronze Age until the last century of our era. As a practice, then, its sheer historical persistence attests to its efficiency. Foot pressing continues today in some rural regions around the Mediterranean and elsewhere as an expression of traditional viticulture.

[79]Burford, Land and Labor in the Greek World, 61; J. H. Young, "Studies in South Attica: Country Estates at Sounion," Hesperia 25 (1956): 122-46.

[80]M. Grant, Cities of Vesuvius: Pompeii and Herculaneum (Weidenfeld & Nicolson, 1971), 192; D. S. Robertson, A Handbook of Greek and Roman Architecture (Cambridge, Eng.: Cambridge University Press, 1929), 310: at Villa Boscoreale, a large estate outside Pompeii, the winepresses were inside room P, while the wine jars for fermentation were in an open court, room R.

[81]Cato Agriculture 3, 11, 12-13.

[82]Columella De re rustica 1.6, 9, 11, 18.

[83]Weinhold, Vivat Bacchus, 154.

[84]Unwin, Wine and the Vine, 72, fig. a; P. Newberry, Beni Hasan I (London: Kegan Paul, Trench, Trübner, 1893), plate 12; Mazar, Archaeology of the Land of the Bible, 166.

Biblical descriptions of the grape harvest and archaeological remains of presses suggest that ancient Israel shared in this tradition of grape pressing by foot. A screw-weight press was developed in the Hellenistic or Roman period, in which wooden or stone weights were screwed down onto a cylinder containing grapes.[85] Examples of the screw press have been unearthed in Israel from the Byzantine period.[86] However, even the screw presses of these later periods likely only supplemented foot treading as a practice. They were situated in the middle of large pressing floors, which suggests that foot treading might have continued.[87] It is possible, of course, that these presses were simply put on the place of the older presses as their replacement. In this case, then, their position marks only an innovation in treading and otherwise bears no relation to the floors. However, with the ready availability of two methods of pressing, more wine could have been produced and quickly. These vintners with these screw presses, then, would have enjoyed the benefits of both tradition and innovation at their grape harvests. For the Iron Age period, though, grape pressing was done exclusively by foot. If there were bag presses or wooden presses, they neither survive in the archaeological record nor are apparent in the literary (i.e., biblical and epigraphic) sources.[88]

דָּרַךְ, "march," or "tread," is the verb for grape pressing in the Bible. It is used as a transitive verb for grapes in Judg 9:27; Jer 25:30; Amos 9:13; and Mic 6:15, and in association with winepresses in Job 24:11; Isa 16:10; 63:2-3; Jer 48:33; and Lam 1:15. Foot pressing, then, is the biblically attested Israelite practice for wine production and no other means of pressing are mentioned.

In Isa 63:3, 6, Yahweh is the treader and people become the grapes as they were in Isa 5:1-7. Here, people are trodden as grapes

---

[85]Broshi suggests its development in the late Roman, early Byzantine periods: "Wine in Ancient Palestine," 23; Dar considers it to be as early as Hellenistic: *Landscape and Pattern*, 1:150.

[86]R. Frankel, "Screw Weights from Israel," in *Oil and Wine Production in the Mediterranean Area, Bulletin de Correspondance Hellenique*: Supplement 26 (ed. M.-C. Amouretti and J.-P. Brun; Paris: Boccard, 1993), 106-18; Roll and Ayalon, "Two Large Winepresses in the Red Soil Regions of Israel," 111-25; S. Saller and E. Testa, *The Archaeological Setting of the Shrine of Bethphage* (Jerusalem: Franciscan Press, 1961), 27-36; Y. Israel, "Ashqelon," *Excavations and Surveys in Israel* 13 (1995): 102-3; E. Shavit, "Rishon Leziyyon," *Excavations and Surveys in Israel 13* (1995): 57.

[87]The presses have a treading floor with a cut square depression in their center for the beam and screw. A mosaic from Mount Nebo depicts treaders holding on to the screw press: Saller and Testa, *Archaeological Setting*, 38-40.

[88]No winepresses have yet been unearthed for Egypt, though they are depicted in tomb paintings. Darby et al. suggest that wood vats might have been used: *Food: The Gift of Osiris*, 2:557; Forbes, *Studies in Ancient Technology*, 3:110; Lutz, *Viticulture and Brewing in the Ancient Orient*, 53; Lesko suggests instead that Egyptian presses were made of baked mud: *King Tut's Wine Cellar*, 17.

and Third Isaiah does nothing to soften the analogy. Isaiah's analogy, we recall, began in chapter 5, when Yahweh planted his vineyard, expecting a good harvest of fruit. He was disappointed by the yield, the unfaithful Judeans. In chapter 63 of Third Isaiah, the analogy turns on the Edomites and reaches its gruesome conclusion with Yahweh in the winepress:

> I have trod them in my anger,
> and trampled them in my wrath,
> their juice spattered on my garments,
> and stained all my robes.
>
> (Isa 63:3)

The gentleness foot pressing afforded grape pips apparently is not extended to people in Third Isaiah's metaphor. Lamentations 1:15 too images the horror of a winepress turned into a human press, with Yahweh treading the virgin daughter Judah. These prophetic passages enlist the Israelite practice of treading grapes to provide a startling depiction of divine wrath.

The splattering and staining mentioned for the winepress indicate both an occupational inconvenience of the vintners and that the wine is dark.[89] When Yahweh treads the Edomites in Isa 63:3, 6, the resulting juice that stains his clothes is red, חָמוּץ (63:1) and אָדֹם (63:2). In Gen 49:11, a color approximating that of blood causes stains as Judah washes his garments in the דַם־עֲנָבִים, "blood of grapes." Asaph Goor interpreted this passage as evidence that ancient Israel used wine for dyeing textiles.[90] However, it seems unlikely that such a valued and enjoyable product would be used in this way, even were there excess in a given vintage. In addition, washing clothes in wine in this passage is extraordinary; it is part of the description of Judah's preeminence in Gen 49:8-12, rather than a daily custom.[91] These passages indicate the messiness and color involved in grape treading. For this reason as well, outdoor pressing remained the traditional

---

[89]Virgil advised disrobing as a way around the problem of staining: "Come hither, O Lenaean sire, strip off thy buskins and with me plunge thy naked legs in the new must": *Georgics* 2.7-8. Nude grape treading apparently made its way to Palestine too. A mosaic from Beth Shean, dating to the sixth century CE, shows three ithyphallic men in the winepress: M. Broshi, "The Diet of Palestine in the Roman Period—Introductory Notes," *IMJ* 5 (1986): 46.

[90]He provided, however, no additional evidence for his suggestion: Goor, "The History of the Grape-vine," 48.

[91]There is a legend that Hannibal washed his horses in wine during the Second Punic War: Forbes, *Studies in Ancient Technology,* 3:115. If true, it was no doubt a gesture of supremacy, rather than daily Carthaginian practice.

practice in ancient Israel. The grape treaders could wash their feet and garment edges with nearby cisterns or jars of water or oil[92] before returning to their dwellings.[93]

The evidence for Israelite winepresses is archaeological and textual. A discussion of the archaeological examples of presses follows below. Next, the archaeological evidence for a wine production center at Gibeon is discussed. Finally, these archaeological materials are brought to bear on a study of the biblical terms for "winepress," which concludes the chapter.

a. Archaeology of the Winepress

Based on the archaeological evidence, the design of the hewn Israelite winepress varied somewhat even in its simplicity. It could be an oval or rectangular basin with small depressions for scooping up the juice and putting it into skins or jars. Another style of winepress had a flat treading floor and a collecting vat where the juice ran via a carved channel. For this press, the channel and vat would be cut lower than the level of the treading floor, so that gravity transported the juice. A small depression as well was often hollowed in the vat, again for scooping up the juice. Ahlström's study of 117 winepresses in the Jenin-Megiddo region is instructive for delineating the types of hewn presses in ancient Israel.[94] The presses Ahlström examined vary in size, in number of accompanying vats, in shape, and in associated features such as cupmarks,[95] mortars, or benches. While he discovered a variety of styles and sizes, 85 of the 117 were rectangular and had rectangular vat(s) accompanying them (Fig. 2).[96] The figure below represents the basic structure of the most common type of winepress found in Israel. Variations would occur particularly with the smaller depression, which could be either rectangular or circular in shape. The larger area, however, tended to be rectangular in design in almost all of the known examples.

---

[92]Ahlström, "Wine Presses and Cup-Marks," 44; Orni and Efrat, *Geography of Israel*, 304.
[93]Social aspects of the vintage are discussed in chapter 5, pp. 179-86.
[94]Ahlström, "Wine Presses and Cup-Marks," 19-49.
[95]Small, circular depressions carved out next to wine and oil presses.
[96]Ahlström, "Wine Presses and Cup-Marks," 20.

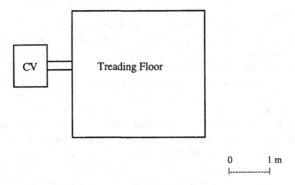

Fig. 2. Press with CV(= Collecting Vat) and Channel[97]

Rafael Frankel, in his study of presses in Galilee, discovered that the earliest, simple winepress tended to be a rectangular treading floor plus a vat(s).[98]   In addition, he notes that there was no significant regional variation of this simple, early press throughout the country. As a general rule, then, we can consider the press with a rectangular treading floor and vat to be typical of ancient Israelite viticulture.

Installations such as the winepress are notoriously difficult to date, since by definition they are cut into the bedrock of an area, and thus lack stratigraphy.[99]   Often they are found in field areas where no building remains offer contemporary evidence for their date. Ahlström has further cautioned that tool marks may not offer clues for dating, since the same kinds of tools for rock cutting may have been in use for hundreds of years.   Also, the soil around a press may offer no help, since it might well have been disturbed by farming over the centuries.[100]   Nevertheless, as a general rubric, winepresses from the later periods—Hellenistic, Roman, Byzantine—tend to be larger and show signs of more working, e.g., masonry, plastering, tesserae,[101] etc.   Yet dating winepresses remains tentative since both

---

[97]Redrawn from Ahlström, "Wine Presses and Cup-Marks," fig. 3, p. 21.
[98]Frankel, *The History of the Processing of Wine and Oil in Galilee in the Period of the Bible, the Mishna, and the Talmud*, iv.
[99]Ahlström, "Wine Presses and Cup-Marks," 21; Borowski, *Agriculture in Iron Age Israel*, 111.
[100]Ahlström, "Wine Presses and Cup-Marks," 21.
[101]Small tiles arranged on the floor.

size and workings could be later expansions of and additions to existent presses.[102]

In the current section, I have selected a variety of winepresses whose origin in Iron Age Israel is fairly well argued by either their excavators or subsequent archaeologists. My goal here is to demonstrate that wine production is evident at sites and in field areas throughout ancient Israel in the Iron II period and to show the variety of the Israelite winepress. Such variety reflects the differing production practices of wine for the Iron Age.

The winepresses of Samaria, Hirbet Jemein, Tell en-Nasbeh, Beth-Shemesh, and Gibeon[103] represent winepresses found within the areas of settlement, rather than out in the fields. They are considered to be Iron Age, as they occur within settlements whose material culture is primarily from the Iron II. The winepresses of Tel Michal, Mevasseret Yerushalayim, and Khirbet er-Ras[104] occur in vineyard areas, and so we rely on the pottery and surrounding installations for an Iron II dating. The many winepresses surveyed by Dar and Ahlström may have originated in different periods or been used throughout historical periods. Because of the difficulties of dating such a simple installation, the authors can only narrow the periods of construction and use to a span of time from the Iron II through to Roman or even Byzantine periods. Still, once the vine was cultivated in an area, it was likely to persist throughout historical periods, since grapes were such a valuable crop. Hence, presses with Iron II pottery near them were likely used during the Iron II period. Some of these presses, too, may have been hewn from earlier presses.[105] The large Hellenistic presses found throughout the area of western Samaria, for example, may well be expansions made from Iron Age ones.[106] If so, the many presses found for this period also reflect significant viticultural activity for ancient Israel. Let us now proceed to a description of the various winepresses representative of Iron Age viticulture.

On the summit of Samaria, rock-cut presses for olive oil and wine production were found on the bedrock, underneath the Building I period of the early ninth century BCE. Stager estimates the date of their origin to be anywhere from the late eleventh century to the early ninth century BCE.[107] The majority of these rock-cut installations

---

[102]Dar, *Landscape and Pattern*, 1:150; Ahlström, "Wine Presses and Cup-Marks," 46.
[103]See below for a discussion of these presses.
[104]See below.
[105]Ahlström, "Wine Presses and Cup-Marks," 23.
[106]Ahlström, "Wine Presses and Cup-Marks," 45; cf. Sallers and Testa, *The Archaeological Setting*, 36-38.
[107]Stager, "Shemer's Estate," 103-4.

were for oil production, but there was one winepress as well. Taken together, the installations are evidence of significant horticultural activity on the site. The winepress is a simple oval design, with one floor for treading. It measured 1.70 m x 0.80 m, 1.20 m deep,[108] and so would have fit 4-5 people comfortably. It had two small depressions in the bottom, presumably for collecting the juice. It was a simple press and differed from the oil presses nearby, which were circular in design, and were in some cases surrounded by a channel.

Hirbet Jemein is a village site in western Samaria, 1.5 km southeast of ʿAzun,[109] which has evidence of terracing and a large winepress[110] and an oil press atop the mound, that is, within the settlement sphere.[111] The entire plan of this small village could be reconstructed because of its good state of preservation.[112] The majority of the houses are of the four-room type characteristic of the Iron Age, and most of the pottery is from the eighth and seventh centuries BCE with a small amount from the Hellenistic period, during which the site was eventually abandoned.[113] The winepress and oil press situated on the mound are the only ones recovered for this village, and so seem to indicate communal use, as does the Samaria winepress above. This arrangement suggests, then, a fair degree of social cohesion, where the villagers could share presses for food production.[114]

Hirbet Hamad, another site in Samaria, further south of ʿAzun,[115] is located on a hillock with an agricultural area. About a kilometer southeast of the settlement site there is a rectangular winepress with a rectangular collecting vat and another rectangular bin cut over the treading floor and sloping into it (Fig. 3).

[108] Stager, "Shemer's Estate" 93, table 1; 96; Reisner et al., *Harvard Excavations at Samaria 1908-1910*, 1:66-67.

[109] For a layout of the site, see Dar, *Landscape and Pattern*, 2, fig. 135.

[110] Dar does not provide the dimensions.

[111] Dar, *Landscape and Pattern*, 1:36; for a schematic of the presses: 2, fig 30.

[112] Dar, *Landscape and Pattern*, 1:37.

[113] Dar, *Landscape and Pattern*, 1:36.

[114] Presses that had Iron Age through Herodian period pottery, Dar asserts, likely had their origin in the Iron Age. These are: Qarawat Bene Hassan, Tower 48, southeast of ʿAzun Tower 229; press 252 in Tzur Nathan, south of Tulkarem; a large press at Tower 1 at Jebel Carson. The Jebel Carson press is distinctive for its wall of dressed stones and a step into the press; so in this case, stones are evident: fig. 87: Dar, *Landscape and Pattern*, 1:41, 148; 2, figs. 87, 134.

[115] Dar, *Landscape and Pattern*, 2, fig. 135.

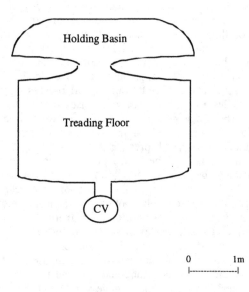

Fig. 3. Press with Holding Basin[116]

The flat raised surface slopes into the treading floor and is connected
to it by a small channel. Dar concludes that it was an area for holding
the grapes before they were pressed.[117] He dates the site by its pottery
and architecture to the Iron Age through to the Byzantine period.[118]
The agricultural area where the winepress occurs he considers to be
Iron Age, with subsequent use and expansion evident from the
Hellenistic period.[119] The winepress was cut in the rock opposite an
oil press, which is of a type common in the Iron Age.[120] From this
association, Dar suggests an Iron II date as well for the winepress.

---

[116]Redrawn from Dar, *Landscape and Pattern*, 2, fig. 89: Hirbet Hamad Press.
[117]Dar, *Landscape and Pattern*, 2, fig. 89.
[118]Dar, *Landscape and Pattern*, 1:38, 40-41.
[119]Dar, *Landscape and Pattern*, 1:41.
[120]It consisted of simple cupmarks, ca. .60 to .80 m in diameter, with depressions
above in the rock, presumably for a beam with stone weights for crushing: Dar,
Landscape and Pattern, 1:148, 166-67, 180; 2, photo 76. For further discussion on
the Israelite olive oil press, see M. Heltzer and D. Eitam, eds., *Olive Oil in Antiquity:
Israel and Neighbouring Countries from Neolith to Early Arab Period* (Haifa:
University of Haifa Press, 1987); R. Frankel, S. Avitsur and E. Ayalon, *History and
Technology of Olive Oil in the Holy Land* (Tel Aviv: Eretz Israel Museum, 1994);
and Frankel, *The History of the Processing of Wine and Oil.*

Winepresses, it should be noted, often occur in association with olive presses, as cultivation of grapes and olives was fundamental to Israelite and neighboring economies. All of the winepresses listed here had olive oil presses nearby except for those at Tel Michal and Gibeon. Additional Iron Age examples of wine and olive oil presses together occur on terraces outside of Jerusalem, at Mevasseret Yerushalayim and Khirbet er-Ras,[121] and at Tel Batash.[122]

The presses for oil and wine production on the terraces outside of Jerusalem are found in association with houses, which is true at Tel Batash, as well. They may represent the small production of the inhabitants working these terraces. The press at Khirbet er-Ras was part of a farmstead with two houses, one a four-room house.[123] Nearby were a bell-shaped vat and a cave, both of which could have been used for storage of the wine.[124]   A winepress at Ras Abu Ma'aruf, 4.4 km northwest of Jerusalem,[125] is interesting in several respects. It is found in association with a single farmhouse that is a four-roomed dwelling, and nearby terraces. The treading floor is small, measuring 1.0 m x .7 m, and is slightly tilted toward a channel that runs along its southern side. The press also had a depression next to the treading floor, presumably for holding the grapes before processing.[126]   The uniqueness of this press, however, lay in its collecting vat. For here, in a partially quarried depression next to the treading floor, was a four-handled jar found in situ with its neck and rim removed.[127] The channel outlet from the treading floor was located exactly over the opening to the jar.[128] Here, then, pottery was planted into the ground to serve as a collecting vat.

Three winepresses judged to be from the Iron Age were found at Tell en-Nasbeh. They are rectangular with a smaller depression, either rectangular or circular, next to them. The first press is located within the settlement area. It is 5.4 m long, 3.7 m wide and has a circular cutting of .50 m wide in its southwest corner, presumably for collecting juice.[129] This press, Jeffrey Zorn argues, is likely from Iron I (Stratum IV), because a casemate wall dating from the

---

[121]Edelstein and Kislev, "Mevasserat Yerushalayim," 54; Edelstein and Gibson, "Ancient Jerusalem's Rural Food Basket," 49.
[122]Kelm and Mazar, "Three Seasons of Excavations at Tel Batash—Biblical Timnah," 32.
[123]Edelstein and Gibson, "Ancient Jerusalem's Rural Food Basket," 48-49.
[124]Edelstein and Gibson, "Ancient Jerusalem's Rural Food Basket," 49. See below, part c, on Gibeon, which also had such bell-shaped vats, probably for wine storage.
[125]J. Seligman, "A Late Iron Age Farmhouse at Ras Abu Ma'aruf, Pisgat Ze'ev A," 'Atiqot 25 (1994): 63.
[126]Seligman, "A Late Iron Age Farmhouse," 66-67.
[127]Seligman, "A Late Iron Age Farmhouse," 66.
[128] Seligman, "A Late Iron Age Farmhouse," 66.
[129]Zorn, "Tell en-Nasbeh: A Re-evaluation of the Architecture and Stratigraphy," 1:107-8, 234; 3, plan 74, p. 996.

beginning of Iron II (Stratum IIIc = 1000-900 BCE) cuts across its northwest corner.[130] This type of press, it will be recalled, is similar to the Samaria press in that its collecting vat is a depression within the treading floor itself, rather than a separate feature. The absence of a separate collecting vat may be a characteristic of the early Israelite winepress.[131] Juice in these presses, then, would have been scooped from the depression within the treading floor.

The two presses from Iron II[132] are also rectangular. These occur outside the walls of settlement and so might have had vine plants near them. Each of these presses has three discrete sections to it. One press is composed of three rectangular parts: a treading floor measuring 4 m x 4 m, with two additional depressions of 4.5 m x 3.5 m and 6.2 m x 4.5 m linked by channels.[133] The second press, though less well defined, is also composed of three parts. It is considerably smaller, measuring 2.1 m x 1.4 m x .20 m, 2.3m x 1.9 m x .30 m, 2.0 m x ? x .80 m.[134] These sections form some sort of wine complex, where treading would occur in one area, and juice would run into the smaller collecting vats. The inhabitants of the village could have shared these presses either by staggered use or by pooling labor resources. With the two collecting basins, one farmer could have pressed his grapes, and once the juice was in a collecting vat, another could begin pressing his grapes, after blocking the first vintner's vat.

Beth-Shemesh has nine winepresses from the Iron II period (Stratum II: ninth to sixth centuries BCE).[135] Several of them occur within building structures from Stratum II and are dated accordingly.

---

[130]Zorn, "Tell en-Nasbeh: A Re-evaluation of the Architecture and Stratigraphy," 1:107-8, 114.

[131]Though the two Late Bronze winepresses from Aphek each have a separate, circular collecting vat: Kochavi "The History and Archaeology of Aphek-Antipatris," 80. The tenth-century presses from Tel Michal also have separate, circular vats.

[132]Zorn's dating is tentative, arrived at from a sequence of winepresses at Tell en-Nasbeh. These presses are more complex than the press from Stratum IV (Iron I), with their separate collecting vats, yet they lack the stone wall, plaster, and tower of the Hellenistic presses at the site. Hellenistic pottery around those two latter presses helped to date them. See Zorn, "Tell en-Nasbeh: A Re-evaluation of the Architecture and Stratigraphy," 2:1.412-13.

[133]Zorn, "Tell en-Nasbeh: A Re-evaluation of the Architecture and Stratigraphy," 1: 232-33; 3, plan 163, p. 991.

[134]One side of a vat has crumbled away: Zorn, "Tell en-Nasbeh: A Re-evaluation of the Architecture and Stratigraphy," 1:232. He notes the existence of two more probable winepresses on the mound that the original excavators identified. Zorn dates them to the Hellenistic period on the basis of a masonry wall, the earliest datable materials being Hellenistic, and notes that one is near a tower, thought to be a Hellenistic feature: 2:1:534; C. C. McCown, *Excavations at Tell en-Nasbeh I: Archaeological and Historical Results* (Berkeley: The Palestine Institute of the Pacific School of Religion and the American Schools of Oriental Research, 1971), 257-58.

[135]Grant and Wright, *Ain Shems*, 5:75-77.

By association with the indoor winepresses from Stratum II, Elihu Grant argued that the outdoor ones as well are evidence of Iron Age wine production at Beth-Shemesh. In addition, the site was not reoccupied after its destruction in the sixth century BCE,[136] so the use of the presses in the Iron II period seems reasonable. Five presses occur within rooms of houses.[137] Unfortunately, the authors do not describe the makeup of the indoor presses, but they do mention that the inner faces of some of the presses were plastered.[138] Some of these presses are circular, while others are rectangular, and all have a circular collecting vat. Most of the treading floors are slanted toward a collecting vat, and this design would have quickened the flow of liquid.[139]

In the region south of Beth-Shemesh, in a triangle with Khirbet Qeiyafa and Tel Beit Naṭṭif, 32 farmsteads were surveyed with 12 wine and oil press complexes.[140] Twenty of the farmsteads were Iron II.[141] The survey indicates, then, that five-eighths of the farmsteads were from the Iron II period. If we assume the same ratio for the presses, then roughly seven of them would be from the Iron II period. Since nine Iron II winepresses were also within the settlement of Beth-Shemesh, it is evident that wine production was significant in the Iron II period.[142]

Tel Michal has four Iron Age presses 180 m east of the tell.[143] This area of the site was only in use for a single period, so the dating of these presses is fairly conclusive. Ze'ev Herzog, the excavator, notes that "although very few pottery shards were found in and around the winepresses, all are from the Iron Age."[144] He considers this finding all the more significant since "the dominant pottery collected from the surface on all parts of the hill dates from the Persian, Hellenistic, and Roman periods."[145] The presses contained none of this later material, indicating that they were covered over and in disuse from the Persian period on. The earliest shards from the winepresses are from the tenth-ninth centuries BCE, with the Tel Michal settlement itself being dated to the tenth century and having a

---

[136]Except for a Byzantine monastery on the northeast corner of the tell: Grant and Wright, *Ain Shems*, 5:85, 249.
[137]Grant and Wright, *Ain Shems*, 5: 75-76; 4, plates 18: 2-4; 19: 5; 20: 2.
[138]Grant and Wright, *Ain Shems*, 5: 76; *Rumeilah*, 3, map 1.
[139]Grant and Wright, *Ain Shems*, 5: 76; Grant, *Rumeilah*, 3, map 1.
[140]Dagan does not list the number of wine and oil presses separately: "Bet Shemesh and Nes Harim Maps, Survey," 94-95.
[141]Dagan, "Bet Shemesh and Nes Harim Maps, Survey," 94-95.
[142]Grant and Wright, *Ain Shems*, 5:76.
[143]Herzog, "A Complex of Iron Age Winepresses (Strata XIV-XIII)," 64.
[144]Herzog, "A Complex of Iron Age Winepresses (Strata XIV-XIII)," 74.
[145]Herzog, "A Complex of Iron Age Winepresses (Strata XIV-XIII)," 74

gap in the ninth. As a result, Herzog reasonably assigns a tenth-century date for the presses.[146]

The four winepresses each had two circular vats. They were arranged in pairs (Fig. 4), indicating some organization for their construction and for wine production.[147]

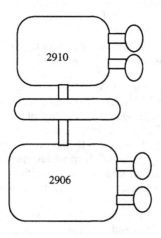

Fig. 4. Winepress Complex[148]

The presses were similar in size. The treading floor of one, labeled 2910, was 2.10 m x 2.7 m, while the floor of the other, 2906, was 2.3 m x 3.5 m, and so was slightly more rectangular in appearance. Each treading floor had channels connecting it to the two circular vats. For each treading floor, one vat is larger and one smaller. So, for complex 2910, its vats are 1.5 m x 1.10 m and .9 m x .75 m. deep.

---

[146]Herzog, "A Complex of Iron Age Winepresses (Strata XIV-XIII)," 75.

[147]The LB presses at Aphek are also aligned together, indicating some organization in construction and probably wine production as well: Kochavi, "The History and Archaeology of Aphek-Antipatris," 80.

[148]Redrawn from Herzog, "A Complex of Iron Age Winepresses (Strata XIV-XIII)," 74, fig. 6.9.

Each vat contained a small hollow in the bottom for collecting the last drops of wine.[149] The winepresses were all plastered. Plastering was necessary, as the soil of the sandy plains was permeable.[150] Complex 2910 was unique in that its plaster contained lots of shell fragments.[151]

Between two of the paired winepresses, there was a rectangular basin (Fig. 4). It measured 1 m x 3.5 m and so was as long as the large treading floor. A plastered channel connected the basin to both the treading floors. Herzog suggests that the rectangular basin acted as a bin to store grapes brought in from the harvest. Any juice expressed from the weight of the grapes would then flow into the treading floor. Similar holding bins were found with the winepresses of Hirbet Hamad[152] and Tzur Nathan in Samaria.[153]

From the evidence of a holding bin, the layout of the presses, and the different-sized collecting vats, Herzog infers a versatility of operation for wine production.[154] He argues that the planners of the complexes preferred several different-sized vats for each press rather than one large press and one large vat. In this way, multiple treadings and fermentations in the vats could occur in quick succession. This design, he notes, favored the small farmer who processed his wine individually.[155] At these presses, one farmer could have trod his grapes and then let the juice ferment in one of the vats, while another came to tread his grapes. When the first used vat was stopped off, the juice for the second farmer would go into the second vat. The eight vats meant that eight farmers could be served by the wine complexes before a period of waiting would result. The construction of these presses implies, as well, a social arrangement during the grape harvest. With the capability for staggered pressing and fermentation, social interaction and mutual guarding would be additional benefits for the farmers. These four Iron Age presses indicate for Herzog an "intensity of viticulture" in the region,[156] further evident by the Persian and Hellenistic winepresses uncovered on the northern hill[157] and south of Tel Michal.[158]

---

[149]Herzog, "A Complex of Iron Age Winepresses (Strata XIV-XIII)," 73.

[150]The LB winepresses at Aphek are also plastered for this reason as well.

[151]Herzog, "A Complex of Iron Age Winepresses (Strata XIV-XIII)," 73. Note too that the plaster of the Ashkelon winepresses also contained shells: Stager, "The Fury of Babylon," 64.

[152]Dar, *Landscape and Pattern*, 2, fig. 89.

[153]Dar, *Landscape and Pattern*, 1:148; 2, figs. 88, 134.

[154]Herzog, "A Complex of Iron Age Winepresses (Strata XIV-XIII)," 73.

[155]Herzog, "A Complex of Iron Age Winepresses (Strata XIV-XIII)," 74.

[156]Herzog, "A Complex of Iron Age Winepresses (Strata XIV-XIII)," 74. A silo from Iron II contained large quantities of charred grape seeds, 64.

[157]The small Persian period press had a plastered treading floor .80 m x .80 m and a circular vat, .70 m in diameter: Herzog, "Persian Period Stratigraphy and Architecture (Strata XI-VI)," *Excavations at Tel Michal*, 102. The Hellenistic press is large, measuring 5.25 m x 6.05 m with vats 1-2 m deep. One vat had steps leading

b. Gibeon

Another site with an "intensity of viticulture" was Gibeon. Indeed, the excavator, J. B. Pritchard, termed the site the "ancient 'Bordeaux' of Palestine."[159] Ten winepresses were discovered at Gibeon.[160] Most are circular in shape with a flat bottom for pressing. Pritchard suggested that these rock cuttings were treading basins for grapes. Only four of the ten presses have some sort of an area carved out from the treading floor that could have served as a collecting vat.[161] The rest simply consist of a treading floor. I stated earlier[162] that the lack of a separate collecting vat might indicate an early Iron Age winepress. Examples were evident at Samaria from 1000-800 BCE and Tell en-Nasbeh from 1200-1000 BCE (Stratum IV). The lack of collecting vats at Gibeon, however, does not indicate an early date for the presses. Instead, it seems to be a function of the production technique practiced at Gibeon. Simple presses, i.e., ones without separate collecting vats and channels, were used for grape treading, while fermentation occurred elsewhere, in large cellars.

The winepresses are found within areas 8 and 17, together with 63 rock-cut cellars, which Pritchard argues were the storage cellars for wine.[163] Together, these installations comprised a significant winery at Gibeon.[164] The pottery from the cellars is chiefly from Iron II.[165] Pritchard estimates that the cellars were mainly in use in the eighth and seventh centuries, and possibly as late as the sixth century.[166]

The cellars were bell shaped and hewn from rock to an average depth of 2.2 m. The opening was narrow, averaging .67 m in diameter (Fig. 5).[167]

---

down into it: Herzog, "Hellenistic Period Stratigraphy and Architecture (Strata V-III)," *Excavations at Tel Michal*, 168-70, figs. 12.4, 12.5.

[158]This Hellenistic press is small, composed of three rectangular parts: Herzog, "Hellenistic Period Stratigraphy and Architecture (Strata V-III)," 176, fig. 12.11.

[159]Pritchard, *Gibeon: Where the Sun Stood Still*, 79.

[160]Pritchard, *Winery, Defenses, and Soundings at Gibeon*, 10.

[161]Pritchard, *Winery, Defenses, and Soundings at Gibeon*, 105a, fig. 4; 126, fig. 12; 127, fig. 12; 205, fig. 5.

[162] See above, p. 154.

[163]Pritchard, *Winery, Defenses, and Soundings at Gibeon*, 13-15.

[164]Pritchard, *Winery*, 1.

[165]Pritchard, *Winery*, 13-15.

[166]Pritchard, *Winery*, 23.

[167]Pritchard, *Winery*, 1.

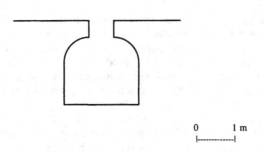

0   1 m
|------------|

Fig. 5. Wine Cellar[168]

These cellars undoubtedly were used to store a commodity. Grains were unlikely, since the damp walls and floors would induce mildew.[169] Water too was unlikely, since Gibeon's elaborate water system included tunnel access to a spring and underground water source, and had a cylindrical construction for storing water that was over ten meters deep and wide.[170] Given the considerable evidence of winepresses and jars at the site, wine becomes the most likely commodity to have been stored within these cellars.   Other bell-shaped cellars occur in association with winepresses, e.g., at Samaria, Beth-Shemesh, Hirbet Jemein, Khirbet er-Ras, Deir Daqleh, and Qarawat Bene Hassan.[171]   At the modern-day winery at Latrun near the site of ancient Gibeon, Pritchard noted, wine was still stored in rock-cut cellars.[172]

Pritchard tested the temperature of these cellars and found them to be almost 20 degrees cooler than the shaded area outside.[173] Wine would have benefited from the cooler temperatures.   Though only one complete storage jar was found in a cellar,[174] Pritchard states that two rows of such jars could be easily stacked in the average cellar.[175]   He adds that the cellars are cut to a depth convenient for filling and emptying them, because one person could stand within and

---

[168]Redrawn from Pritchard, *Winery*, fig. 6, no. 109.
[169]Pritchard, *Winery*, 25.
[170]Pritchard, "Gibeon," in *NEAEHL*, 2:512.
[171]Samaria has 6 bell-shaped pits, ca. 3.50 m³ with a capacity of ca. 21,000 liters: Stager, "Shemer's Estate," 96-98; Beth-Shemesh: Grant and Wright, *Ains Shems* 5:70;   Deir Daqleh, Hirbet Jemein, Qarawat Bene Hassan: Dar, *Landscape and Pattern*, 1:149, 151; 2, plates 73, 84, 85, 90, 171; Khirbet er-Ras, Edelstein and Gibson, "Ancient Jerusalem's Rural Food Basket," 49.
[172]Pritchard, *Winery*, 26.
[173]65° F compared to 83.5° F: Pritchard, *Gibeon: Where the Sun*, 84.
[174]Pritchard, *Winery*, fig. 32, no. 8.
[175]Pritchard, *Winery*, 25.

hand the heavy jars up to another individual.[176]　　The large storage jars would be removed and the contents either consumed or transferred to smaller jars.

The capacity for wine production suggested by these 63 cellars (95,000 liters),[177] if ever met, would have required considerable grape treading. And, if the ten discovered winepresses were the only ones in use, then they were most likely shared by vintners. The juice was probably put in jars after treading and allowed to ferment in the cellars. The lack of vats meant that there was no other place for the juice to begin to ferment except in the press itself. But letting the juice begin to ferment within the treading floor is a practice best exercised when there is not an immediate need for the press. Such would be the case if only a few vintners with small harvests, say a treading floor's worth of grapes, were using the presses. If there were more farmers, collecting vats offered a way to maximize the use of a treading floor. If there were many farmers, all with their own harvests, then a different technological strategy was needed.

This strategy is perhaps evident at Gibeon. Many large cellars were built. These cellars stored the jars of wine, but they also probably functioned in the fermentation process itself. As we shall see in the next chapter, fermentation rooms filled with jars was a known Roman practice. It might also have been the practice of Gibeonite vintners. If so, then the presses could quickly be emptied and reused while the juice fermented in cellars. The jars would need to be left open and room left in them as the carbon dioxide escaped. The simple press without a separate vat could only contain one batch of juice at a time. The use of a simple press, then, implies that this level of production was adequate for the grape harvests involved, or that the juice fermented in jars away from the press, perhaps in the home or in the vineyard tower. At Gibeon, simple presses were sufficient, because the cellars were extraordinary; they constituted the heart of the winery.

Gibeon itself was a medium-sized site, 6.4 hectares.[178] As a rough estimate, using the density coefficient of sites of 250 inhabitants per hectare,[179] approximately 1,600 people lived in Gibeon. The capacity of the cellars was 95,000 l.[180] This means that each person

---

[176]Pritchard, *Winery*, 25-26.

[177] Pritchard, "Gibeon," in *NEAEHL*, 2:512.

[178] A hectare is just under two and a half acres.

[179]This seems rather high. M. Broshi and I. Finkelstein estimate approximately 270 people per hectare: "The Population of Palestine in Iron Age II," *BASOR* 287 (1992): 48; Stager cautiously estimates between 250 and 300 people per hectare as an upper limit: "The Archaeology of the Family in Ancient Israel," table 3, p. 21. At any rate, the estimate is far below Pritchard's own unrealistic estimate of 4,000-6,000 people: *Gibeon: Where the Sun*, 104.

[180]Pritchard, "Gibeon," in *NEAEHL*, 2:512.

would then have had approximately 60 l of wine for a yearly ration. This is a fairly modest amount by any reckoning and could easily have been stored within the individual dwellings of the citizens.[181] The Gibeon cellars could have been a measure to ease the domestic storage space of its citizens. It was, however, a lot of initial labor to carve out the cellars and required an ongoing cooperative effort for sharing storage. Since this amount of wine could have been stored in the individual dwellings, and since there were no other communal storage facilities found for other staples, such as grain and oil, it is reasonable to assume that the wine cellars were constructed with trade in mind. Their locations are confined to two areas, nos. 8 and 17, located on the northeast section of the mound near the winepresses.[182] While this layout could simply reflect the collaborative efforts of the Gibeon citizens, it seems to imply some sort of citywide trade activity in wine. Perhaps the cellars represent not the total wine production of Gibeon, but only that portion that was saved for trade.

Trade must have been an important facet of the winery at Gibeon, and this is suggested both by the capacity of the cellars and by the inscribed jar handles, twenty-seven with the name Gibeon on them.[183] The jars themselves, Pritchard estimated, held 15-22 l, and, with an almost 4 cm diameter opening, proved impractical for anything but a liquid commodity.[184] The inscriptions conform for the most part to a set pattern of *gbʿn gdr* + personal name.[185] If *gdr* is a "vineyard wall,"[186] it could perhaps denote the vineyard area of the wine producer. Pritchard demonstrated that 68 percent of the inscriptions mention one of four personal names. These four are, for him, the vintners in the Gibeon area.[187] He suggested that the existence of personal names with the addition of a town name on the jar handles was either a kind of quality control for the wine production, or a means to facilitate return of the empties.[188] The cellars and the inscribed jars suggest the possibility of organized

[181]See chapter 2, pp. 47-51.

[182]Pritchard, *Winery*, 1.

[183]J. B. Pritchard, *Hebrew Inscriptions and Stamps from Gibeon* (Philadelphia: University Museum, 1959), v.

[184]Pritchard, *Gibeon: Where the Sun*, 49; R. Amiran, "A Note on the 'Gibeon Jar,'" *PEQ* 107 (1975): 131-32.

[185]Pritchard, *Hebrew Inscriptions and Stamps from Gibeon*, 7; F. M. Cross argues for a reading of *gdd* rather than *gdr*, as the third letter is similar in form to the middle letter, *dalet*, in approximately half of the cases, "Epigraphical Notes on Hebrew Documents of the eighth-sixth centuries B.C.," *BASOR* 168 (1962): 20.

[186]chapter 3, p. 96.

[187]Pritchard, *Hebrew Inscriptions and Stamps from Gibeon*, 7. Cross notes that the names probably were of the vintners rather than potters since they were inscribed after firing: "Jar Inscriptions from Shiqmona," *IEJ* 18 (1968): 232, n. 40; see also N. Avigad, "Two Hebrew Inscriptions on Wine Jars," *IEJ* 22 (1972): 1-9.

[188]Pritchard, *Gibeon: Where the Sun*, 51.

communal wine production at Gibeon. The presses themselves are not remarkable, but the capacity for wine storage in the cellars suggests that it was a center of Israelite wine production.[189]

I have demonstrated in this section that winepresses are both present and varied in Iron II sites in Israel. These presses, nevertheless, may represent only a fraction of the Iron Age presses, since there are hundreds more in fields whose date cannot be fixed. Still, the winepresses described here show the range of types for the Israelite winepress. Some of that range may be hinted at as well in the biblical texts where there exist three words for "winepress." We turn now to a discussion of the biblical terminology for this installation.

c. Biblical Terms for "Winepress"

There are three biblical terms for "winepress": יֶקֶב, גַּת, and פּוּרָה. יֶקֶב is the most frequent, occurring sixteen times. גַּת is used five times, occurring also in place names, such as the Philistine city of Gath (e.g. 1 Sam 5:8; 7:14), Gath-(Ha)hepher (2 Kgs 14:25; cf. Josh 19:13), and Gath-Rimmon (Josh 19:45, 21:24). The third term, פּוּרָה, is used only twice, both times in postexilic texts: Isa 63:3 and Hag 2:16. Four of the five occurrences of גַּת occur in exilic and postexilic texts: Neh 13:15; Isa 63:2; Lam 1:5; Joel 4:13 [ET 3:13]. Also, in these four uses, the verb דָּרַךְ, "tread," accompanies גַּת. The fifth occurrence of גַּת occurs in Judg 6:11, where Gideon is beating out wheat in the winepress. What all five occurrences of גַּת have in common is that an agricultural *process* is described as taking place in the press.

The range of uses differs somewhat for יֶקֶב. יֶקֶב is used in both pre-exilic and exilic texts, but in its sixteen occurrences, the agricultural process of treading appears only three times: Job 24:11; Isa 16:10; and Jer 48:33. In eleven of its occurrences, the product or yield from a winepress is noted with יֶקֶב. That is, wine is either mentioned (Hos 9:2; Jer 48:33; Isa 16:10) or somehow implied in the description (e.g., a thirst in Job 24:11; an overflow in Prov 3:10 and Joel 4:13 [ET 3:13]). In these eleven cases, then, the press is described in terms of the *product* it yields, rather than the agricultural process involved as it was for גַּת. The difference, it seems, is one of viewpoint towards the press, i.e., whether its function or its consequent bounty is highlighted.

Since דָּרַךְ is used so often with גַּת, it is reasonable to suggest that the treading floor is most likely meant. This is probable as well for the fifth reference, Judg 6:11, since the flat surface of a winepress could serve as a makeshift threshing floor in times of duress.[190] Since the *product* of the winepress, either as liquid or as offering, is stressed in eleven of sixteen occurrences of יֶקֶב, I suggest that יֶקֶב signifies the installation in general.[191] In other words, גַּת is best translated "treading floor,"[192] while יֶקֶב denotes the more general term, "winepress." Thus, if a winepress as envisioned by a biblical author was simply a treading floor and a circular depression within for scooping up the juice, such as the one at Samaria, for example, then either term, גַּת or יֶקֶב, sufficed. If, however, a press was a more complex installation, comprising a treading floor, collecting vat(s), and channels, then יֶקֶב proved to be the more comprehensive term. Most of the Iron Age presses and the majority of the undated presses found through survey work had, as we saw above, a rectangular treading floor and at least one collecting vat. These, then, were the installations, יְקָבִים. If this distinction is accurate, then it explains why יֶקֶב is the much more frequent term for winepress than גַּת.

The third term, פּוּרָה, is used, as we have seen, only twice, both in postexilic texts: Isa 63:3 and Hag 2:16. In Isa 63:3, פּוּרָה is in parallel with גַּת, as the press where Yahweh will tread people.[193] פּוּרָה is not simply Yahweh's special press for human destruction, since for that he uses a גַּת in Lam 1:15, and indirectly threatens to use the יֶקֶב in Isa 5:2b for the same purpose. In 63:3, Third Isaiah contrasts Yahweh's vintage with that of the Edomites. The latter are all spattered red from treading in a יֶקֶב. Yahweh then treads people in his anger:

---

[190]For Borowski, the three biblical terms denote three kinds of winepress. The גַּת, he states, was within the city and built of stones and mortar. He bases this on the story of Gideon hiding from the Midianites in the winepress, but offers no evidence for why he considers it to be built of stones and mortar: *Agriculture of Iron Age Israel*, 111. The archaeological evidence so far does not bear out this assumption, though there is one possible stone winepress at Tel Batash: Kelm and Mazar, "Tel Batash (Timnah) 1987-1988," 108-10.

[191]Ahlström considers the יֶקֶב to be the collecting vat, while the פּוּרָה was the installation as a complex of parts: treading floor and vat(s). See the discussion of פּוּרָה below. Ahlström, "Wine Presses and Cup-Marks," 19-49; Ross also understands the יֶקֶב to be the collecting vat: "Wine," 850. However, יֶקֶב cannot be restricted to mean only the collecting vat, since דָּרַךְ is used with it in Isa 16:10, Jer 48:33, and Job 24:11. For Borowski, יֶקֶב denotes the rock-hewn winepress near a vineyard. He argues that יֶקֶב, which occurs seven times with גֹּרֶן, "threshing floor," another food-processing installation, is located outside the city walls. By contrast, he notes, גֹּרֶן is never paired with גַּת: Borowski, *Agriculture in Iron Age Israel*, 111.

[192]Ross, "Wine," 850.

[193]See above, pp. 146-47.

פּוּרָה ׀ דָּרַכְתִּי לְבַדִּי
וּמֵעַמִּים אֵין־אִישׁ אִתִּי
וְאֶדְרְכֵם בְּאַפִּי וְאֶרְמְסֵם בַּחֲמָתִי

(In) a press, I trod by myself,
and from peoples, none was with me;
I trod them in my anger and
trampled them in my rage.

(Isa 63:3)

Yahweh's robes too get spattered, so the פּוּרָה is not superior to the יֶקֶב in terms of cleanliness. Two features of Yahweh's treading here receive contrastive stress through syntax[194] and repetitive parallelism.[195] The first feature is that Yahweh works alone: לְבַדִּי and אֵין־אִישׁ אִתִּי. The second feature is that his rage fuels the task: בְּאַפִּי and בַּחֲמָתִי. The uniqueness of a פּוּרָה press in distinction to a יֶקֶב or a גַּת, I suggest, is in its volume. It is, for Third Isaiah, a large press that one person would not ordinarily work alone. Yet in Isaiah's view, Yahweh does just that, and the פּוּרָה is even large enough to contain the deity's rage.

The Haggai passage (2:16) as well intimates that capacity is somehow the discernible feature of the פּוּרָה, but here it is expressed in terms of measures. The יֶקֶב that is expected to yield fifty פּוּרָה measures only nets twenty. As a measure here, the term פּוּרָה makes sense as a collecting vat of a winepress, a יֶקֶב, though it is also the treading surface itself in the Third Isaiah passage above. *Capacity*, in both texts, is stressed. The פּוּרָה in the Haggai verse signifies, in other words, a collecting vat's worth of wine, but the יֶקֶב, as installation, is disappointingly low in its output. However, since פּוּרָה is used only in these two passages, it is not possible to say more than that capacity is its highlighted feature.[196]

---

[194]פּוּרָה comes first in the verse for emphasis: Joüon, *A Grammar of Biblical Hebrew*, § 155o.

[195]Three such parallels are drawn in v 3.

[196]Borowski suggests that the פּוּרָה might have been a "portable winepress," though he adds that he knows of no Iron Age examples: *Agriculture of Iron Age Israel*, 111; For Ahlström, the פּוּרָה was the winepress installation, inclusive of a treading floor and collecting vat(s). He offers, however, no explanation for how he arrived at this conclusion, nor does he discuss the use of both פּוּרָה and יֶקֶב for treading in Isa 63:3: "Wine Presses and Cup-Marks," 19.

יֶקֶב, גַּת, and פּוּרָה are thus the three biblical terms for winepress, and I have suggested more precise interpretations based on their use in biblical passages and with the knowledge of the variety of presses evident in archaeological remains. It is time now to turn to wine production itself and the social aspects involved in harvesting and treading the grapes. Isaiah's Song of the Vineyard has thus far proven to be an instructive rubric for discussion of the tasks required to start a vineyard in ancient Israel. Construction of the two installations—tower and winepress—concludes the list of preparations in the Song. A discussion of the grape harvest and wine production comes next as the payoff to all that planning and care, but since Isaiah's vintner never saw a successful harvest, his Song is of limited value for further historical reconstruction. We turn instead to other biblical and comparative materials for illumination of ancient Israel's autumn grape harvest and wine making.

Chapter 5

The Grape Harvest

This chapter is devoted to a discussion of the grape harvest, from picking grapes to wine production and storage. Since chapter 4 dealt with the actual process of grape treading in a winepress,[1] the discussion in the present chapter focuses on the additional practical and social activities involved in the Israelite vintage season. The first section (I) details the harvesting and collection of grapes from the vineyard, and section II focuses on the festive aspects of the vintage. The last two sections, III and IV, are concerned with the wine produced at the vintage. Section III details the chemical process involved in fermentation for the ancient Israelite vintner, while section IV is an analysis of all the different terms for wine in biblical and epigraphic materials.

I. Practical Aspects of the Vintage

a. Peak Ripening of the Grapes

The grape harvest in modern Israel occurs in the fall, sometime in August to September.[2] Grapes begin their growth in the spring.

---

[1] See especially, pp. 145-48.
[2] Turkowski, "Peasant Agriculture in the Judaean Hills," 27; Ahlström, "Wine Presses and Cup-Marks," 41; Moldenke and Moldenke, *Plants of the Bible*, 244; Dar, *Landscape and Pattern*, 1:154; Ross, "Wine," 850. W. Dommershausen, "יָיִן," in *Theological Dictionary of the Old Testament* (ed. G. J. Botterweck and H. Ringgren; Grand Rapids, Mich.: Eerdmans, 1985) 6: 60. The ancient Greek and Roman vintages also apparently took place in September: Hesiod *Works and Days* 610; Columella *De re rustica* 11.2.63-64; White, "Farming and Animal Husbandry," 217; Unwin, *Wine and the Vine*, 94. In modern Greece, too, the vintage occurs in September, Friedl, *Vasilika: A Village in Modern Greece*, 30, 276.

Then, in a period of five to seven weeks in the summer, the berries reach their highest acidity and begin to accumulate sugar.[3]   In the last half of the summer, they begin to lose the chlorophyll that gives them their green color.[4]   In the next five to eight weeks grapes change color and grow four to five times in size.[5] Oenidin, the anthocyanin present in *Vitis vinifera*, and the tannins on the grape's skin account for the berry's color change during ripening.[6] Indeed, the onset of the harvest is determined, then and now, on the basis of the grapes turning a dark red.   This color change, known today as *véraison*, French for "ripening," signals the beginning of the vintage season. The leaves turn color slightly later, usually after the harvest, and then are shed for winter's dormancy.

Grapes at the height of their ripening are 70-80 percent water and from 10 to 25 percent sugar.[7]   This is the optimal time for harvesting because the maximal amount of juice and sugar would yield the most wine.   Before this point the grapes are smaller and sour. Grapes in this phase are denoted in the Bible by the term בֹּסֶר (Isa 18:5).   That sour grapes were not generally desirable is evident from their use in a biblical proverb:

אָבוֹת יֹאכְלוּ בֹסֶר וְשִׁנֵּי הַבָּנִים תִּקְהֶינָה

Fathers will eat sour grapes
and the children's teeth will be dulled.
(Ezek 18:2, cf. Jer 31:29-30)

In this proverb, sour grapes symbolize sin, in a metaphor about who is beset with its consequences.   Sour grapes, then, were not desired because of their presumed effect on teeth.   They were probably also avoided because their low sugar levels would have resulted in a wine of minimal alcohol.   The proverb hints at an additional viticultural wisdom, viz., that an unskilled vintner, one given to tasting grapes before they were ripe, or worse, preferring them, would teach his sons similar bad habits.   Sour grapes were not the desired object of the harvest in ancient Israel.   Instead, "grapes," עֲנָבִים, (Gen 40:11; Lev 25:5; Num 6:3-4; Deut 32:32; Jer 8:13; Amos 9:13) and "grape

---

[3]Winkler et al., *General Viticulture*, 138.
[4]Unwin, *Wine and the Vine*, 36.
[5]Winkler, *General Viticulture*, 138; Olmo, "The Origin and Domestication of the *Vinifera* Grape," in *OAHW*, 41.
[6]Unwin, *Wine and the Vine*, 36; Winkler et al., *General Viticulture*, 159.
[7]Unwin, *Vine and the Wine Trade*, 34; de Blij, *Wine: A Geographic Appreciation*, 22; White, "Farming and Animal Husbandry," 256; Olmo, "Grapes," 294.

cluster," אֶשְׁכּוֹל, (Num 13:24; Deut 32:33; Cant 7:8, 9; Isa 65:8; Mic 7:1) are the more frequent terms for the fruit of the vine.

Both terms, עֲנָבִים and אֶשְׁכּוֹל, are used together in the harvests of Gen 40:10 and Num 13:23. In Gen 40:10, Pharaoh's cupbearer takes grapes from clusters where they have ripened:

$$\text{הִבְשִׁילוּ}^8 \text{ אַשְׁכְּלֹתֶיהָ עֲנָבִים}$$

Its clusters ripened into grapes.

In Num 13:23, Moses's spies are able to harvest "one cluster of grapes," אֶשְׁכּוֹל עֲנָבִים אֶחָד, from the Wadi Eshkol. This particular cluster is so large that it requires two men[9] to carry it back to Moses on a pole. Hence, with this grape cluster, some pomegranates, and figs, the spies handily show Moses that the land "flows with milk and honey, and this is its fruit" (v 27). After forty days of spying (v 25), of course, the cluster would have become a (large) bunch of raisins.

This story of the people's first glance at what the land could yield clearly illustrates the value of viticulture to ancient Israel. To be sure, pomegranates, figs, and all products under the rubric of "milk and honey" get displayed as well, but it is the grape cluster that is uniquely esteemed in this passage by its exaggerated size.[10] The grape harvest of this land, in other words, is so fruitful that one cluster requires two men to lift it. Such bounty overall bodes well for this wandering camp's relocation.

Today's vintner can calibrate the peak ripening moment for the grape with a hydrometer, an instrument that measures the sugar

---

[8]הִבְשִׁילוּ , "boiled," may connote fermentation, i.e., boiling, as well as ripening since the entire viticultural process from harvest to serving the liquid is telescoped in the cupbearer's dream sequence. However, the liquid in Pharaoh's cup is not mentioned as wine (40:11), so the description may simply represent juice from ripened grapes. Only juice drunk soon after crushing would still be unfermented, since the yeasts from the skins begin to work on the juice within a few hours.

[9]Contrary to subsequent popularizing traditions in art, the two men are not named in the biblical text as Joshua and Caleb.

[10]This camp, in fact, is swayed by size. Upon seeing the fruit, the people want to take possession of the land, undaunted by reports of the strong people dwelling there. They are dissuaded only when some of the spies start a "rumor about the land," דִּבַּת הָאָרֶץ, that the people there are also exaggerated, i.e., they are giants, הַנְּפִילִים (vv 32-33). Num 13:22 and 13:33 specify that one tribe, the Anakites, was taller than most people, having a long "neck," עֲנָק, a tradition also reflected in Deuteronomy (Deut 2:10-11, 21; 9:2). The spies exaggerate by claiming that all the people of the land are such giants that the Israelites become mere "grasshoppers." Surely, such a report would have deflated even the most stalwart and eager of the warriors in the camp.

percentage in liquid samples taken from several grapes.[11]   Additional
kits also test for tartaric and malic acids.[12]   A certain amount of these
acids in wine is desired today for a crisp taste and also because they
protect the grape juice from various spoilage organisms.[13]   The
modern vintner can simply purchase these tools for testing grapes and
then know when the best time for the grape harvest has arrived. Even
with the scientific advance of the hydrometer and acid kits, however,
the modern vintner still must go into the vineyards every day to test
the grapes. And, not all modern vintners bother with the hydrometer.

In the anthropological study of a modern Tuscany vineyard,
mentioned earlier, Calabresi records that the husband and wife,
Adamo and Maria, assess the peak moment for the grape harvest by
walking about the vineyard sampling their grapes by taste. The peak
moment for the harvest becomes between them a daily debate.[14]
Adamo's wisdom is almost masked by its simplicity when he states:
"An experienced farmer knows when a grape is ripe by the taste."[15]
Daily attention to the crop as it goes into the final swell of its ripening
is, for the modern vintner, necessary as well as exciting.

The daily attention and excitement were likely not all that
different for the ancient vintner. He could have visited the vineyard
daily, and even slept over in it as the harvest approached. Columella
advised first pressing one grape in the vineyard to see if its pip was
colored.[16] This test complemented simple tasting and enabled the
ancient vintner to proceed with confidence to the harvest.   The
Israelite vintner might have had a similar test.   There were
undoubtedly procedures for testing grape ripeness in ancient Israel
that simply go unrecorded in the Hebrew Bible. A kind of finesse
with grapes would have separated the wise from the foolish vintner.
This subtle wisdom and skill of the vintner is underrepresented in the
Bible. Biblical descriptions of grape handling usually involve Yahweh
as vintner; and his "skill" as deity, of course, is both beyond reproach
and never subtle. As deity, he need not test, gauge, or ponder—still
less, sample or discuss with a wife the ripeness of his crop.

As a crop, grapes were much more vulnerable than olives, the
other major fall crop of ancient Israel. Grape berries quickly become
overripe and drop to the ground and rot. Olives, by contrast, could
be harvested any time in a two-month period, as they remain stable on

---

[11]Cox, *From Vines to Wines*, 113-16.
[12]Cox, *From Vines to Wines*, 116.
[13]Cox, *From Vines to Wines*, 116.
[14]Calabresi, "*Vin Santo* and Wine," 122.
[15]Calabresi, "*Vin Santo* and Wine," 123.
[16]*De re rustica* 11.2.68-69.

the tree when they ripen, and the ones that do fall do not easily rot.[17] Grapes reach their peak and have to be harvested within several days. The Israelite's grape test was likely as simple and skilled as that of Adamo and Maria, involving taste, sight, and perhaps debate between husband and wife. Once the peak time arrived, the harvesting of the fruit could begin.

b. Cutting Grapes from the Vine

The story of the spies in Num 13 describes the actual procedure for grape harvesting:

וַיִּכְרְתוּ מִשָּׁם זְמוֹרָה
וְאֶשְׁכּוֹל עֲנָבִים אֶחָד

They cut from there a shoot
and one cluster of grapes.
(v 23)

The description of grape harvesting in Gen 40:11 is less specific, stating only that "I," that is, the cupbearer, "took the grapes," וָאֶקַּח אֶת־הָעֲנָבִים. Grape harvesting was done by cutting the cluster from the vine in ancient Israel.[18] This method is evident from the biblical term for the grape harvest itself, בָּצִיר[19] (Lev 26:5; Judg 8:2; Isa 24:13; 32:10; Jer 48:32; Mic 7:1), from the root בָּצַר, "cut off."[20] "Gather" (אסיף/אסף), as used in the Gezer and biblical calendars (Exod 23:16; 34:22; Lev 23:39; Deut 16:13), is a more general term for the collection of grapes off the vine and into containers. It is, more precisely, the *second* task of grape harvesting after first cutting the cluster from the vine. It is not surprising that the biblical calendars would reflect the fruit's collection since those texts focus on the festival offerings of the harvests. Its use in the Gezer calendar is

---

[17]King, *Amos, Hosea, Micah*, 159; Borowski, *Agriculture in Iron Age Israel*, 119; Turkowski, "Peasant Agriculture," 27; Hopkins, *The Highlands of Canaan*, 230; Baly, *Geography of the Bible*, 86; du Boulay, *Portrait of a Greek Mountain Village*, 276-77; Calabresi, "*Vin Santo* and Wine," 131.

[18]No tools, however, are evident in the many harvest scenes of Egyptian New Kingdom tomb paintings: Lesko, *King Tut's Wine Cellar*, 17.

[19]Waltke and O'Connor, *An Introduction to Biblical Hebrew Syntax*, 88: בָּצִיר is one of five nouns used with a *qātîl* pattern for agricultural periods. The others are: קָצִיר, זָמִיר, אָסִיף, and חָרִישׁ.

[20]BDB, 131; KB, 1:148.

explained simply by the nature of the inscription: it describes the harvests of an entire agricultural year in seven terse lines.

The בָּצִיר, as the time of the grape cluster cutting, is, at times, contrasted with the specific techniques involved in the other harvest periods, namely the "threshing" of grain, דַּיִשׁ (Lev 26:5), the "beating" of olive trees, וְנֹקֶף (Isa 24:13), and the "gathering" of summer fruit, אֹסֶף (Mic 7:1). The first two terms describe procedures unique to the grain and olive harvests, respectively. The third, gathering, was, as noted above, a task subsequent to first getting the fruit off its vines and trees. Collection of the yield, of course, would apply to all harvests. The two practices, of cutting the fruit from the vine and then gathering, undoubtedly reflect Iron Age viticultural practice.

The grape cluster is connected to the vine's branch or shoot by a smaller, woody stem termed a peduncle.[21] While severing this peduncle can be accomplished by repeated twisting, it is much easier and faster to clip it with a cutting tool.[22] Other ancient texts describe the grape harvest with cutting. A text from Ugarit, for example, describes a grape harvest with "cut off a bunch of grapes": *šmtr. [uṭkl . . .].*[23] Hesiod mentions cutting grape clusters for harvesting as well.[24] Cato and Columella describe the cutting of grape clusters off the vine and mention special tools for the purpose. Cato recommends a *falcula viniatica*, "small grape sickle."[25] Columella advises the use of these *falculae* too and warns that hand picking loses too much of the fruit to the ground.[26]

The Bible refers to vineyard cutting tools as מַזְמֵרוֹת, "pruning hooks." Judging by the term's use in Isa 2:4; 18:5; Mic 4:3; and Joel 4:10 [ET 3:10], these were metal tools used in vineyard cultivation in ancient Israel.[27] In Isa 2:4; Mic 4:3; and Joel 4:10 [ET 3:10], we recall, spears are turned into pruning hooks, or, in the case of Joel 4:10 [ET 3:10], vice versa, and so are clearly made of metal. In Isa 18:5, they are clearly cutting tools, used to clip and clear away the tendrils, before the grapes are ripe:[28]

---

[21]Cox, *From Vines to Wines*, 31.

[22]Calabresi, "*Vin Santo* and Wine," 129.

[23]*KTU* 1.41, lines 1-2; for a discussion of this passage, see J. de Moor, *An Anthology of Religious Texts from Ugarit* (Nisaba; Leiden: Brill, 1987), 157-58.

[24]*Works and Days* 676.

[25]Cato *Agriculture* 11.5.

[26]He recommends as many sickles and iron hooks, *falculae et ungues ferrei*, as possible, "so that the vintager may not strip off the bunches of grapes with his hand, which causes no small part of the fruit to fall to the ground": Columella *De re rustica* 12.18.2. Elsewhere he mentions a "vintner's knife," *vinitoria falx*, 4.25.1.

[27]See chapter 3, pp. 120-22.

[28]Perhaps this is the summer pruning, likely indicated in the Gezer calendar. See chapter 1, pp. 34-38.

כִּי־לִפְנֵי קָצִיר כְּתָם־פֶּרַח וּבֹסֶר גֹּמֵל יִהְיֶה נִצָּה
וְכָרַת הַזַּלְזַלִּים בַּמַּזְמֵרוֹת וְאֶת־הַנְּטִישׁוֹת הֵסִיר הֵתַז

> For before the harvest, when the blossom is over
> And the flower becomes a ripening grape,
> He will cut the vine branches with pruning hooks,
> And the tendrils he will cut away.

Vine wood is thick and tough and would have needed metal tools for pruning.

The term itself is derived from זָמַר, "to prune," but there is no reason to assume that the tool would have been restricted in use to pruning woody shoots and buds, when it could have just as effectively cut the cluster peduncles at the harvest. Indeed, in Num 13:23, the grape cluster is cut precisely at its peduncle, termed זְמוֹרָה, "shoot." Here, and elsewhere, I suggest, the occurrence of זְמוֹרָה, implies the use of מַזְמֵרוֹת, because pruning involved cutting with a metal tool. The wood piece denoted by זְמוֹרָה, in other words, indicates not any kind of tree branch, but rather a shoot or branch cut with a pruning tool. In the biblical texts, this "pruned (off) branch" is usually from the vine, though other fruit trees were pruned for greater yields as well (Num 13:23; Ezek 15:2; also Ezek 8:17; Nah 2:3).

While any sharp blade could have cut grape clusters, the pruning hook was best for cutting and gathering them toward the laborer. In shape, the pruning hook was similar to a sickle and it is possible that the ancient farmer used a sickle for both the grain and grape harvests.[29] Among the five farm tools listed in a Ugaritic text, sickle (ḥmrṭṭ) is the only small cutting tool,[30] and we know from other texts that farms outside of Ugarit were cultivating vines and olive trees.[31] It is difficult to gauge how many such tools a typical Israelite vintner would have had at his disposal for the vintage, though the archaeological remains of tools thus far suggest that it was not many. The biblical tradition of converting spears to pruning hooks and back

---

[29]For a comparison with forms of modern sickles and "garden tools," see Turkowski, "Peasant Agriculture," 102.

[30]An axe (mʿṣd) is mentioned, but would be unwieldy for cutting the vine's small peduncle. The other tools listed are the hammer (mqb), pick (nit), and shovel (krk).

[31]PRU V, 48; M. Liverani, "Economy of Ugaritic Royal Farms," in Production and Consumption (ed. C. Zaccagnini; Budapest: University of Budapest Press, 1989), 135-36; Heltzer, "Vineyards and Wine in Ugarit (Property and Distribution)," 119-35: see especially a table detailing texts with vineyard plots on 122-23.

again (Isa 2:4; Mic 4:3; Joel 4:10 [ET 3:10]) also suggests the rarity and value of Iron Age tools, since the metal was being recycled.

c. Gleanings from a Vineyard

In six of the twelve times that בצר is used for the vintage[32] (Mic 7:1; Jer 6:9; 49:9; Obad 5; Deut 24:21; Judg 8:2), grape harvesting (בצר) is associated with gleanings (עוֹלֵלוֹת). Gleanings were the grapes that were *not* cut from the vine during the harvest, either through oversight or in some charitable gesture toward the poor who would come into the vineyards later. The task of grape harvesting in these passages entails then, oddly, *not* harvesting all the fruit. The custom of gleaning seems to have applied to all the major crop yields (e.g., Isa 17:5-6 for grain, olives, and fruit; Isa 24:13 for olives and grapes; Mic 7:1 for grapes and figs). עָלַל is the verb used for grape gleaning, while לָקַט is used for the gleaning of grain (e.g., Lev 19:9; Ruth 2:2, 7-8; Isa 17:5).[33]

Jeremiah 49:9 and Obad 5 are similar, and pair grape harvesters, בֹּצְרִים, with thieves:

אִם־בֹּצְרִים בָּאוּ לְךָ

לֹא יַשְׁאִרוּ עוֹלֵלוֹת

אִם־גַּנָּבִים בַּלַּיְלָה

הִשְׁחִיתוּ דַיָּם

If grape-gatherers came to you,
would they not leave gleanings?
If thieves came in the night, they
would pillage only what is sufficient.
(Jer 49:9)

The Obadiah verse is essentially the same, though expanded, and in a reversed order, with thieves coming before grape harvesters:

---

[32] בָּצִיר in Lev 26:5; Judg 8:2; Isa 24:13; 32:10; Jer 48:32; Mic 7:1; in verb forms, including participles, in Lev 25:5,11; Deut 24:21; Judg 9:27; Jer 49:9; Obad 5.
[33] Olive gleanings apparently were gotten in the same manner as olive harvesting, through beating the trees (e.g., Isa 17:6; 24:13).

אִם־גַּנָּבִים בָּאוּ־לְךָ אִם־שׁוֹדְדֵי לַיְלָה
אֵיךְ נִדְמֵיתָה הֲלוֹא יִגְנְבוּ דַיָּם
אִם־בֹּצְרִים בָּאוּ לָךְ הֲלוֹא יַשְׁאִירוּ עֹלֵלוֹת

> If thieves came to you, if plunderers at night—
> How you have been destroyed—
> would they not steal only what is sufficient?
> If grape-harvesters came to you, would they
> not leave behind gleanings?
>
> (Obad 5)

The practice of grape harvesting as given rhetorically in these two passages, then, included leaving some grapes behind. And, the prophets' view of thieves may not be simply naive.

The premise for comparison between grape harvesters and thieves seems to be that their tasks both entailed carrying only what one can, while leaving the rest behind. The grape harvester and the thief alike had an imposed limit on how thorough they could be in their efforts. The job, in other words, be it in the vineyard or at a targeted home, was always larger than the task force, so the grape harvester and thief took only what they could carry. Something was always left behind, presumably even if the thieves had donkeys to carry the loot for their getaway. In these biblical passages, gleanings, i.e., fruit left on the vines, were factored in as a part of the Israelite grape harvest.

Two biblical laws provide social or ethical rationales for leaving gleanings during the grape harvest:

> When you harvest your vineyard,
> do not glean afterwards; it shall be for
> the alien, the orphan, and the widow.
> (Deut 24:21)

> Your vineyard you shall not glean,
> the fallen berries of your vineyard you shall not glean;[34]
> for the poor and the sojourner, you shall leave them.
> (Lev 19:10)

---

[34]Here, the second verb for gleaning is לְקֵט, though it seems to refer to berries broken off, פֶּרֶט, i.e., on the ground as grain would be.

These laws reflect concern for people who were either landless or impoverished, and so sanction a kind of workfare system for ancient Israel. They also detail, then, essentially two grape yields, one harvested by grape gatherers, and another by the poor who were permitted to enter these vineyards, not their own, after the grape harvest. The laws simply mandate that the first harvesters not go back over the vines for the remainder of the fruit.

Jeremiah, via metaphor, is to violate this custom and be both harvester and gleaner. Yahweh commands him to glean thoroughly to see if he can find a worthy remnant of Israel. Jeremiah as the בּוֹצֵר is commanded to harvest twice and so get the gleanings by returning his hand upon the vine shoots:

> Glean thoroughly as (with) a vine the remnant of Israel;
> return your hand, like a grape harvester upon shoots.[35]
>
> (Jer 6:9)

The theological implication is clear: Yahweh is certain in his anger that even with the unusual practice of both harvesting and gleaning, Jeremiah will not be able to find even one batch of righteous Israelites.

At some point in Israel's history, gleanings were left intentionally on the part of the grape harvesters for charity. The grape harvester, then, attained a certain finesse, working quickly to harvest the grapes at their peak of ripening, but also leaving some behind as charity toward nonlandowners. Varro mentions as well an additional skill of harvesting that might have been present in Israel too: that of selecting grapes used to make wine and those that were for eating.[36] The amount of fruit left for gleanings was perhaps negligible. Nevertheless, it was a sanctioned or protected part of the crop that was not harvested.

In Judg 8, Gideon calls the Ephraimites to battle late, and the latter become angry with him. Gideon was able to pacify them with a rhetorical question drawn from grape harvest traditions:

> Are not the gleanings of Ephraim
> better than the grape harvest of Abiezer?[37]
>
> (Judg 8:2)

---

[35]סַלְסִלּוֹת, "shoots:" KB, 3:715; BDB, 700.
[36]Varro *On Farming* 1.54.2.
[37]Gideon was an Abiezrite (Judg 6:11), so he is using self-deprecation to appease the Ephraimites.

Gideon's flattery contrasts the *quality* of their grapes over *quantity* and mollifies the Ephraimites. Their vanity is clearly evident here. For it is highly improbable that any vintner would prefer the few grapes left for gleaning to the full baskets of a harvest, no matter how extensive a vineyard might have been. Greed, it shall be recalled, had motivated the Ephraimites to confront Gideon in the first place. They were angry because a late mustering meant they could not reap as much of the spoils of war.[38] It was a group, then, that thought in terms of quantity over quality, until vanity blinded them.

d. Baskets for Grape Gathering

Once the clusters were cut from the vine, some container was needed to "gather" (אסף) the fruit. Most likely, baskets were used for this purpose in ancient Israel, as they were in other ancient societies practicing viticulture. Baskets for grape harvesting are depicted on the ancient Egyptian tombs of Khaemwese, Nakht, and Petosiris.[39] In the *Iliad*, boys and girls harvest grapes with woven baskets.[40] Cato and Columella both mention baskets for the grape harvest.[41] Modern small vineyards, such as Adamo's in Tuscany and those in Vasilika, Greece, continue to use baskets for fruit collection and transport.[42] We can infer from these parallels that baskets served these purposes for the Israelite grape pickers as well.[43]

The Hebrew Bible contains several instances of fruit collection in baskets. In Deut 26:2, 4, a basket, טֶנֶא, is filled with fruit from the harvest. Two baskets of figs, דּוּדִים, one good and one bad, illustrate for Yahweh the division of the people of Judah (Jer 24:2).[44] A "basket of summer fruit," כְּלוּב קָיִץ, is used in Amos as Yahweh's visual aid and devastating word play of the end (קֵץ) of the people (Amos 8:1-2).[45]

---

[38]The metaphor breaks down when Gideon explains that the gleanings were the princes of Midian, and these were more valuable than more regular gains in battle (v 3). It is improbable that gleanings were ever the very best of one's grape crop.

[39]Johnson, *Vintage*, 31; Darby et al., *Food: The Gift of Osiris*, 2:558, fig. 14.15.

[40]Homer *Iliad* 18.537.

[41]Cato *Agriculture* 23.1; Columella *De re rustica* 12.18.2; 12.52.8.

[42]Friedl, *Vasilika*, 20; Calabresi, "*Vin Santo* and Wine," 129.

[43]In Baytin (biblical Bethel), women gather straw and make baskets: Antoun, *Arab Village*, 119.

[44]Jehu used these kinds of baskets to send the heads of Ahab's sons to the king (2 Kgs 10:7).

[45]Baskets for grain products had yet another term, סַל, e.g., Gen 40:17; Exod 29:3, 23, 32; Lev 8:2, 26, 31; Num 6:15, 17, 19.

The baskets were either carried to the winepress by individuals or brought on the backs of farm animals.[46] A vineyard path for grape pickers is mentioned in the *Iliad*[47] with no mention of animals. Numbers 22:24 notes a vineyard path that can fit both Balaam and his ass, although the animal is there for travel—travel that has been halted—and not for the grape harvest. Most biblical descriptions of field labor (e.g., Ruth 2:3-18) seem to entail only a human work force, yet farm animals are present on occasion, mentioned especially in the legal codes (e.g., 1 Kgs 19:19; Exod 21:28-37; 22:2, 9-15), so we cannot rule out their use in the vineyard. A donkey, for whatever symbolic reasons,[48] is tied to the vine in Gen 49:11, where an abundance of wine is also detailed, and so it is possible that he was used to carry that wine.[49]

Traditional grape harvesting has persisted in the rural Greek villages of Messenia, and descriptions of the vintage offer an analogy for reconstructing the Israelite harvest. In those villages, the grape harvest

> involves cutting clusters with a small knife, then
> carrying baskets of these on the shoulder to load
> the larger pair of baskets, which a donkey will
> transport to the village or field house where the
> wine is made.[50]

Workers probably filled their baskets with clusters, brought them to the winepress, and returned to the vines for more harvesting. It should be noted that during all facets of the grape harvest, the grapes would be handled with care. In antiquity and today, "if picked grapes are handled roughly and held for even a short time, organisms begin to develop."[51] Some skin breakage would occur in the baskets simply from the grapes' own weight. This would begin the process of fermentation, but it also would have attracted insects.

---

[46]Ross notes an intermediate stage of drying grapes in the sun, but there is no biblical evidence for this practice, "Wine," 850; Hesiod mentions sun-drying the grapes: *Works and Days* 609-17. Since drying grapes would have greatly diminished wine volume, the practice was probably not widespread among poor farmers.

[47]Homer *Iliad* 18.534-5.

[48]The donkey is associated with the king (1 Kgs 1:5, 33, 38), as the beast upon which he rides, and is part of a messianic image in Zech 9:9 (cf. Matt 21:5, 7; John 12:14-15). See W. T. in der Smitten, "חמור," in *Theological Dictionary of the Old Testament*, 4:465-70; C. Westermann, *Genesis 37-50: A Commentary* (trans. J. J. Scullion, S. J.: Minneapolis: Augsburg, 1986), 231.

[49]Or, was the source of manure for the soil: Weinhold, *Vivat Bacchus*, 81.

[50]Aschenbrenner, "A Contemporary Community," 55.

[51]Amerine and Singleton, *Wine: An Introduction*, 95.

At some point, the stems and leaves would have to be removed from the fruit. Virgil notes that those in charge of larger baskets were the ones who did this chore, and that it fell to those treading the grapes to pick out what had been missed.[52] He indicates, then, a differentiation in basket sizes and a chore division. In modern viticulture, removing stems is the first step in processing the grapes, after they arrive from the vineyards.[53] The Israelite family might have had some tradition not reflected in biblical texts for when to remove the stems and leaves, and who would do so. The stems are high in tannins, ash, and acid, and so lend to the wine an additional bitter taste, unwanted by modern palettes. They would also hurt bare feet during treading, so their removal before treading is probable for the Israelite vintner, even if his palette might have differed from the modern one.

Once the treading floor of the press was filled or the vines were stripped of (most of) their grapes, treading could begin.[54] In fact, it should. The time between picking and the start of processing in the press would have been kept short since grape skins break easily.[55] Virgil gives advice for hygiene at the winepress, and it is unknown whether ancient Israelites abided by a similar wisdom: "The men that tread must get into the press, having scrupulously cleaned their feet, and none of them must eat and drink while in the press, nor must they climb in and out frequently."[56]

II. Social Aspects of the Vintage

Because of the short period of *veráison*, small vintners were likely to enlist their families and perhaps friends and neighbors in the grape harvest. Hired hands are solicited for a grape harvest in the New Testament, where their staggered shifts result in resentment (Matt 20:1-16). But in the Hebrew Bible, no evidence of hired hands or slaves remains. Neither is any division of labor mentioned, say, between the family and hired laborers or by gender. Columella stressed that at no time should the winepresses be left unattended during the harvest, because of the threat of pests and theft.[57] This period of heightened activity would likely have engendered family closeness, with all its joys and irritations. In Tuscany, Adamo called

---

[52]Virgil *Georgics* 6.2.
[53]Machines called "crusher-stemmers" are used in modern wineries: Amerine and Singleton, *Wine: An Introduction*, 96.
[54]Chapter 4, pp. 142-65.
[55]Amerine and Singleton, *Wine: An Introduction*, 95.
[56]Virgil *Georgics* 6.2; Friedl, *Vasilika*, 20.
[57]Columella *De re rustica* 12.18.4.

on his wife, his two sons, and their wives and children to get the grape harvest in quickly.[58] In Vasilika, a rural village of Greece, for example, the nuclear family harvests the typical family vineyard of two dunams and can do so in a few hours.[59] As Friedl notes for Vasilika, the grape harvest is not at all a rueful exploitation of a farmer's children. It is instead an "occasion much enjoyed, since men, women, and children all pick the grapes together, talk and joke as they do so, and eat a little of the produce besides."[60]

Calabresi captures this same family joy when she describes the younger son, his wife, and their two adolescent girls all harvesting with Adamo and Maria on their Tuscany vineyard: "The girls trotted up and down behind them chatting and laughing to each other in high voices."[61] In Homer's idealized vineyard on Achilles' shield, boys and girls are depicted harvesting the grapes.[62] Turkowski notes that for the Judean hill farmers of this century, September was enjoyable for everyone, as it was the time for the "crowning of their efforts and the gathering of grapes and other fruits."[63] Sounds and actions of joy are also evident in biblical traditions of the vintage. The vulnerability of the fruit and the anticipation of what it yielded, viz., wine that would cheer both gods and mortals (Judg 9:13), undoubtedly lent a tense, expectant mood to the Israelite grape harvest. Hence, it is not surprising that emotional expression through sound, dancing, and prayer is evident in the Bible.[64] As Braudel noted for France the "...hard toil of harvest and grape picking could also be a time of festivity: the countryside resounded with merrymaking and near-continuous feasting."[65] A similar collective expression was probably practiced in ancient Israel.

A terse description of the grape harvest occurs in Judges when Gaal moves to Shechem:

וַיֵּצְאוּ הַשָּׂדֶה וַיִּבְצְרוּ אֶת־כַּרְמֵיהֶם
יִדְרְכוּ וַיַּעֲשׂוּ הִלּוּלִים

They went out to the field and
harvested their vineyards; they
trod (grapes), and gave praise.
(Judg 9:27)

---

[58]Calabresi, "*Vin Santo* and Wine," 123.
[59]Friedl, *Vasilika*, 19-20.
[60]Friedl, *Vasilika*, 20.
[61]Calabresi, "*Vin Santo* and Wine," 129.
[62]Homer *Iliad* 18.537.
[63]Turkowski, "Peasant Agriculture in the Judaean Hills," 27.
[64]Broshi, "Wine in Ancient Palestine," 24; Ross, "Wine," 850.
[65]Braudel, *Identity of France*, 2:244.

The grape harvest itself is described only in its barest essentials. The men could all be harvesting their own vineyards or acting as pooled labor for each other. While the nuclear family could harvest a two-dunam vineyard easily, larger vineyards probably required the labor of extended family members and perhaps anyone else the vintner trusted. Gaal himself may be one of these trusted laborers. He has not lived in Shechem long enough to cultivate his own vineyard, since he does not yet know who Abimelech, the ruler, is (v 28), yet he presumably helps with the grape harvest. Verse 26 notes that the Shechemite vintners "trusted" Gaal (v 26: וַיִּבְטְחוּ־בוֹ בַּעֲלֵי שְׁכֶם), and his participation in their vintage may well be an expression of that trust.

After they are done treading their grapes, the laborers offer praise, וַיַּעֲשׂוּ הִלּוּלִים (v 27). Isaiah contains a similar tradition of harvest praise:

$$\text{כִּי מְאַסְפָיו יֹאכְלֻהוּ וְהִלְלוּ אֶת־יְהוָה}$$
$$\text{וּמְקַבְּצָיו יִשְׁתֻּהוּ בְּחַצְרוֹת קָדְשִׁי}$$

For those who gather it will eat it and praise Yahweh.
And those who collect it will drink it in my holy courts.
(Isa 62:9)

Although praise is listed between eating and drinking, it is clearly a response to the harvests just completed. In Joel 2:24-26, a bountiful harvest of full winepresses and threshing floors is predicted with its attendant praise (וְהִלַּלְתֶּם).

The biblical traditions of praise for agricultural plenty may have their origins in the worked fields. Peak work periods of the harvests, of barley, wheat, figs, olives, and grapes, would have stressed the labor pool and engendered among the workers some anxious hope for a good yield, as the harvesters' lives were depending on it. Judges 9:27 might reflect one such response, that of praise upon a successful harvest.

a. Shouting and Singing

Several biblical scenes offer glimpses into the joy and celebration of an Israelite harvest. Biblical texts indicate that shouts, singing, and dancing were all aspects of the vintage. While these features may well have accompanied the grape harvesting and other field chores throughout the year, they are specifically mentioned in the Bible for wine making. Often, the festive facets of grape treading are evident in prophetic threats of their removal, but we glimpse in this *via negativa* something of what was pleasant in Israelite daily life. For prophetic threat presumably proved most effective by holding up what was dear to the audience. For First Isaiah, often what was dear was in a vineyard,[66] whether Israelite or, as in the verse below, Moabite:

וְנֶאֱסַף שִׂמְחָה וָגִיל מִן־הַכַּרְמֶל
וּבַכְּרָמִים לֹא־יְרֻנָּן לֹא יְרֹעָע
יַיִן בַּיְקָבִים לֹא־יִדְרֹךְ הַדֹּרֵךְ
הֵידָד הִשְׁבַּתִּי

Gladness and joy are gathered[67] away
from the fruitful field, and in the
vineyards are no triumphant or ringing
shouts. The grape treader treads no wine
in the presses.
I stopped the vineyard shout.

(Isa 16:10)

Jeremiah 48:33 closely parallels Isa 16:10 and was probably borrowed from Isaiah or is a shared tradition against Moab. It is worth quoting in full:

וְנֶאֶסְפָה שִׂמְחָה וָגִיל
מִכַּרְמֶל וּמֵאֶרֶץ מוֹאָב
וְיַיִן מִיקָבִים הִשְׁבַּתִּי
לֹא־יִדְרֹךְ הֵידָד
הֵידָד לֹא הֵידָד

---

[66] Isa 1:8; 3:14; 5:1-7, 10; 7:23; 16:8-9; 24:7; 27:2; 32:12; 36:16; see also 34:4, a postexilic addition.
[67] Isaiah alludes to a harvest gathering; this time, gladness and joy are stripped from the land.

> Gladness and joy have been gathered
> from the fruitful land of Moab;
> I stopped the wine from the winepresses;
> they will not tread with shouting;
> vintage shouts are vintage shouts no more.
>
> (Jer 48:33)

Three different words denote the vocal noises of the vintage in the Isaiah passage: הֵידָד,יְרֹעָע,יְרֵנֻּ. The Jeremiah passage dispenses with the first two terms altogether and instead gives הֵידָד three times. The noise described by these terms contributed to or expressed the gladness and joy that were gathered up from these vineyards. Grape treading itself is not a loud activity, so the sound Isaiah recounts with these three terms must be vocal. The workers used their voices in some sort of group expression, described here and repeatedly as "shouting." Since these sounds constitute a communal celebration at the vintage, all three terms deserve some attention.

Both רָנַן and רוּעַ are verbs for vocal expression in a variety of contexts in the Bible and are often translated as "cry out" or "shout" respectively.[68] רָנַן is used almost exclusively for shouts of celebration and joy, the exception being when Wisdom summons through shouting in Prov 1:20 and 8:3. It is associated with joy and with singing primarily in the Psalms and in prophetic texts. It accompanies rejoicing in Pss 5:12 [ET 5:11]; 67:5; 90:14; 92:5 (with שׂמח); and in Isa 35:2 (with גִּיל). רָנַן is often coupled with other vocal celebration in the Bible (e.g., Isa 12:6; 24:14; 54:1; Jer 31:7,12; Zeph 3:14; Zech 2:14 [ET 2:10]). Isaiah 24:14 and Jer 31:12 both detail the start-up of the vintage after a period of destruction. In Isa 24:14, the remnant will once again enjoy these pleasures of the grape and olive harvests, and in Jer 31:12, vintage festivities will resume for the people as part of their homecoming. In other words, both pleasures, of festivity and of viticulture itself, are esteemed in the promise that the people will one day know them again. These contexts demonstrate that the shouting was most often joyous and therefore appropriate for the vineyard scene depicted in Isa 16:10, where it is part of the gladness taken away from the Moabite vineyard. Singing might naturally have been an aspect to this vocal noise, even though no texts explicitly state this.

The second term for noise making, רוּעַ, is unusual for a harvest and has distinctive militaristic connotations. רוּעַ sounds a war cry or

---

[68]BDB, 943, 929.

alarm before battle, as when the Israelites fell the walls of Jericho (Josh 6:20), or as a triumph over vanquished enemies (e.g., Job 38:7; Ps 41:12; Isa 10:24). The military sense of רוּעַ is well established in the Bible, and even prodigious.[69] Its use for a vintage in Isa 16:10, however, requires examination.

We can reasonably infer from Isaiah's combination of these two verbs, רָנַן and רוּעַ, in 16:10 that the noise, whatever its precise makeup, was produced by multiple voices present at the grape treading, in the case of רוּעַ, like that of an army. The coupling of רוּעַ with רָנַן accomplishes at least three things for the description. First, it "pumps up the volume" considerably and quickly, since רוּעַ was so often the sound of war heard from afar and made by an entire force. Some anxiety over the outcome surrounds any military venture. The shouting of triumph in battle acted as a kind of psychological strategy to steel the Israelite army and unnerve the opposing force. Second, it adds alliteration to Isaiah's verse, and third, the combination expresses a triumphant relief and joy that the incoming harvest is successful. Isaiah would seem to be mixing metaphors, one agricultural, one militaristic; but by doing so he gains volume, alliteration, and some allusion to the emotional release of a successful vintage. While military endeavors can on occasion be portrayed in viticultural terms (Judg 8:12; 20:45), the reverse is not the case, except here. Militaristic shouting may stem simply from the vintner's knowing that vines can fail even after he has waited, worked, and watched for a year for this important harvest.

Yet one more verb is used to describe the vocal expression of the grape treaders. This one, הֵידָד, is restricted to harvest scenes and found only in Isaiah and Jeremiah.[70] In Isa 16:9, הֵידָד is the shout for the summer fruit harvest and the cereal harvest and is not mentioned for the grape harvest, as it is in its other uses. Jeremiah uses הֵידָד not only in 48:33, but earlier in 25:30-31, in which the treaders' shout occurs with שָׁאַג, "to roar," and both sounds are heard as a "clamor,"

---

[69]Besides its notorious role at Jericho, shouting was a military practice at many Israelite battles. Its militaristic sense is so prevalent that Joshua mistakes this sound (רוּעַ) of shouting golden calf worshippers for a war cry (Exod 32:17). In their readying for battle against the Philistines, the Israelites shout when the ark is brought into their camp. The noise travels and spooks their opponent (1 Sam 4:5). Later, the Israelites rush the Philistines with this shouting before and after Goliath is slain by David (1 Sam 17:20, 52). Defeated, though still readied, Moabite soldiers continue to shout (Isa 15:4). Yahweh goes forth to battle as a warrior shouting (Isa 42:13).

[70]In Jer 51:14, it is shouted by soldiers. Still, they are compared in their numbers to locusts, notoriously aggressive pests particularly at fruit harvests, and so Jeremiah might well be drawing on הֵידָד's agricultural meaning here. See Joel 2:24-25, where wine vats will flow again after years of locust/army ravages. Locusts are an apt metaphor for invading armies, for their devastation of harvests is quick, owing to their multitude, and complete.

שְׁאוֹן, "to the ends of the earth." One feature of הֵידָד, then, apparently was its volume. It derived not from a few straggling voices or from some timid chanting, but was a notable sound. These shouts, then, constitute part of the communal celebration—"gladness and joy"—that Yahweh stripped from the Moabite vineyard.

b. Dances

Two texts mention dancing at the vineyard. Judges 21:21 places women dancing in a vineyard. This "festival of Yahweh," חַג־יְהֹוָה, takes place at Shiloh and the Israelites assume that women dancers will be in the vineyard. Since a vineyard is the locale for at least the dancing portion of the festival, it is likely that the yearly festival is that of the grape harvest. Dancing women may have added to the overall gladness and joy and been a facet of the Israelite vintage. Women need not have been present at the vintage solely for the sake of entertainment, as a kind of ancient (treading) floorshow for the tiring vintners. Rather, dancing could have formed one aspect of the celebration, which fluidly moved throughout the day in treading, shouts, music, and dances. Women could have worked and sung in the presses and vineyards, with dancing comprising only one of the activities in which they participated. Custom may have dictated that only women and not men danced, but of this we cannot be sure.[71]

In the Judg 21:21 passage, "the daughters of Shiloh" come "dancing the dances," בְנוֹת־שִׁילוֹ לָחוּל בַּמְּחֹלוֹת, as the Benjaminites lie in wait in the vineyards. These women are the Israelite solution to get all the bachelor Benjaminites married. During the performance, each Benjaminite is instructed to take a dancer for a wife. These women, "daughters," are indeed unmarried, for the Israelites anticipate only that their fathers and brothers would come to complain about the theft. As bizarre and rapacious as the plan is, the passage nevertheless

---

[71]Forbes mentions both men and women dancing: *Studies in Ancient Technology*, 3:110. King David is chastised by Michal for dancing, but defends himself by saying that it was for Yahweh (2 Sam 6:5, 16, 20-21). Michal might well have been angry either because David broke with some tradition wherein the men, or at least the king, did not typically dance, or because he exposed himself while dancing. The latter scenario is unfortunate for Yahweh, who in Exod 20:26 took precautions with human altar climbing precisely to avoid just such a scene. Queen Vashti refused to display her beauty before the king and his eunuchs (Esth 1:11-12). Though dancing is not specified, her refusal was intolerable to the men because it thwarted a custom of displaying women.

contains perhaps a hint of the social function of a grape harvest to ancient Israel.[72]

In the festivity and communal participation of the vintage, the Israelite harvest may have served the double purpose of encouraging marriage arrangements among neighbors. The only textual evidence for this is indirect, yet with the excitement and celebration of this important harvest, and with unmarried daughters on hand harvesting and dancing, the time was certainly ripe with opportunity. Judges 21:21 cannot be taken as representative of the Israelite strategy for marriage. Instead, it approximates the taking of spoils in war, as indeed the Israelites will remind any complaining Shiloh men, in what must be a veiled threat: "Allow us to keep them, because we did not capture them in battle" (v 22). Women are part of the spoils of war, and that is why their removal from Shiloh is more theft than it is kidnapping. In this passage, the Israelites want only the spoils of war, the women, while foregoing the battle from which they would earn the right to plunder. In effect, they tell Shiloh to count its blessings.

Jeremiah 31:13 describes a general agricultural setting of grain, wine, oil, flocks, and gardens when the exiled people will return to their land:

אָז תִּשְׂמַח בְּתוּלָה בְּמָחוֹל
וּבַחֻרִים וּזְקֵנִים יַחְדָּו

Then a virgin will rejoice in dance,
with young men and old together.
(Jer 31:13)

The dancer is an unmarried woman, but young men and elders will join in the dancing too. Was mixed dancing, then, an Israelite harvest custom or Jeremiah's expansion of one signaling that the homecoming of the exiles would be even greater than their original harvest times? These cryptic passages are far too meager a basis on which to reconstruct a social practice of Israelite mating, yet they offer the possibility that the vintage was remembered as a time to find wives.

---

[72]The rape of the Sabine women also occurs at an agricultural festival of some sort, the Consualia: Livy *The History of Rome* 1.9.6-14. T. H. Gaster considers the Judges story to be an etiological legend for "mass-mating" at seasonal festivals. He adds, by way of (unsettling) explanation, "a country carnival is an obvious place for sex-starved males to go stalking mates": *Myth, Legend, and Custom in the Old Testament* (New York: Harper & Row, 1969), 444-45.

III. Wine Production

a. The Chemistry of Fermentation

Wine production is a fairly simple process that requires only two elements: grapes and yeast. The yeasts convert the sugar in the grape juice into alcohol.[73]   Since these elements occur naturally together, the technological contribution to the process is minimal, merely that it break the grapes' skin.   The accidental discovery of wine, then, must have occurred fairly early in history. It could have been a simple discovery, made by having left some grapes in a jar. Once the skins were broken, the yeast would have worked on the liquid, with wine being the result. Some appreciation of the chemistry of wine production is in order at this point, since it enables us to understand how the ancient farmers could enjoy an alcoholic drink from their vineyards.

The yeast, called *Saccharomyces cerevisiae*,[74] is found naturally in sufficient quantity on grape skins to ferment the juice.[75] When fruit falls to the ground in a vineyard, insects bring in additional yeasts, which then multiply in the fallen fruit.   Yeasts can survive from season to season in the soil itself.[76]   In this way, a vineyard attracts what it needs. Grapes also have additional wild yeast strains on them called "apiculate" because their shape is pointed on one end. These apiculate yeasts multiply rapidly in fermentation until the juice reaches 4-6 percent alcohol.   Then they are inhibited, while the *Saccharomyces cerevisiae* continues to ferment. Six to twelve hours after treading, fermentation reaches its peak and then proceeds at a slower rate for another two to five days.[77] Roman fermentation, done inside, in jars, lasted three to nine days.[78]

---

[73]Unwin, *Wine and the Vine*, 46; Weinhold, *Vivat Bacchus*, 161.

[74]In older texts, like that of Forbes, *Studies in Ancient Technology*, 3:65, the yeast is termed *Saccharomyces ellipsoideus* due to its elongated oval shape. See Renfrew, *Palaeoethnobotany*, 131; Amerine and Singleton, *Wine: An Introduction*, 65.

[75]Amerine and Singleton, *Wine: An Introduction*, 52. It is as well the same yeast commonly used in bread making.

[76]Amerine and Singleton, *Wine: An Introduction*, 54.

[77]Dar, *Landscape and Pattern*, 1:155; Forbes, *Ancient Technology*, 3: 66.

[78]Grant, *Cities of Vesuvius*, 191-202; Pliny *Natural History* 14.124; Columella *De re rustica* 12.28; Cato mentions a thirty-day fermentation: *Agriculture* 26. Dayagi-Mendels notes that fermentation could last up to six months: *Drink and Be Merry*, 30; see also Forbes, *Studies in Ancient Technology*, 3:112; White mentions no specific time frame, but shows the storage facilities that allowed for longer fermentation before the wine was decanted into amphorae for storage: *Roman Farming*, 427.

Yeasts need a carbon and nitrogen source for their multiplication. Grape juice has carbon in equal proportions of the six carbon sugars (hexoses), viz., glucose and fructose.[79] The juice also has sufficient amino acids to provide a nitrogen source for the yeast cells, and in this is almost unique among fruits.[80] Given the natural ease with which yeast and grape juice combine, it is not surprising that grape wine was the most prominent type of wine in antiquity. Dates were the other major fruit that had sufficient carbon and nitrogen sources in their juice and yeasts on their skin to yield a potent wine. Date wine is attested in Egypt and Mesopotamia and, as we shall see later,[81] probably in ancient Israel as well.[82] Honey wine or mead from grain, by contrast, would have needed an extra nitrogen source to ferment. In antiquity, raisins could have provided this source.

An additional chemical asset to yeasts is the manner of their cell multiplication. Most bacterial and other one-cell microorganisms multiply by splitting into two new cells, a process called binary fission. Yeast cells, by contrast, form a bud, which grows and then splits off from the "older" cell. After a cell does this a few dozen times, the "older" cell dies. The breakdown of the older cells then becomes an additional nutrient source to the remaining yeast cells.[83] What this means is that yeast cells thrive (up to a point) in the grapes' juice. The combination of these chemical features meant for the ancient vintner that wine fermentation would rarely fail. While the grape harvest had its risks and uncertainties, once the vintner got the grapes to the press, fermentation was a hardy and therefore dependable process.

Yeast cells work on the simple sugar in order to multiply their own cells. Alcohol is a byproduct of this process, and one that eventually kills the yeast cells. The wild yeast strains will die once the alcohol in the must, that is, the fermenting juice, reaches 4-6 percent,[84] and at 14 percent the *Saccharomyces cerevisiae* die off.[85] Hence, the naturally produced wine of ancient Israel did not get higher than 14 percent alcohol. Carbon dioxide is the other byproduct. The basic chemical formula is: 1 glucose > 2 ethanol + 2 CO_2 + about 56 kilocalories of energy.[86]

Fermentation is an anaerobic process. In the process of fermentation, carbon dioxide is given off and is at times rather

---

[79] Amerine and Singleton, *Wine: An Introduction*, 55.
[80] Amerine and Singleton, *Wine: An Introduction*, 56.
[81] Chapter 6, pp. 200-202.
[82] Forbes, *Studies in Ancient Technology*, 3: 61; Moldenke and Moldenke, *Plants of the Bible*, 170; Borowski, *Agriculture in Iron Age Israel*, 127.
[83] Amerine and Singleton, *Wine: An Introduction*, 58.
[84] Amerine and Singleton, *Wine: An Introduction*, 53.
[85] Lucas and Harris, *Ancient Egyptian Materials and Industries*, 18.
[86] Known as the Gay-Lussac equation: Unwin, *Wine and the Vine*, 49.

potent.[87] Adamo in Tuscany knew stories of people passing out from the fumes and drowning in the fermenting wine.[88] The carbon dioxide drives oxygen from the surface of the must, so that the yeast cells can continue the process of converting sugar into alcohol. The liquid during this phase is bubbling and foaming. The wine, at this point, is described as "boiling" today, as it was, also, in ancient Israel (Gen 40:10).[89] The yeast cells consume any oxygen present in the must for their initial (aerobic) multiplication. After that is achieved, fermentation occurs without oxygen. In fact, the presence of additional oxygen is detrimental to the must.

The major concern in wine making is to allow the carbon dioxide to escape without prolonged contact with the air, since this risks contamination from a fungus called *Mycoderma aceti* that converts the ethanol to acetic acid (vinegar).[90] The ancient vintner, in other words, would want to leave his grape juice exposed to the air long enough to allow the carbon dioxide to escape, but before vinegar resulted. Again, vintner skill entered into the picture. Stirring is important to keep the surface juice from turning to vinegar.[91] Adamo's procedure was to let the juice steam and froth for about ten days, during which time he would stir daily. Since he made wine by traditional methods—treading, no additional yeasts, outdoor fermentation—his practice is illustrative for reconstructing the Israelite vintage. We can reasonably infer that the Israelite vintner learned to stir the must to prevent his wine from becoming vinegar.

There are enough examples of collecting vats in Israelite winepresses to indicate that the juice would have spent some time fermenting in them. Also, given the heat and hours it would take to tread grapes, fermentation probably began in the treading floor itself. For those presses without collecting vats, fermentation would simply begin within the treading floor itself. Within six to twelve hours, fermentation reached its peak, and then tapered into a slower process lasting two to five days.[92] Fermentation is accomplished in any suitable container where the must takes up only one-half to three-quarters of its capacity. Room must be left to allow for the

---

[87]Calabresi, "*Vin Santo* and Wine," 123.

[88]Calabresi, "*Vin Santo* and Wine," 123.

[89]See above, p. 169n8; Calabresi, "*Vin Santo* and Wine," 126.

[90]Lesko, *King Tut's Wine Cellar*, 20; Broshi, "Wine in Ancient Palestine," 33; Unwin, *Wine and the Vine*, 46-48; Lucas and Harris, *Ancient Egyptian Materials and Industries*, 19; Badler, et al., "Drink and Be Merry! Infrared Spectroscopy and Ancient Near Eastern Wine," 32.

[91]Calabresi, "*Vin Santo* and Wine," 130; Unwin, *Wine and the Vine*, 48.

[92]Dar, *Landscape and Pattern*, 1:155.

considerable volume of carbon dioxide produced.[93]

The slower phase of fermentation could have occurred in two places for the Israelite vintner. The juice could have remained in the collecting vats for that period of time or have been transferred into jars for storage and completion of the fermentation. Fat-bellied jars dating to Iron II were used at Ashkelon for the fermenting and storage of wine.[94] The wine jars from King Tutankhamon's tomb seemed to have been both for fermentation and for storage. One of the jars apparently had been sealed too soon and ruptured from the carbon dioxide.[95] Some of the other jars have small holes in their necks or in the stoppers, presumably to allow the carbon dioxide to escape.[96] Greco-Roman fermentation seems to have taken place primarily in earthenware jars.[97]

Collecting vats had two advantages for the fermentation process over the treading floor itself. They were smaller, which meant that the amount of surface juice exposed to the air was lessened. And, since they were deeper, scooping juice for transfer into jars was easier. The ancient vintner would not have known the chemistry of fermentation and acetic acid. But through experimentation with the flavor and potency of his wines, he likely learned at some point to transfer some or all of the must from the treading floor to the smaller venues of either a collecting vat or jars. Most probably the jars were then stored in the house or a cellar. Fermentation in the winepress itself, either the treading floor or collecting vats, meant that some precautions were probably necessary, not only against theft, but also against debris or animals falling into the must. Columella, for example, offered advice for when a mouse or snake fell into the must and drowned. He recommended that the little intruder be taken out, burned, and its ashes then stirred back into the must as a curative measure.[98]

b. Collecting Wine from the Vats

The fermenting wine would either remain in the vats or be scooped into jars. The ancient farmer could have transported the

---

[93]Lucas and Harris, *Ancient Egyptian Materials and Industries,* 18; Amerine and Singleton, *Wine: An Introduction,* 102.
[94]Stager, "The Fury of Babylon," 66, with photo.
[95]Lesko, *King Tut's Wine Cellar,* 21.
[96]Lesko, *King Tut's Wine Cellar,* 20; Darby et al., *Food: The Gift of Osiris,* 2:569, fig. 14.9.
[97]Cato *Agriculture* 26; Dar, *Landscape and Pattern,* 1:156.
[98]Columella *De re rustica* 12.31.

wine by bringing small jars or wineskins to the vats, filling them, and then returning to the house by foot, donkey, or cart.

Travelers are depicted in the Bible using both skins and jars for their wine provisions. David and the Gibeonites (1 Sam 16:20; Josh 9:4, 13),[99] for example, carried wineskins. A passage in Job suggests that wineskins were used for the transport of new wine:

הִנֵּה־בִטְנִי כְּיַיִן לֹא־יִפָּתֵחַ
כְּאֹבוֹת חֲדָשִׁים יִבָּקֵעַ

> My belly, like wine, will not be opened,
> like new skins, will (not) be burst.
> (Job 32:19)

The context for this passage is Elihu's first speech in Job, and so the image is meant to convey that Elihu has thus far managed to hold his words back (v 4), but can no longer. His words, like new wine, are ready to burst through their constraints, i.e., his silence up until now. He contrasts his words with those of Job and the three friends precisely in that they are new. Old wine could stay in new skins because its fermentation is completed. Elihu thus implies here that the discussion has become old and needs both the urgency and freshness of his perspective. At this point in the discussion, Elihu can keep quiet no more than new wine can stay in new skins.

Wine that is not finished fermenting, as we have seen, still needs an opening for the release of carbon dioxide. In the New Testament, the Synoptic Gospels also use wine and wineskins, ἀσκοί, for a parable:

> No one puts new wine into old wineskins,
> Otherwise the wine will burst the skins.
> (Mark 2:22)[100]

Such a mistake was costly, for both the wine and the skins would be lost. Parables are effective, pithy metaphors of wisdom and occur frequently in the Synoptic Gospels. Apart from the theological freight that the parables carry, they also provide details of daily,

---

[99]This is perhaps appropriate, since Gibeon yielded a winery from the Iron II period.
[100]Matt 9:17; Luke 5:37-38.

Palestinian life, since they must be drawn from such to be understood. These are the givens about life that everyone knows or can be reminded of. For implicit in an assertion of what "no one" does is the custom which "everyone" does do. The parable preceding the wineskin parable, for example, is on sewing: "no one" sews a piece of new unshrunken cloth onto an old coat, for the new patch would pull away from the coat cloth as it shrank (Mark 2:21).

The custom evident in the wineskin parable is that new wine, i.e., wine from the press and still fermenting to some degree, was poured into new wineskins. "No one" put new wine into old skins, because the ongoing fermentation of new wine would give off carbon dioxide and expand the skin. An old bag has little stretch left to give. A new skin, on the other hand, one relatively new off the animal, would stretch some. A custom of skin transport of new wine seems evident in this parable and in the simile of Job 32:19. Jars were probably also used, but simply are not recorded in the biblical texts for this purpose. In addition, skins had the advantage over jars of being lighter and easier to transport. The vintner may have used whatever empty containers he had to transport the wine from the press to his house. At home, he would then pour the wine into larger jars for storage.

Once the wine was in jars, some sort of stopper became necessary to prevent conversion of the wine into vinegar. Unfired clay was probably the best solution for stoppering because it could be shaped to the jar's opening. The seal would become tighter still as the wine moistened the clay and caused it to expand. Unfired clay balls or stoppers were found at Ashkelon and at Godin Tepe, Iran, along with wine jars.[101] Such clay balls are found elsewhere in ancient Israel and have long been thought to be loom weights.[102] Stager discovered that the clay balls at Ashkelon fit into the mouth of amphorae. He suggests that these may be the "corks" of ancient wine jars. The small holes of the balls could have allowed carbon dioxide to escape and then been stopped with some sort of material, such as mud and straw. Different stoppering devices are evident in Egypt.[103] Olive oil was a possible sealant too.[104] If oil were used, it would prevent air from reaching the wine and lend a not unpleasant taste to the wine. A jar sealed with olive oil, of course, could not be used for transport without some additional form of stoppering.

---

[101]Badler, et al., "Drink and Be Merry! Infrared Spectroscopy and Ancient Near Eastern Wine," 28; Stager, "The Fury of Babylon," 64, 68, with photo.

[102]Z. Gal, "Loom Weights or Jar Stoppers?" *IEJ* 39 (1989): 281-83; Stager, "The Fury of Babylon," 64, 68, photo.

[103]C. Hope, *Jar Sealings and Amphorae of the 1st Dynasty: A Technological Study* (Warminster: Aris and Phillips, 1977); Dayagi-Mendels, *Drink and Be Merry: Wine and Beer in Ancient Times*, 24.

[104]de Blij, *Wine: A Geographic Appreciation*, 50.

IV. Biblical Terms for Wine

Once the fermentation process was complete, the desired product was, of course, wine.[105] Unfermented grape juice was most likely rare in ancient Israel, because the yeasts of the skins would have begun to work on the juice soon after the grape skins were broken. In the Hebrew Bible, four terms denote "grape wine."[106] They are: יַיִן, תִּירוֹשׁ, עָסִיס, and חֶמֶר. A fifth term, שֵׁכָר, probably also denotes a wine, though one made from dates, and so is included in the present section. יַיִן is the most frequent term, occurring 141 times in a variety of biblical sources. תִּירוֹשׁ occurs 38 times, also in a variety of biblical material. עָסִיס occurs five times, and חֶמֶר only once in the Masoretic text. שֵׁכָר, "strong drink," occurs a total of 23 times.

a. יַיִן

The precise etymology of יַיִן is unknown.[107] The origin of יַיִן is considered to be non-Semitic.[108] The Hebrew term is clearly related to the Greek οἶνος and the Latin *vinum*,[109] but how so remains obscure. The Greek term had an original –w-, represented by a *digamma* (no longer written).[110] In Ugaritic, the primary term for wine is *yn*. This initial -y- from the Late Bronze Ugaritic term cannot produce the original -w- of the Greek term.[111] Initial –w- shifts to –y- in Northwest Semitic languages, including Ugaritic and Hebrew. Hence, the development of this term for wine did not occur directly

---

[105]Moldenke and Moldenke, *Plants of the Bible*, 214; A. van Selms, "The Etymology of Yayin, 'Wine'," *JNES* 3 (1974): 76; Stager, "Firstfruits," 173.

[106]The LXX uses one term for wine, οἶνος.

[107]Dommershausen, "יַיִן," 60; Stager, "Firstfruits," 173; A. van Selms, "The Etymology of Yayin, 'Wine'," *JNES* 3 (1974): 76. van Selms suggests that it comes from the Hebrew verb, יָנָה, to "oppress," and so connects it to the method of production, 82; However, his hypothesis cannot be correct, since, as we shall see presently, the presence of an initial –w- in the Greek and Hittite terms for wine means that the term originally had an initial –w- that shifted to a –y- in Hebrew.

[108]J. Brown, "The Mediterranean Vocabulary of the Vine," *VT* 19 (1969): 146-70; Ross, "Wine," 849; Stager, "First Fruits," 173; Jastrow, "Wine in the Pentateuchal Codes," 184; The primary Akkadian term for "wine" is *karānu*.

[109]A Mediterranean word *wīn* from a *woin* is postulated: *American Heritage Dictionary of Indo-European Roots* (ed., C. Watkins; Boston: Houghton Mifflin, 1985), 73.

[110]I am indebted to Michael Coogan for his clarification of this complex etymological issue.

[111]van Selms, "The Etymology of Yayin," 76, 84.

from Ugaritic into Hebrew and Greek. If it had, the Greek term would have an initial -y- as well. Instead, the term must be an Indo-European loanword into Semitic that probably developed in the Levant and in the Mediterranean regions from the Hittite term for wine, *wa-āna-as*.[112] This would account for the different initial phonemes and reinforce the historical hypothesis of the Anatolian origins of viticulture.[113] The development of the term, then, would have followed the vine's cultivation.

יין/יי is also the term found in epigraphic evidence from the Israelite Iron Age. The notation in the Samaria ostraca is *yn*, יי.[114] These ostraca bear the characteristic feature of the northern dialect with its collapse of the diphthong, -*ay*-. Ostraca from Arad, by contrast, have *yyn*, יין.[115] Two Iron II decanters have the term in inscriptions: one found near Hebron (ליחזיהו יין כחל, "belonging to/for Yaḥaziyahu *khl* wine");[116] and the other of unknown provenience (למתניהו יין נסך רבעת), "belonging to/for Matanyahu, wine libation, a fourth").[117] The frequency of this term in the Bible and in the epigraphic evidence demonstrates that יין/יי was undoubtedly the most common term for wine in Iron Age Israel.

b. תִּירוֹשׁ

A second term for wine in the Hebrew Bible is תִּירוֹשׁ, whose etymology is also unknown. It is often rendered in English as "new wine."[118] In Mic 6:15, תִּירוֹשׁ is associated with grape treading, and in Isa 65:8 with the grape cluster itself, so its sense of newness, i.e., right out of the vineyard, may arise from these associations. Often, it is understood to be a more archaic term for wine than יין.[119]   A

---

[112]Brown, "The Mediterranean Vocabulary of the Vine," 148.

[113]See chapter 1, pp. 12-15.

[114]See chapter 2, pp. 51-59.

[115]*Inscriptions Reveal* (ed. M. Tadmor; Jerusalem: Israel Museum, 1973), 62-65; D. Pardee, *Handbook of Ancient Hebrew Letters* (Missoula, Mo.: Scholars, 1980), 42-44. See chapter 6, pp. 218-19.

[116]Its precise provenience is unknown since it was acquired from an antiquities dealer, but the uncontracted diphthong is typical of Judah: Avigad, "Two Hebrew Inscriptions on Wine Jars," 3-5; *Inscriptions Reveal*, 53, 114-15, with photo. See below, p. 206.

[117]R. Deutsch and M. Heltzer, "A Wine Decanter with a Hebrew Inscription," in *Forty New Ancient West Semitic Inscriptions* (Tel Aviv-Jaffa: Archaeological Center, 1994), 23-26, photo, fig. 6.

[118]BDB, 440; KB, 4: 1591.

[119]W. Holladay, *A Concise Hebrew and Aramaic Lexicon of the Old Testament* (Grand Rapids, Mich.: Eerdmans, 1978), 389; Ross, "Wine," 849; Jastrow, "Wine in the Pentateuchal Codes," 183.

corresponding term occurs five times in Ugaritic as *trt*,[120] once in parallel to *yn*, "wine" (2 Aqht VI:7).[121] But the Ugaritic evidence can hardly establish this term as archaic, since *yn* also occurs and more frequently (133 times) in the extant texts.[122] If it is from a Semitic root, then the most likely candidate is יָרַשׁ. Van Selms, for example, argues that תִּירוֹשׁ evolved from יָרַשׁ, to "dispossess" or to "trample" because of its process of production.[123] He adds that it was a more "sophisticated word" used mainly in poetry. Van Selms is correct to note that תִּירוֹשׁ usually occurs in poetry, but it is uncertain whether it stems from a Semitic root or is a loanward. Before we examine the biblical contexts of its use, let us first assess whether production techniques might have given rise to a distinction in wine types.

In the Greco-Roman world, wine pressed from its own weight was valued and distinguished. It was termed *protopum* or *mustum lixivum*.[124] This is the first or new wine that was produced before treading, and perhaps denoted in Hebrew by the term תִּירוֹשׁ.[125] Some of the Israelite winepresses, it will be recalled, had channels connecting a wait station to the treading floor.[126] Perhaps the wine that flowed from the harvested grapes before treading was collected and differentiated from יַיִן, the "bulk" juice gotten through treading. If so, תִּירוֹשׁ might have had additional value for having come naturally, that is, without human labor. Its pristine quality and rarity—recall that it appears far less frequently than יַיִן—then make it a candidate for special offerings.

The term תִּירוֹשׁ occurs in a variety of biblical texts: narrative and poetry (Gen 27:28; Num 18:12; Deut 7:13; 11:14; 12:17; 14:23; 18:4; 28:51; 33:28; Judg 9:13; 2 Kgs 18:32; Neh 5:11; 10:37, 39; 13:5, 12); prophecy (Isa 24:7; 37:16; Hos 2:10-11 [ET 2:8-9]; 2:24 [ET 2:22]; 4:11; 7:14; 9:2; Joel 1:10; 2:19; 2:24; Mic 6:15; Hag 1:11;

---

[120]The Proto-Semitic *t* remains in Ugaritic and appears as a *š* in Hebrew: S. Segert, *A Grammar of Phoenician and Punic* (Munich: Beck, 1980), 34; R. E. Whitaker, *A Concordance of the Ugaritic Literature* (Cambridge, Mass.: Harvard University Press, 1972) 621-22.

[121]A. Herdner, *Corpus des Tablettes en Cunéiformes Alphabétiques* X (Paris: Librairie Orientaliste de Paul Geunthner, 1963), no. 17, VI:7, p. 82; C. Gordon, *Ugaritic Literature* (Rome: Pontifical Biblical Institute, 1949), 89.

[122]Whitaker, *A Concordance of the Ugaritic Literature*, 311-13.

[123]van Selms, "The Etymology of Yayin, 'Wine,'" 83. It came, he argues, from a newer type of process, treading over pressing, but never succeeded in supplanting יַיִן. See also Ross, "Wine," 849; Dommershausen, "יַיִן," 61.

[124]Forbes, *Studies in Ancient Technology*, 3:110; Columella *Agriculture* 12. 41.

[125]Moldenke and Moldenke, *Plants of the Bible*, 213; Jastrow, "Wine in the Pentateuchal Codes," 183.

[126]See chapter 4, pp. 148-65.

Zech 9:17); and, to a lesser extent, writings (Ps 4:7; Prov 3:10).[127] Of
its 38 occurrences, תִּירוֹשׁ appears 32 times in association with grain, or
grain and oil, where it is a tithe or offering from the land. It would
seem from context, then, that תִּירוֹשׁ was valued for gift giving and as a
symbol of the land's bounty. In 35 of its 38 occurrences, תִּירוֹשׁ
denotes a product of the land. Its agricultural origin is stressed over
its quality as a drink. It has this sense too in a Phoenician inscription
from Karatepe (ca. 725 BCE[128]), where *trš* is paired with *šbʿ*,
"plenty," to denote a town's wealth.[129]

There are three passages that do describe תִּירוֹשׁ as a drink,
rather than as a product from the farm. In Hos 4:11, תִּירוֹשׁ is parallel
with יַיִן in its effects, so it is clearly a potent drink. It is also drunk in
Isa 24:7 and Judg 9:13. Since תִּירוֹשׁ is so often the wine described in
reference to its agricultural origins, I suggest that it is the new wine
of the fall vintage. This interpretation is similar to the Greek and
Roman traditions of a new wine noted above, but is not restricted in
definition only to the wine produced from the grapes' own weight.
For that wine amount would still be minimal in comparison with the
rest of the yield, and few would get to drink it. Instead, I suggest that
תִּירוֹשׁ denotes *any* batch of new wine, available to more drinkers and
so highly anticipated. The vintage occurred only once a year, in the
fall, and it provided the wine for the entire upcoming year. By the
end of the year, well before the next vintage, the wine supplies might
have been scarce and, by then, sour. As the vintage approached, new,
fresh wine would undoubtedly have been anticipated. In such a case, a
new term for this fresh new wine of the year could have arisen quite
naturally, and then been retained in these biblical and other traditions.
In this understanding, *seasonal freshness*, rather than technology,
would have led to תִּירוֹשׁ as an additional term for wine. This fresh
wine of the season would likely have been esteemed and so became the
wine described in the biblical traditions of tithes and offerings.

תִּירוֹשׁ is the wine mentioned to the Jerusalem citizens in the
Rabshakeh's speech. The Rabshakeh tempts the citizens by offering to
bring them to a land that yields grain and תִּירוֹשׁ wine (2 Kgs 18:32/Isa
37:16). The grain and the wine here signify what the new land would
bring them. It is not a meal, but a vision of prosperity (see also Gen
27:28). The Rabshakeh's speech is tempting because the citizens are

---

[127] תִּירוֹשׁ becomes the dominant term for wine in the Dead Sea Scrolls, 1QS6:4ff.;
1QS2:17ff., 20; 1QH10:24: D. Barthélemy and J. T. Milik, *Discoveries in the
Judaean Desert I: Qumran Cave I* (Oxford: Clarendon, 1955), 118.
[128] Segert, *A Grammar of Phoenician and Punic*, 34.
[129] C. Gordon, "Azitawadd's Phoenician Inscription," *JNES* 8 (1949): 115; R.
Marcus and I. J. Gelb, "The Phoenician Inscription from Cilicia," *JNES* 8 (1949):
119-20. For the text itself, see H. Donner and W. Röllig, *Kanaanäische und
Aramäische Inschriften* (Wiesbaden: Harrassowitz, 1966) 1: no. 26, pp. 5-6.

promised not only wine and grain, but also the vineyards and fields to produce them. He is offering them nothing less than the chance to be vintners.

c. עָסִיס

The term עָסִיס is used five times in the Bible for a grape product, all in exilic or postexilic prophetic texts: Isa 49:26; Amos 9:13;[130] Joel 1:5; 4:18 [ET 3:18]; and Cant 8:2. Hence, עָסִיס may simply be a later term for wine, in addition to יַיִן and תִּירוֹשׁ.[131] In Amos 9:13 and Joel 4:18, עָסִיס is used to describe the wine that will flow from mountains in a future of prosperity. Both passages highlight the bounty of the land. In Amos 9:13, the grain and grape harvests will yield plenty. In Joel 4:18, water and milk will also abound. This image of wine flowing from mountains, as prosperity evident on Israel's very topography, is also an allusion to agricultural terracing, where, as we have seen, Israelite vineyards often were located.[132] In these two contexts, עָסִיס denotes an agricultural yield more than it does a type of wine. In the other two verses, Joel 1:5 and Isa 49:26, drunkenness is the named effect of drinking עָסִיס, and so it is clear that the product was alcoholic, though some render the term "grape juice."[133] Again, because of the speed of the process of fermentation,[134] עָסִיס, along with the other three wine terms, was most likely an alcoholic liquid,[135] since drinking it, in these two verses, results in drunkenness.

Other translations render עָסִיס "sweet wine"[136] or "new wine."[137] In the process of fermentation, all the sugar is converted into alcohol.[138] Once the alcohol reaches 14 percent, the remaining yeasts would die. At that point, it would have been possible to make a sweet wine by adding raisins, honey, or some other fruit juice as a sweetening agent. I suggest that "my pomegranate wine," עָסִיס רִמֹּנִי, in Cant 8:2, besides representing a potent sexual metaphor, was a

---

[130]Verses 9-15 being widely considered to be a later addition to the book of Amos: S. M. Paul, *Amos* (Hermeneia: Minneapolis: Fortress, 1991), 288-89.

[131]And so the JPS renders the term simply "wine," though not in Joel 1:5, where it is "new wine." So too, the NRSV of Isa 49:26 has simply "wine."

[132]See chapter 3, pp. 94-97.

[133]KB 2: 860; NAB, except for Joel 4:18; Dalman, *Arbeit und Sitte in Palästina*, 4:372.

[134]See pp. 187-90 above.

[135]Ross, "Wine," 849.

[136]BDB, 779; NRSV, except Isa 49:26.

[137]BDB, 779. JB; NJB; JPS Joel 1:5; NEB "fresh wine."

[138]See pp. 187-90 above.

grape wine flavored with pomegranate juice, rather than a wine made wholly from pomegranates.[139] With their tough rinds, pomegranates would have required a technology other than foot treading; and without rinds, picking the juice pellets for even one liter's worth of liquid would have been extraordinarily labor-intensive. Another method of making sweet wine would have been to stop the fermentation process before all of the grapes' sugar was converted to alcohol. Columella gives recipes for making sweet wine and is careful to state that the procedure occurs after fermentation of the grapes is complete. Then, he advises the vintner to add crushed iris.[140] Only in Amos 9:13 may a quickened process of production be at all possible. That is, the passage describes a kind of prosperity where the grape treader will overtake the one who sows. Such heightened and hurried activity could apply to the wine's chemical fermentation as well, so that it retains some of its sugars. But this is at best a tenuous inference.

"New wine" is probably the best rendering for עָסִיס, on the basis of both its etymology and its biblical contexts, if it is to be taken as distinct in meaning from יַיִן and תִּירוֹשׁ. Its etymology might indicate its newness, for עָסִיס comes from the root עָסַס, to "press, crush."[141] Since all wine was pressed from grapes in ancient Israel, this עָסִיס might have signified a "just-pressed" wine. If so, that newness is reinforced in Amos 9:13 and Joel 4:18, when this wine will simply drip from the mountains.[142] The freshness is (all too) evident in its use in Isa 49:26. In a metaphor of revenge, Yahweh asserts:

וְהַאֲכַלְתִּי אֶת־מוֹנַיִךְ אֶת־בְּשָׂרָם וְכֶעָסִיס דָּמָם יִשְׁכָּרוּן

> I will feed your oppressors their flesh,
> and they will be drunk with their blood
> as with wine.
>
> (Isa 49:26)

עָסִיס may serve as an apt simile to blood in this verse, not only for its shared color, but also because both are freshly made. We know the blood is fresh, since the oppressors would only have so long to drink

---

[139] It is attested only here, but thought to be a brand of wine interred with King Tutankhamon: Lesko, *King Tut's Wine Cellar*, 35.
[140] Columella *De re rustica* 12.27.
[141] BDB, 779.
[142] See chapter 3, pp. 93n17, 96.

their own before it dried.[143] In addition, it is potent, since drinking
blood like wine has induced a frenzy of energy. The oppressors are
"drunk" and wild for more of what made them so: they are literally
bloodthirsty, and so, bent on their own destruction.

d. חֶמֶר

The final biblical term for grape wine is חֶמֶר. It is only used
once in the Masoretic Text, in Deuteronomy:

וְדַם־עֵנָב תִּשְׁתֶּה־חָמֶר

And from the blood of a grape,
you drank wine.
(Deut 32:14)

Isaiah 27:2 mentions a כֶּרֶם חֶמֶד, "pleasant vineyard" that Yahweh
would tend day and night in a vision of a future deliverance. The text
follows the LXX, Ben Asher Masoretic texts, Targumim, and Syriac.
The Leningradensis Masoretic text has, instead חמר,[144] and the Isaiah
Scroll from the Dead Sea has חומר.[145] A confusion between *resh* and
daleth is understandable in the Isaiah verse, given the similarity of the
two letters. The Deut 32:14 passage clearly indicates wine. But, since
a vineyard is both pleasant and the locale for wine production, either
rendition of the Isaiah verse is possible. *Ḥmr* is used for wine in one
ritual text from Ugarit, as a part of a feast of bread and wine.[146]
Johannes de Moor has argued that this text is of a fall festival and
therefore at the time of the vintage, when new wine would be in
production.   Hence he translates *Ḥmr* as the "freshly, fermenting
wine," right from the vats.[147] De Moor's suggestion is plausible as
well for the Hebrew texts Deut 32:14 and Isa 27:2. For the Hebrew
verb חָמַר denotes "ferment" or "boil."[148] Both texts associate the wine

---

[143]This omophagic meal by definition cannot be finished.
[144]*BHK*.
[145]E. Kutscher, *The Language and Linguistic Background of the Isaiah Scroll*
(Leiden: Brill, 1974), 375.
[146]*KTU* 1.23, line 6.
[147]de Moor, *An Anthology of Religious Texts from Ugarit*, 119.
[148]BDB, 330.

with its agricultural origin, viz., grape and vineyard respectively. Psalm 75:9 describes a wine, יַיִן חָמָר, "fermenting (i.e., fresh) wine."

חֲמַר/חַמְרָא is the general Aramaic term for wine (Ezra 6:9; 7:22; Dan 5:1-2, 4, 23), with no apparent nuance of freshness. And in a text from Palmyra, a quality of freshness is ruled out for the wine described as "old wine," חמרא עתיקא.[149] Two Persian-period jar inscriptions from Shiqmona contain חמר as a term for wine as well; there it is a product from the "press of Karmel," *gt. krml*. *Yn* occurs as well on one of the jars.[150] Since חָמָר is only used once in the Bible, possibly twice, it is difficult to gain further clarity about its meaning from contexts.

e. שֵׁכָר

שֵׁכָר, "strong drink," is another biblical term for an intoxicating drink and merits some discussion. The verb, שָׁכַר, means "be, or become, drunk,"[151] and so, undoubtedly the drink was potent. שֵׁכָר occurs in the Bible most often with יַיִן, being paired with יַיִן in 20 of its 23 uses (e.g., Deut 29:5; Lev 10:9; Judg 13:4, 7: 1 Sam 1:15; Prov 31:4; Isa 5:11, 22; 28:7; 56:12; Mic 2:11).[152] The term *škr* (שכר) is also paired with *yn* in an ostracon from Ashkelon dated to the late seventh century BCE.[153] שֵׁכָר in the Bible is often translated as "strong drink" without further specification.[154] The question is whether שֵׁכָר is a term for wine or denotes another kind of intoxicant, viz., beer.

Beer was the other major alcoholic beverage in antiquity and is frequently attested in ancient Egypt and Mesopotamia.[155] The production of beer required water, barley, and some sort of fermentation agent to stimulate the low levels of yeast present in the grain. Beer had a practical advantage over wine in that it could be brewed throughout the year as it was needed, as long as grain and water were available. The yeast levels are increased in production by first allowing the barley to germinate in water. Since barley does not have much natural sugar in it, enzymes[156] first convert the starch to maltose, and, with additional soaking, the maltose is then converted to

---

[149]H. Ingholt, "Un nouveau thiase à Palmyré," *Syria* 7 (1926): 134.
[150]Cross, "Jar Inscriptions from Shiqmona," 227.
[151]BDB, 1016.
[152]Num 28:7 and Ps 69:13 are the exceptions.
[153]Stager, "The Fury of Babylon," 66.
[154]BDB, 1016; NRSV.
[155]See chapter 1, pp. 21-27.
[156]Called diastase: Lucas and Harris, *Ancient Egyptian Materials and Industries*, 5.

alcohol by another enzyme.[157] The different enzyme strands are not always present in sufficient quantities for successful fermentation.[158]

Since there is so little natural sugar present in grain, the resulting alcohol content of the beer is low, approximately 2-6 percent.[159] Ancient beer,[160] then, was much weaker in alcohol than wine; it had 2-6 percent compared to the 14 percent of wine. Beer, in other words, was *not* strong, and so if שֵׁכָר does denote beer in the Bible, "strong drink" is inaccurate. In addition, if שֵׁכָר is beer, then its association with the verb, שָׁכַר, is unexpected, since wine could be as much as seven times more potent than beer. Since beer was so much weaker than wine, and it required water, a scarce resource for the Israelite farmer, it likely did not play a large role in Israelite daily life. Beer production is nowhere described in biblical texts, and שֵׁכָר is never mentioned as one of the products of the grain and bread of the farm. Biblical silence alone does not rule out beer as a possible translation for שֵׁכָר, but in combination with these other factors of potency and water scarcity, beer is unlikely.

Because שֵׁכָר is so often paired with יַיִן in the Bible, and is associated with the verb for becoming intoxicated, it is reasonable to assume that it was a drink that differed from יַיִן, yet was at least as potent. Of course, שֵׁכָר could be a simple synonym of יַיִן, or a generic term like "drink." However, the presence of *škr* on the Ashkelon ostracon is suggestive of another possible interpretation. The ostracon lists units of wine, *yn* *'dm*,[161] and *škr*; and we presume that these are discrete commodities.

Several scholars have suggested that שֵׁכָר is date palm wine, and I concur.[162] Date palms did grow in ancient Israel, especially in the drier areas. Remains of dates, for example, were found at Jericho from the Middle Bronze Age,[163] and are evident for the Iron Age as well.[164] Date palm wine is attested from ancient times, particularly in Egypt and Mesopotamia.[165] It could have been made in much the same way as grape wine.[166] That is, the fruit had sufficient sugar and juice

---

[157]Called zymase: Lucas and Harris, *Ancient Egyptian Materials and Industries*, 6.
[158]Lucas and Harris, *Ancient Egyptian Materials and Industries*, 7.
[159]Lucas and Harris, *Ancient Egyptian Materials and Industries*, 5-7.
[160]Without the addition of any sweetening agents.
[161]See below.
[162]Forbes, *Studies in Ancient Technology*, 3:77; Stager, "The Fury of Babylon," 66.
[163]Hopf, "Jericho Plant Remains," 589.
[164]Stager, "Farming in the Judean Desert during the Iron Age," *BASOR* 221 (1976): 157 n. 22; Borowski, *Agriculture in Iron Age Israel*, 127; Pritchard, *The Ancient Near East in Pictures Relating to the Old Testament,* plate 374.
[165]Forbes, *Studies in Ancient Technology*, 3:61; Zohary, *Plants of the Bible*, 60; Borowski, *Agriculture in Iron Age Israel*, 127; Moldenke and Moldenke, *Plants of the Bible*, 170. Stager notes the existence of a date wine known as *šakra* in Syriac; see C. Brockelmann, *Lexicon Syriacum* (Halle: Niemeyer, 1928), 801.
[166]Lucas and Harris, *Ancient Egyptian Materials and Industries*, 22.

within itself to produce an alcoholic drink, and could be trod like grapes.[167] In fact, dates are 60 percent sugar,[168] which is double the sugar content of grapes (25-30 percent).[169] Wild yeasts are also found on the dates' skins in sufficient quantities for successful fermentation.[170] One biblical text suggests that שֵׁכָר was a wine rather than beer. The Nazirite vow (Num 6:3) prohibits the use of any יַיִן and שֵׁכָר and the vinegar resulting from these two fermented products: חֹמֶץ יַיִן וְחֹמֶץ שֵׁכָר. This would seem to indicate that both liquids could undergo the "unwanted" acetic fermentation, and this is true only of wines. And finally, one date palm tree yields approximately 45 kg of fruit.[171] By comparison, one grape vine will average 3.64 kg of fruit.[172] Hence, the farmer did not have to have many trees in order to make some date wine.

The alcoholic potency of date wine, however, could not have been higher than that of grape wine. Dates had double the sugar content of grapes, but yeasts die off once the alcohol reaches about 14 percent.[173] Date wine would have been just as potent as grape wine and, since all its sugar could not be converted to alcohol, it would have tasted sweeter than grape wine. A potent drink that was also sweet might easily have led to drunkenness and then earned a name that suggested its potency, שֵׁכָר. In this understanding of שֵׁכָר as date wine, the Ashkelon ostracon becomes clearer, for it lists both *škr* and *yn ʾdm*, "red wine." Since grape wine was almost all red because of the genetic makeup of *Vitis vinifera* and the chemical processes of fermentation,[174] the qualifier *ʾdm* seems redundant. However, if two kinds of wine are listed in the ostracon, then *ʾdm* could differentiate the "red" grape wine from שֵׁכָר, the brown date wine.

---

[167]A date palm wine made from the sap of the tree was known in ancient Egypt. It was used at times to wash out the abdominal cavity during mummification: Lucas and Harris, *Ancient Egyptian Materials and Industries*, 21; Herodotus *Histories* 2.86. Tapping the date palm tree for sap usually rendered it worthless for subsequent fruit bearing, and it generally died. Fruit trees were highly valued in ancient Israel, as the Deuteronomic proscription against killing them illustrates (20:19). A palm sap wine, then, is unlikely for ancient Israel. Perhaps the Egyptians reserved this wine for mummification rituals, as a symbolic association between the inner cavity of the deceased and the tree.

[168]Renfrew, *Palaeoethnobotany*, 152.

[169]Unwin, *Vine and the Wine Trade*, 34; de Blij, *Wine: A Geographic Appreciation*, 22; White, "Farming and Animal Husbandry," 256; Olmo, "Grapes," 294.

[170]Renfrew, *Palaeoethnobotany*, 152; Forbes, *Studies in Ancient Technology*, 3:62.

[171]Renfrew, *Palaeoethnobotany*, 152.

[172]Cox, *From Vines to Wines*, 32.

[173]See above, pp. 187-90. It was not until distillation was mastered in the seventeenth century CE that alcohol drinks could be stronger than the 14 percent characteristic of wine.

[174]See above, pp.187-90, and chapter 3, pp. 106-9.

f. Descriptive Terms for Wine

There are several additional biblical terms for an alcoholic drink, and these, for the most part, describe a specific *quality* of the liquid, e.g., "mixed" or "old." They are then drink descriptions, rather than synonyms for the term "wine." The first term is מֶמְסָךְ, "mixed drink,"[175] used twice: Prov 23:30 and Isa 65:11.[176] In Prov 23:30, מֶמְסָךְ occurs parallel with יַיִן. The root מסך, "mix,"[177] is elsewhere parallel with יַיִן too (Isa 5:22; Ps 75:9), indicating that mixing of some kind was part of the preparation of this drink.

Various elements used in antiquity for mixing with wine include water, spices, and sweetening agents, such as honey or raisins. A common Greco-Roman practice was to mix wine with water.[178] Hesiod mentions a ratio of three parts water to one part wine.[179] In the *Odyssey*, a wine is diluted with twenty parts water to one part wine.[180] The only biblical reference to mixing water with wine is Isa 1:22, where the practice is a sign of the overall degradation of Jerusalem and so presumably does not reflect an Israelite drinking preference:

כַּסְפֵּךְ הָיָה לְסִיגִים סָבְאֵךְ מָהוּל בַּמָּיִם

Your silver has become dross,
Your drink is cut with water.

That the writer sees this practice as unfortunate is evident by its comparison with unrefined silver. Both products, silver and wine, apparently were desired without their impurities or additional elements. Isaiah uses the term סֹבֶא for "drink" (cf. Hos 4:18). The verb סָבָא means to "imbibe, drink largely."[181] In Deut 21:20, סֹבֵא describes a drunken son. In the latter text, as well as in Prov 23:20-21, excess is indicated as סֹבֵא is paired with זוֹלֵל, "glutton." Proverbs 23:20 also pairs סֹבֵא with יַיִן, while in Isa 56:12, the object of

---

[175]BDB, 587.
[176]Both are postexilic texts.
[177]BDB, 587.
[178]Homer *Iliad* 9.203; *Odyssey* 9.209; Pliny *Natural History* 14.9; Cato *Agriculture* 24.
[179]*Works and Days* 660.
[180]Homer *Odyssey* 9.209. This wine was particularly sweet unmixed, 205. Its owner, Maron, keeps it a secret, and Odysseus will later use it to get Polyphemos drunk, 347, 360.
[181]BDB, 684.

imbibing is שֵׁכָר. The term, סֹבֵא, seems to refer to the *use* of alcohol, i.e., to the drinker's habits, rather than the liquid itself, and so the general term, "drink," is the best rendering, as long as alcoholic content is assumed.[182] סֹבֵא refers, then, to an alcoholic drink that drinkers use to excess. In Isa 1:22, cutting such a drink with water rendered it less potent, less of a סֹבֵא.

מִמְסָךְ is often translated "spiced wine,"[183] though no spices are specified for wine mixing in any of the contexts in which wine or מֶסֶךְ is used.[184] Columella mentions cardamom and saffron in wine recipes.[185] If similar spices were added to Israelite wine, the practice is unrecorded or else assumed by the use of the term מִמְסָךְ.

Sweetening agents, such as honey, raisins, or dates, are another possibility for mixing agents in the Israelite מִמְסָךְ. Honey-sweetened wine is mentioned in classical sources.[186] I mentioned earlier that pomegranate juice added to grape wine would have sweetened the drink.[187] Raisins would also add sweetness to wine and could have been dried from the same harvest and put directly into the must or the wine-filled jars. Hesiod mentions the processing of raisin wine by first drying grape clusters in the sun for ten days, and then leaving them another five days in the sun before making the wine.[188] It should be noted that Hesiod's description is of raisin wine rather than *raisin-flavored* wine. That is, the wine was made from treading raisins, rather than adding raisins to pressed grape juice.

It is improbable that raisin wine was common, simply because drying ripe grapes for fifteen days drastically reduces the volume of wine from a grape harvest. Also, it meant additional work and security for the grape-drying phases. However, it is possible that, since raisin processing would be going on at the time of wine production, a vintner might apportion a certain amount of his wine to be sweetened with raisins from the vintage. The addition of raisins or other fruits, again, would not add substantially to the potency of the wine, but it was a fairly easy process and could enhance a wine's taste.

Flowers served as other possible agents for ancient mixing with wine. Columella's recipes for mixing sweet wines include one with the addition of irises to the must. This recipe, in fact, he recommends

---

[182]BDB, 685.
[183]NEB and REB have "spiced wine" for Isa 65:11 and "spiced liquor" for Prov 23:30.
[184]NRSV, RSV are therefore more exact. They have "mixed wine" for both verses. The JB has "well-blended" for Isa 65:11, but omits the term completely for Prov 23:30.
[185]*De re rustica* 12.20.5.
[186]Homer *Odyssey* 9.208; Dayagi-Mendels, *Drink and Be Merry*, 33.
[187]See above, pp. 197–98.
[188]*Works and Days* 677–80.

over the others.[189]   I suggest that a flower-sweetened wine is evident
in ancient Israel as well.   The biblical term סְמָדַר, "blossom of
grape,"[190] occurs only three times, all in the Song of Songs.[191]   It is
associated with vines and vineyards in all three verses (2:13; 2:15;
7:13).[192] It apparently does refer to the vine's flower blossom, for it
both has a scent and opens (2:13; 7:13).[193]   The term סמדר is also
inscribed on an eighth-century BCE jar from Hazor, dating to the
second half of the eighth century BCE:[194]

לפכח סמדר

Semadar, belonging to Pekaḥ

The jar, belonging to or a gift to a certain Pekaḥ, may have contained
a wine sweetened with flowers from the same vineyard from which
the grapes were harvested.[195] A nice touch, surely.

   To conclude, מִמְסָךְ refers to a "mixed wine," but precisely what
additional element was used or understood to be a part of this drink is
unknown.   Proclivities of taste undoubtedly played a role, but these
are not known to us.   If Isa 1:22 reflects a basic preference for
undiluted wine, then water is perhaps the least likely candidate for
mixing with wine.   On the other hand, water stretched wine and
lessened potency, which might have had a practical purpose for at
least some Israelite dining occasions.[196]

   One other possibility exists for the meaning of מִמְסָךְ.   In Ps
75:9, מסך is parallel with יַיִן חָמַר, "fermenting (i.e., fresh) wine.[197] If
this parallel is meant to be exact, then perhaps מִמְסָךְ refers to a wine

---

[189]Columella De re rustica 12.27.
[190]BDB, 701.
[191]Thus far, it is unattested in Ugaritic sources.
[192]Vine blossoms begin to form in the spring, usually in April, when mean
temperatures reach 20 degrees C: Unwin, Wine and the Vine, 34.
[193]Of the vine blossom, Pliny states, there is "not a perfume known which in
exquisite sweetness can surpass it": Natural History 14.2.
[194]Stratum Va: Y. Yadin, Hazor II (Jerusalem: Magnes, 1960), 73-74,   plate 95.4;
Avigad, "Two Hebrew Inscriptions on Wine Jars," 3; Inscriptions Reveal, 109, 55,
120.
[195]Vine flowers blossom in the spring, so the flowers would have been added then to
wine stored from the fall vintage, or they would have been dried for a later mixing
with wine. Yadin suggests only that it was a drink made from the blossom: Hazor II,
74.
[196]In chapter 6, I estimate that a small, two-dunam vineyard could produce 694 liters
of wine, so stretching wine might not have been necessary, depending, of course, on
the intake levels. See pp. 211-12.
[197]See pp. 199-200 above for a discussion of חָמַר.

that is still fermenting, i.e., still "mixing." The meanings of חמר and
מסך in wine contexts both require clarification. Hence, the weakness
of this proposal is that their interpretations are mutually dependent.

Three additional terms qualifying wine in epigraphic sources
are *khl, yšn,* and *'dm.* All three occur on inscriptions along with the
term *yn/yyn.* כחל occurs only once in the Bible, in Ezek 23:40, as a
verb for painting the eyes. Its appearance in an inscription is, as we
have seen, from an Iron II decanter found near Hebron.[198] It
mentions a certain wine, יין כחל.[199] It is possible that this phrase
described a wine of a hue dark enough to be comparable to eye
makeup. Alternately, כחל could denote a geographic region. Avigad
suggests a place name, Beit Kaḥil, 4 km northwest of Hebron.[200]

Yšn occurs on Samaria ostraca with *yn* apparently to describe
"old wine" (ישן יי). It occurs on ostraca nos. 1, 3, 5-10, 12-14, and
36.[201] יי without this qualifier occurs only twice, in nos. 11 and 44.
The other commodity noted by these ostraca also occurs with a
qualifying term, in the phrase שמנ רחצ.[202] This oil, as Stager has
convincingly argued, was the finest oil, the oil produced by crushing
the olives, immersing them in water, and then skimming the "washed
oil," *šmn rḥṣ,*[203] off the top.[204] By analogy, we can reasonably infer
that ישן יי was too the finer quality wine than that noted without the
qualifier, ישו. In the Palmyra inscription mentioned above,[205] "old
wine," חמרא עתיקא,[206] does seem to indicate a finer quality, as it was
the wine for a banquet of priests.[207] With the wine commodity of the
Samaria ostraca, quality was apparently marked by its age rather than
by its process of production, as it was for oil.

It is difficult to establish how long wine could have lasted in the
Iron Age before either being consumed or turned to vinegar. Varro
mentions an old wine as being a year old.[208] If containers were

---

[198]Avigad, "Two Hebrew Inscriptions on Wine Jars," 3-5; *Inscriptions Reveal,* 53.
[199]*Inscriptions Reveal,* 114-15.
[200]Avigad, "Two Hebrew Inscriptions on Wine Jars," 4.
[201]For nos. 3, 6, 8, and 36, one of the three graphemes for ישן is not discernible. For
no. 8, two are not legible: Kaufman, "The Samaria Ostraca," 141-44; Aharoni, *Land
of the Bible,* 358-63. See discussion of the ostraca in chapter 1, pp. 51-59.
[202]In nos. 16-21, 53-55, 59 (נ illegible in 59). שמנ, unlike יי, does not occur without
its qualifying term: Kaufman, "The Samaria Ostraca," 141-44; Aharoni, *Land of the
Bible,* 358-63.
[203]Stager, "The Finest Oil in Samaria," 242; K. F. Vickery, *Food in Early Greece*
(Urbana, Ill.: University of Illinois Press, 1936) 52.
[204]For a discussion of the production of this oil, see Stager, "The Finest Oil in
Samaria," 241-45.
[205]See above, p. 200.
[206]Ingholt, "Un nouveau thiase à Palmyre," 134.
[207]B. Porten, *Archives From Elephantine* (Berkeley: University of California Press,
1968), 179-86.
[208]Varro *On Farming* 1.65.1.

completely airtight, then the wine could last over a year. Biblical texts are silent about any tradition of aging wine. In Ruth 2:14, bread is being eaten with vinegar (חֹמֶץ). There is no hint about whether the diners enjoy this vinegar as a condiment, though it is a sign of Boaz's generosity that he offers it to Ruth. Judging from the other biblical uses of vinegar (Ps 69:22 [ET 69:21]; Prov 10:26; 25:20), vinegar was not generally enjoyed. In Ps 69:22, it slakes thirst in the same way that poison eases hunger, and in Prov 10:26, it is likened to smoke in the eyes. If the Ruth verse indicates that this vinegar is the last of Boaz's wine, it is only spring, the time of the barley harvest, so Boaz would have a long wait for the new wine of the fall vintage. Whether ישן יין was understood in eighth-century BCE Samaria to signify wine of a specific age or, more generally, the oldest wine of a farm, as a mark of its value, cannot be known.

Thus far we have examined the activities of the grape harvest and vintage, the chemical process of fermentation, and the various terms used for wine in biblical texts. Wine was obviously the primary goal of the vintage, and its enjoyment is the subject of the next chapter.

Chapter 6

Wine Consumption

We have discussed thus far the activities involved in the cultivation and maintenance of the Israelite vineyard and the equipment, processes, and labor used in wine production. This chapter is devoted to an analysis of family wine consumption. For the goal of all this viticultural activity was to produce wine to drink. While a percentage of the vintage went to raisins or grapes,[1] the work of the Israelite vineyard had wine as its primary goal, (Num 18:27; Deut 11:14; 15:14; 16:13; Judg 9:13; 1 Chr 27:27; Neh 5:11; Prov 3:10; Isa 16:10; 27:2; 62:8; Jer 40:12; Joel 4:13, 18 [ET 3:13, 18]; Amos 9:13-14).[2] It was a desired product of the farm for several reasons: it was a liquid resource, was storable, and brought levity. To cultivate a vineyard, tread its grapes, and not get to drink the wine became a repeated biblical symbol of misfortune (Deut 28:39; Job 24:11; Amos 5:11; Mic 6:15; Zeph 1:13).

Biblical traditions of drinking are frequent and undoubtedly varied in their meanings. To recall, יַיִן alone occurs 141 times, while תִּירוֹשׁ, עָסִיס, and חֶמֶר occur a total of 44 times. A verb meaning to "get

---

[1]Borowski, *Agriculture of Iron Age Israel*, 113; A. Goor and M. Nurock, *The Fruits of the Holy Land* (Jerusalem: Israel Universities, 1968), 23; King, *Amos, Hosea, Micah*, 118; 1 Sam 25:18; 30:12; 2 Sam 6:19; 16:1; 1 Chr 12:40; 1 Chr 16:3; Cant 2:5). Large quantities of charred grape seeds were found in Stratum XIV (tenth century BCE) at Tel Michal: Herzog, "A Complex of Iron Age Winepresses (Strata XIV-XIII)," 74.
[2]The classification, *vinifera*, "wine-bearing," testifies to the plant's primary historical use: Johnson, *Vintage*, 17.

drunk," שָׁכַר, is used nineteen times in biblical texts, and so it is evident that wine consumption, at times to the point of inebriation, was an inscribed feature of Israelite social life. The effects of alcohol in a culture, especially an ancient one, are difficult to assess. Drinking wine affected, even altered social relations in many ways, some lost to the historian, others evident from biblical traditions. Wine, in those traditions, was used at meals as a part of daily sustenance, and, often in greater quantities, at social celebrations where it led to merrymaking and drunkenness.

The present work has focused on the family as the basic social unit practicing viticulture in ancient Israel, and it remains as the basis for an analysis of wine consumption. The present chapter, then, is devoted to an analysis of biblical scenes of family drinking. Analysis of those texts shows that wine provided or enhanced family life with at least three significant elements: sustenance, enjoyment, and celebration. Sections I and II focus on wine in the everyday life of the farmer. Sections III and IV provide a discussion of the two types of banquet traditions mentioned in the Bible, the *mišteh* and the *marzēaḥ*.

I. Sustenance

Wine was a staple food of ancient Mediterranean life. A product of the farm, from an anticipated harvest to its joyous production in the press, wine naturally made its way to the table as a component of the meal.[3] It was not a luxury or only occasionally present, but was instead commonplace in ancient Israel. With the scarcity of water and its risk of contamination,[4] wine proved to be an important liquid source. In the story of Daniel, even prisoners, at least those groomed for service in the royal court, were afforded a ration of wine with their food (Dan 1:5). It was, according to Cato, part of the farm rations for ancient Rome. Cato advised that for the family farm, each field worker be given as much as ten amphorae[5] of wine a year.[6] This works out to be approximately 260 liters or 347 bottles of the .75 l size (today's standard bottle) a year, almost a bottle

---

[3]Broshi, "Wine in Ancient Palestine," 34; Goor, "The History of the Grape-vine," 48.

[4]King, *Amos, Hosea, Micah*, 118; Ross, "Wine," 851; Amerine and Singleton, *Wine*, 14; Darby et al., *Food: The Gift of Osiris*, 2:533; G. Austin, *Alcohol in Western Society from Antiquity to 1800* (Santa Barbara: ABC-Clio, 1985), xvi.

[5]A Roman amphora typically had a capacity of about 26 liters: Powell, "Weights and Measures," 904; Hooper and Ash, in Cato *Agriculture*, 531; Forbes, *Studies in Ancient Technology*, 3:119.

[6]Cato *Agriculture* 57.

a day.[7] Since the agricultural strategies for ancient Israel were similar, in its mixed farming of grains, grapes, and olives,[8] we reasonably infer from analogy that the Israelite's wine consumption would approach, or even surpass, that of a Roman farmhand.

We saw in chapter 3 that a two-dunam vineyard, eminently within the labor resources of a small farm, was a conservative estimate for the vineyard plot of the ancient Israelite.[9] If this plot were planted according to modern practices of having vines 1.83 m apart, in rows 3 m wide, then a two-dunam vineyard would hold 275 vines. The ancient Israelite vineyard likely varied in spacing and was even at times fragmented, depending on the nature of the farm. Still, this number of vines can act as a rough estimate for calculating wine consumption. One vine yields approximately 8 lb grapes.[10] Hence 275 vines will yield 2,200 lb of grapes. It takes on average 12 lb of grapes to produce a gallon of wine.[11] A two-dunam vineyard, therefore, will yield approximately 183 gallons, or 694 liters of wine. This averages to 1.9 l of wine a day. By today's bottle measures, this amount of wine was roughly 2.5 bottles of wine a day per household, though some would undoubtedly have been reserved for trade and festivity. Storing 694 liters of wine would require 15 jars the size of the Judean royal jars, whose average capacity, it will be recalled, is 45 l.[12]

On a lower estimate, the Talmud mentions fifteen jars of wine for the seventh year.[13] If we calculate this amount of household wine using both Albright's jar estimate of 22 l and the standard measure of a Roman amphora at 26 l,[14] then the "wine cellar" of the ancient Jew in the strapped sabbatical year approaches somewhere between 330-

---

[7]White estimates one liter a day for the Roman citizen as well: *Roman Farming*, 496, n. 55. For ease in comparison, I have calculated wine estimates in liters and number of wine bottles of today's standard capacity (.75 l).

[8]Holladay, "The Kingdoms of Israel and Judah," 386.

[9]Chapter 3, 111-12.

[10]In this instance, retaining the American system of measurement that Cox used is easier. I have converted the end amount into the metric system. See Cox, *From Vines to Wines*, 32.

[11]Cox, *From Vines to Wines*, 32.

[12]Chapter 2, pp. 50-51. Ussishkin averages the capacities from nine jars, 39.75; 43.00; 43.20; 44.25; 44.75; 45.33; 46.67; 46.75; 51; 80: "Excavations at Tel Lachish," 162-163.

[13]*y. Sheb.* 5. 7, 36a. It does not mention measurements or how many people are served by fifteen jars, since the focus is on what the potter can sell in the sabbatical year. A nonsabbatical year might easily require more than fifteen jars. I assume, however, that since they are likely storage jars, they contain the wine for a household rather than an individual.

[14]Albright's widely accepted estimate of the royal bath measurement is 22 l, and the Roman amphora can vary from 22 to 26 liters in capacity. See Albright, *Tell Beit Mirsim* 3: 58-59, n.7; Y. Aharoni, "Hebrew Ostraca from Arad," *IEJ* 16 (1966): 2; de Vaux, *Ancient Israel*, 202; Powell, "Weights and Measures," 904.

390 l. This figure would be the estimate for a year in which no pruning was done, and hence would constitute a bare minimum. These calculations of Israelite wine amounts must remain hypothetical, of course, since archaeology, biblical text, and epigraphy have so far not provided more precise evidence for the Iron Age. Still, the estimates provide at least a range of wine amounts, from 330 to 694 liters a year. Even with this broad range, it is easy to see that wine was available for daily meals as well as special occasions.

The first such use of wine was its most common, as a daily provision. Grain[15] and wine were the basic foodstuffs for the Israelite meal (Judg 19:19; Ruth 2:14; Prov 9:5; Eccl 9:7; Hos 9:2).[16] They were brought to those in mourning when food preparation was forbidden (Jer 16:7-8).[17]   Meals shared in some kind of commemorative purpose, too, had bread and wine as their basis (e.g., Gen 14:18-24; Mark 14:12-25).[18] The bowls, plates, jugs, and chalices present in Iron Age tombs indicate that bread and wine constituted provisions in death as well.[19]

Psalm 128:2-3 contains the only biblical scene of a farmer's table. It includes wine, but only through allusion:

יְגִיעַ כַּפֶּיךָ כִּי תֹאכֵל אַשְׁרֶיךָ וְטוֹב לָךְ
אֶשְׁתְּךָ ׀ כְּגֶפֶן פֹּרִיָּה בְּיַרְכְּתֵי בֵיתֶךָ
בָּנֶיךָ כִּשְׁתִלֵי זֵיתִים סָבִיב לְשֻׁלְחָנֶךָ

For you shall eat of the toil of your hands;
you shall be happy, and it shall go well with you.
Your wife is like a fruitful vine inside your
house; your children will be like olive shoots
around your table.

                                        (Ps 128:2-3)

---

[15]Forbes, *Studies in Ancient Technology*, 3:86; Matt 6:11; Luke 11:3. Bread, לֶחֶם, also meant more generally "food."
[16]Fruits, legumes, vegetables, and animal products of milk and meat supplemented the fare. See Hayes and Miller, *A History of Ancient Israel and Judah*, 52-53; Rosen, "Subsistence Economy in Iron Age I," 342; Holladay, "The Kingdoms of Israel and Judah," 386-87; Broshi, "The Diet of Palestine in the Roman Period," 41-56; E. Vermeule, *Greece in the Bronze Age* (Chicago: University of Chicago Press, 1964), 180, 317-18.
[17]de Vaux, *Ancient Israel*, 60.
[18]See below.
[19]E. Bloch-Smith, *Judahite Burial Practices and Beliefs about the Dead* (Sheffield, Eng.: JSOT Press, 1992), 73, 108.

The passage makes overt what is otherwise assumed in biblical texts, namely, that agricultural productivity had its reward first and foremost at the farmer's table. The wife as fruitful vine would bring joy like wine, and the children as transplanted olive shoots were full of rich promise, like new oil. These are images from the deity's perspective of the usefulness and delight of family members to the farmer. Two pleasures of the table, family and crops, are combined in this passage. To the farmer, in other words, family members were as valued and as basic as having wine and oil at one's table.

The domestic pottery of the Iron Age provides a material complement to the psalmist's description of daily dining. Pottery assemblages show the types and styles of vessels used, but not their function or what they contained.[20] A jug, for example, could have functioned as a serving vessel for both wine and water. Small bowls were the most common ceramic form represented in Iron Age domestic assemblages, and these were likely used for both eating and drinking.[21] The drinkers who incur Amos's disgust drink their wine in bowls: יַיִן מִזְרְקֵי (6:6).[22] מִזְרָק is otherwise restricted to ritual or temple contexts (e.g., Exod 27:3; 38:3; 1 Kgs 7:40; Jer 52:18), and so, Amos's disgust may in part be directed at the kind of bowls the drinkers chose to use. כּוֹס is a more frequent term for drinking vessel and apparently signifies a cup similar to our bowl.[23] Bowls were the standard drinking vessel of the Iron Age.[24] Royalty also often used bowls for drinking in the ancient Near East. Asshurbanipal and (presumably) his queen use them to drink under a trellised grape vine in the Nineveh relief.[25] Bowls were found along with chalices and wine jars in King Tutankhamon's tomb from the New Kingdom

---

[20]For a helpful analysis, see Rast, *Taanach I: Studies in Iron Age Pottery*, 29, 70-78.
[21]N. L. Lapp, "Pottery Chronology of Palestine," in *ABD*, 5: 432. For representative types throughout the Iron Age, and differences between styles from the north and south, see R. Amiran, *Ancient Pottery of the Holy Land: From its Beginnings in the Neolithic Period to the end of the Iron Age* (Jerusalem: Massada, 1963), 192-212. Bowls were typically coarsely made, and carinated (i.e., having a sharp, inward change in direction of the wall) or rounded. In the Iron II period, workmanship improved and a burnished red slip predominates with the bowls, also mostly carinated or rounded in shape: Amiran, *Ancient Pottery of the Holy Land*, 192, 195; W. E. Rast, *Through the Ages in Palestinian Archaeology: An Introductory Handbook* (Philadelphia: Trinity, 1992), 118; J. L. Kelso, "The Ceramic Vocabulary of the Old Testament," *BASOR Supplemental Studies* 5-6 (1948): 11.
[22] About these banquet participants, see below, pp. 243-46.
[23]Kelso, "Ceramic Vocabulary," 19.
[24]Hunt, "The Pottery," in *Tell Qiri: A Village in the Jezreel Valley*, 198, n. 64.
[25]Stronach, "The Imagery of the Wine Bowl: Wine in Assyria in the Early First Millennium, B.C." in *OAHW*, 175-95. Shalmaneser III is shown with a drinking bowl on the Black Obelisk: M. Roaf, *Cultural Atlas of Mesopotamia and the Near East* (New York: Facts on File, 1990), 175.

period.[26] Chalices—vessels with a shallow bowl on a high foot—were another type of drinking vessel and were characteristic of the Iron I period in both the northern and southern regions of the country.[27] They were plain in Iron I and decorated in the Iron II period.[28]

For serving wine, various shapes and styles of jugs—vessels that are typically wider than they are high—were common in the Iron Age.[29] They have one or sometimes two handles and are small enough for one person to lift or carry. This jug may be meant by the biblical term, כַּד, which women carry to the well (Gen 24:14, 17, 43; see also Eccl 12:6) and soldiers carry in one hand (Judg 7:16, 19). Such jugs are to flow at the banquet given by King Ahasuerus for the people of Susa (Esth 1:8). It would seem from these contexts that such a jug served double duty as a transport and serving vessel. The wine (and oil) taxes of Ugarit were measured by a jar amount termed a *kd*.[30] Of special note in the archaeology is a jug with a strainer spout.[31] This jug was likely used to strain wine of its lees when serving. A kind of strained wine is noted in Isaiah:

וְעָשָׂה יְהוָה צְבָאוֹת לְכָל־הָעַמִּים בָּהָר הַזֶּה מִשְׁתֵּה שְׁמָנִים
מִשְׁתֵּה שְׁמָרִים שְׁמָנִים מְמֻחָיִם שְׁמָרִים מְזֻקָּקִים

> And the Lord of hosts will make
> for all peoples on this hill,
> a feast of rich foods, a feast of
> well-aged foods: oils, meats,
> aged, refined wines.
>
> (Isa 25:6)

It does not mention a jug, but some container is implied as this is "kept, strained" wine. A separate ceramic strainer for placing over a cup or bowl in pouring was used in New Kingdom Egypt.[32] The Israelite homes without strainer jugs probably strained with the use of

---

[26]Lesko, *King Tut's Wine Cellar*, 39.

[27]Amiran, *Ancient Pottery of the Holy Land*, 213.

[28]Amiran, *Ancient Pottery of the Holy Land*, 213-15; Rast, *Taanach I: Studies in Iron Age Pottery*, 14, 20-21.

[29]Rast, *Taanach I: Studies in Iron Age Pottery*, 28-29.

[30]Heltzer, *The Rural Community in Ancient Ugarit*, for wine: 40 (*PRU*, V, 4; *CTA*, 67); for oil: 42 (*PRU*, II, 82); Liverani, "Economy of Ugaritic Royal Farms," 144.

[31]Amiran, *Ancient Pottery of the Holy Land*, photo 257, p. 259. These kinds of jugs are found at Philistine sites and were long called "beer jugs." With the significant Iron Age winery at Ashkelon, Stager argues that these are jars for straining wine of its lees: "The Fury of Babylon,"65.

[32]Lesko, *King Tut's Wine Cellar*, 40.

a cloth by pouring the wine through it into a second vessel.[33] Or the wine could have been strained before it was put into jars.[34]

Decanters were characteristic of the Iron II period especially and were probably used to serve wine, as well as other liquids. They typically were well made, with narrow, ridged necks, a handle drawn from the ridge, and a wide body.[35] As already noted, a decanter dating from the late eighth to early seventh centuries BCE and having a height of 22.6 cm was found near Hebron.[36] To recall, the decanter has an inscription with a *lamed*, personal name, and the term "wine" on it: כחל יין ליחזיהו, "belonging to Yaḥaziyahu *kḥl* wine."[37] This decanter, then, was clearly used for wine and has a capacity of ca. 2.5 l.[38] Another Iron II decanter, also discussed earlier, whose exact provenience is unknown, is 19.7 cm, with a capacity of 1 l, 270 cm³.[39] It too had an amount of wine on its inscription: למתניהו יין נסכ רבעת, "belonging to Matanyahu, libation wine, a fourth."[40] The *lamed* affixed to the personal names in these and other jar inscriptions can signify either "of" or "for." If the former, "of," then ownership is meant by affixing the *lamed* to a personal name, as "belonging to" indicates. But, if the latter sense is meant, "for," then the jars denote not the owners, but the intended recipients. This was, we recall, part of the interpretive problem with the Samaria ostraca, which had personal names both with and without a *lamed* affixed to them.[41] A narrow neck, such as the decanter has, gave greater precision in pouring and so indicates that the liquid it contained was valued.[42] However, wine need not have been the only valued liquid in ancient Israel. In fact, for Israel, it is probable that all liquids were scarce enough to be worth protecting by serving with vessels that would not spill easily.

---

[33]Jesus used the idea of straining to mock the Pharisees who, he claimed, strained out a gnat, but would swallow a camel whole: Matt 23:24; Ross, "Wine," 851; Dommershausen, "יַיִן," 61; Forbes, *Studies in Ancient Technology*, 3:112; Moldenke and Moldenke, *Plants of the Bible*, 214. Roman vintners seem to have used strainers made of straw: Columella *De re rustica* 12.19.4.

[34]Badler et al., suggest that this may indeed have been the case at Godin Tepe, late Period V (3100-2900 BCE), where the jars had tartaric acids, but no pips, vegetal matter, or phenolic acids: "Drink and Be Merry! Infrared Spectroscopy and Ancient Near Eastern Wine," 32.

[35]Amiran, *Ancient Pottery of the Holy Land*, photos 255-56, p. 259.

[36]Avigad, "Two Hebrew Inscriptions on Wine Jars," 1; *Inscriptions Reveal*, 53.

[37]*Inscriptions Reveal*, 114-15. See chapter 5, pp. 194, 206.

[38]Avigad, "Two Hebrew Inscriptions on Wine Jars," 1; *Inscriptions Reveal*, 53. Incidentally, this amount approximates the 1.9 l estimate given above.

[39]Deutsch and Heltzer, "A Wine Decanter with a Hebrew Inscription," 23.

[40]Deutsch and Heltzer, "A Wine Decanter with a Hebrew Inscription," 23-26, photo, fig. 6. See chapter 5, p. 194.

[41]See the discussion in chapter 2, pp. 51-59.

[42]P. Rice, *Pottery Analysis: A Sourcebook* (Chicago: University of Chicago Press, 1987), 225.

Dipper juglets—small jugs—were also characteristic of Iron II.[43] These were probably used to "dip" wine out of the large storage jars and put it into jugs for serving. An example of one such storage jar, from Tell Beit Mirsim, has a circular depression off its rim for holding a dipper juglet.[44] A jar from 'Ein-Gev, dating to the ninth century BCE and 41.5 cm high, had the following Aramaic inscription:

לשקיא[45]

belonging to the cupbearers

This jar differs from the decanter above by its greater size, its two handles, and its wider neck. Perhaps wine was brought from storage to the table in this jar, from which the cupbearers then served either by dipping cups into the jar or by pouring directly from it. Or the jar may represent the wine ration that the cupbearers themselves were given. In either case, this inscribed jar indicates a certain degree of wealth, enough to require cupbearers, and so cannot be considered a representative vessel of small-farm life.

Biblical descriptions of provisions for travel also indicate that wine was a part of daily sustenance. In Judg 19:19, a man and his concubine have enough for a meal in the bread and wine they have packed. The Gibeonites pack only bread and wine for their meals for a feigned long journey (Josh 9:12-13).[46] Jesse sends the young David with a gift to Saul consisting of a donkey load of bread, one kid goat, and one skin of wine (1 Sam 16:20). The latter two examples mention the container[47] for the wine as a נאד, a "skin."[48] נאד denotes a container made of animal skin rather than clay, for in Josh 9:12-13, these נאדות have worn down to the point of bursting.[49] And the LXX renders נאד from these passages as ἀσκός. This is the same term as in the Gospel parable about wineskins that burst (Matt 9:17; Mark 2:22; Luke 5:37-38). Hence, the term נאד denotes a skin container (most

---

[43] Amiran, *Ancient Pottery of the Holy Land*, 259.
[44] Amiran, *Ancient Pottery of the Holy Land*, photo 250, p. 243.
[45] *Inscriptions Reveal*, 58, 129.
[46] For a discussion of Gibeon, see chapter 4, pp. 158-62.
[47] Judg 19:19 does not mention a container for the wine.
[48] BDB, 614.
[49] In Ps 119:83, the plaintiff feels like a skin (נאד) in smoke. Smoke would put only a soot residue on a clay vessel. While perhaps an undesired state, it is only surface grime and does not convey the level of discomfort he has experienced.

likely of a goat)[50] for liquid,[51] often used for travel. Wineskins were probably common for transport, as they could be easily strung along upon one's back or side.[52]

One pottery form from the Iron Age seems to have been used primarily for travel. The so-called pilgrim flask was in use throughout the Iron Age and was a development from the Late Bronze period.[53] This flask typically has a spherical body, one handle, and little if any base.[54] Because of this latter feature, it does not stand upright well and so was probably for the carrying of liquids during travel, rather than for daily meals at home.

David, when he traveled, acquired substantial provisions, first from Abigail (1 Sam 25:18), and then from Ziba (2 Sam 16:1). In both texts, נֶבֶל is the term used for wine container. The LXX has "vessel," ἀγγεῖον, for 1 Sam 25:18, and simply νέβελ for the Hebrew נֶבֶל in 2 Sam 16:1. While the amounts require multiple donkeys in both cases, the wine provisions themselves remain small. Abigail gave 200 loaves, 200 fig cakes, 100 raisin bunches, 5 sheep and some grain, yet only 2 containers of wine. Her apparent[55] parsimony is somewhat unexpected, since her husband is inside drinking himself into a stupor (vv 36-37).[56] Ziba gave David only one container, a נֶבֶל, of wine with his 200 loaves, 100 summer fruits, and 100 bunches of raisins, and he remarks that the wine is reserved for those who faint. So, though Abigail's wine gift seems small, it is double that of any of the other rations specified in the Bible.[57]

It is impossible to discern the amount of wine indicated in these texts. In 1 Sam 10:3, where נֶבֶל is also used to carry wine, the LXX has ἀσκός, "skin." If נֶבֶל denotes a skin bottle,[58] then, as in the above passages, not much wine is given, though the exact amount is not indicated.[59] But a skin container would have been easier to carry than

---

[50]Brown, "The Mediterranean Vocabulary of the Vine," 149. Ross remarks that such a skin could have been simply a whole goat hide with the legs and neck tied: "Wine," 851; For a picture of a modern-day wineskin made from a goat, see Dayagi-Mendels, *Drink and Be Merry*, 41.

[51]Elsewhere it holds milk (Judg 4:19), and tears (Ps 56:8).

[52]Odysseus accepts a gift of one skin full of wine and is carrying it on him when he meets up with the Cyclops Polyphemous in the cave: Homer *Odyssey* 9.196, 212.

[53]Lapp, "Pottery," 442.

[54]Amiran, *Ancient Pottery of the Holy Land*, 276.

[55]We cannot know the capacity of the container. Proportionate to the bread and fig cakes at least, it appears to be somewhat stingy.

[56]The pun on her husband's name, Nabal, is undoubtedly intentional and multivalent. Nabal, נָבָל, is a "fool," נָבָל, who is drinking himself into a coma, and he drinks so much wine that he himself becomes, in effect, a human wineskin, a נֶבֶל.

[57]Though Ezra can call in 100 baths of wine from the treasurers: Ezra 7:22.

[58]BDB, 614.

[59]S. M. Paul and W. G. Dever, *Biblical Archaeology* (New York: Quadrangle/New York Times Book Co., 1974), 174.

a jar for travel on foot.[60] נֵבֶל could as well refer to a traveling jar, strapped in a saddle bag over the animal, rather than a skin. The container noted in the Samaria ostraca for wine and oil is a נבל.[61] We can probably assume that for taxation purposes, jars would have been preferred to skins, though the containers themselves were not found.[62] The Lachish relief shows the transport of both pottery and skin containers, with pottery vessels on carts and the softer, bent bags or skins carried over the shoulder.[63]

Wine provisions for the army seem to be somewhat more abundant from the Arad letters. Grain, wine, and oil are the mainstays of the provisions in these letters.[64] In letter 56[65] soldiers get three baths worth of wine[66] along with a donkey load of flour for bread.[67] In letter 57[68] they are to receive two baths worth of wine for four days. But they also receive 300 loaves of bread and a full donkey load of wine.[69] Three baths of wine go to an individual in letter 58A.[70] From this letter, then, we can estimate wine consumption at 66 liters, but we do not know how long this amount of wine is intended to last. Because neither the number of men nor the time periods involved are given, it is impossible to draw conclusions about daily drinking habits. What these requisitions for wine amounts

---

[60]The three men in 1 Sam 10:3 are, as far as we can tell, on foot, and so the wine container is probably a wineskin. The man who is carrying the three loaves of bread could easily be transporting them manually. It is the man charged with carrying the three goat kids who faces the real challenge! Either a cart of some sort goes unmentioned, or he is herding the three animals along the way.

[61]Ostraca 1, 3-19, 21, 53-55, 59: Kaufman, "The Samaria Ostraca," 141-44; Aharoni, *Land of the Bible*, 318-321. Kelso suggests that נֵבֶל evolved in meaning from wineskin to include ceramic wine jar as well: "The Ceramic Vocabulary of the Old Testament," 14.

[62]Chapter 2, pp. 51-59.

[63]For pottery vessels of Lachish captives, see Ussishkin, *The Conquest of Lachish by Sennacherib*, fig. 84, p. 108; fig. 85, p. 110; skins or perhaps bags, fig. 70, pp. 86-87; fig. 80, p. 103; fig. 85, p. 100.

[64]J. Lindenberger, *Ancient Aramaic and Hebrew Letters* (Atlanta: Scholars Press, 1994), 101.

[65]Lindenberger, *Ancient Aramaic and Hebrew Letters*, 107.

[66]Liquid measures for the Iron Age are uncertain, but see Powell, "Weights and Measures," 897-908; Albright, *Tell Beit Mirsim*, 3: 58-59, n.7; Aharoni, "Hebrew Ostraca from Arad," 2, n.3; Paul and Dever, *Biblical Archaeology*, 175.

[67]Interestingly, Eliashib is instructed to take from the old flour and use the wine from the "craters," האגנת; (BDB, 8 translates as "bowls"). Craters would have been for mixing and not storage. Since mixing wine with water was a characteristic Greek practice, mention of this vessel supports the view that the *ktym* (kittim), those to whom the provisions were sent, were Greek mercenaries: Aharoni, "Hebrew Ostraca from Arad," 4-5; Miller and Hayes, *A History of Ancient Israel and Judah*, 389, 417.

[68]Lindenberger, *Ancient Aramaic and Hebrew Letters*, 108.

[69]Lindenberger, *Ancient Aramaic and Hebrew Letters*, 108. On p. 141, he estimates the donkey load at approximately 134-209 liters. This would be, thus, a fair amount of wine, though we do not know how many men this was supposed to serve.

[70]Lindenberger, *Ancient Aramaic and Hebrew Letters*, 108.

do indicate, however, is that wine was used as a staple of the army diet in ancient Israel.

The *lmlk* jars, if they contained wine,[71] indicate the same. In fact, with an average of 45 liters each, they indicate significant consumption. Almost 400 have so far been found at Lachish alone.[72] The numbers of these jars and their distribution throughout Judah suggest that they were for army provisions during Sennacherib's attacks through Judah. They likely contained all needed provisions—wine, water, grain, and oil—rather than just wine. Ussishkin provides no estimate of the number of inhabitants at Lachish at the eve of its destruction by Sennacherib's forces, citing only the mass grave of 1,500 skeletons thought to be from the attack. Taking this 1,500 as a very rough estimate of Lachish's population, the wine amounts are too high. For, 400 *lmlk* jars would hold approximately 18,000 liters of wine.[73] This works out to be six modern-size bottles per person, and, depending on the time remaining between the delivery of the wine and the citizens' death, seems a bit high, even if morale had, by that point, given way to despair.

To conclude, then, the biblical and epigraphic evidence cited above indicates that wine provided daily sustenance for the army, for travelers, and, most basic of all, for the family. Wine use, in these instances, was primarily for sustenance; it was taken along with bread and perhaps some fruit to comprise a meal. It was, in this sense, one facet of a meal along with others. In this use, wine functioned as a staple element of the Israelite meal. It did not stand apart from other elements or get used in any focused sense as it does for celebration and feasting.

II. Enjoyment of Wine

Viticulture gave the Israelite farmer wine, while his orchards and fields gave him olive oil and bread. At its best, Israelite farm life gave

וְיַיִן ׀ יְשַׂמַּח לְבַב־אֱנֹוש
לְהַצְהִיל פָּנִים מִשָּׁמֶן
וְלֶחֶם לְבַב־אֱנֹוש יִסְעָד

---

[71]Cross, "Judean Stamps," 22; Na'aman, "Hezekiah's Fortified Cities," 5-21.

[72]Ussishkin, "Excavations at Tel Lachish 1978-1983: Second Preliminary Report," 164.

[73]Ussishkin, *The Conquest of Lachish by Sennacherib*, 56; Broshi, "Wine in Ancient Palestine," 25; Cross, "Judean Stamps," 20-22.

> wine that gladdens the human heart,
> oil to brighten the face,
> and bread that sustains the human heart[74]
> (Ps 104:15)

The enjoyment of the latter two products, oil and bread, for the ancient Israelite is fairly easy to explicate. Olive oil provided dietary fat, was an emollient, and a lamp fuel; the latter two functions literally did "brighten the face." And bread was the primary foodstuff of the Israelite meal. Man did not live by bread alone (Deut 8:3), but without it, he did not survive. Bread in ancient Israel sustained the human heart that wine then gladdened. The enjoyment of wine was likely deemed to be an essential of life, just as oil and bread were.[75]

Wine is a unique agricultural staple, for it not only added to the sustenance of the farmer, but also, as an intoxicant, affected, even altered him. The biblical scenes of drinking present traditions of wine use, of how it could "gladden the human heart" as well as madden, trick, and subdue it. Overwhelmingly, these effects are staged in social contexts. This is not surprising, for alcohol, with its ability to lower inhibitions, was early on valued as a social stimulant.[76] Hence analysis of wine consumption in biblical scenes involves its social dimension. With one notable exception—Noah—no one drinks alone in the Bible. Behaviors and values of social interaction, i.e., cultural mores, are caught up in the not-so-simple act of drinking wine.

Commensality, the sharing of food and wine, was an important social component in biblical scenes of daily life. Sharing meals was a natural extension of the work of the farm. Just as families worked their cultivation and harvests together, so they shared in meals from the produce. The main meal was probably taken in the evening after a day of hard work and when there was no longer enough light for

---

[74]Wine, oil, and bread typified Mediterranean culture. For Mesopotamia, beer, oil, and bread were the cultural markers. The wild Enkidu becomes human after eating bread, rubbing oil on his skin, and drinking beer until he was merry. "Gilgamesh (Old Babylonian Version)" (S. Dalley, *Myths from Mesopotamia* [Oxford: Oxford University Press, 1991], tablet II. iii. lines 15-25, p. 138. The products in these literary traditions are agricultural staples and markers of the culture that produced them.

[75]Ross, "Wine," 851.

[76]W. Burkert, "Oriental Symposia: Contrasts and Parallels," in *Dining in a Classical Context* (ed. W. Slater; Ann Arbor, Mich.: University of Michigan Press, 1991), 7; P. Michalowski, "The Drinking Gods: Alcohol in Mesopotamian Ritual and Mythology," in *Drinking in Ancient Societies*, ed. L. Milano, 29; Austin, *Alcohol in Western Society*, xvi; Amerine and Singleton, *Wine*, 325; Johnson, *Vintage*, 12.

more work.[77] Meals or feasting sometimes occurred in daylight as well: Gen 18:1-3 without wine; Gen 43:16; Job 1:1; Isa 5:11.

Wine enhances a meal in two primary ways; it helps to ease bodily weariness and aids in communication. First, it provides liquid refreshment and aids in digestion by stimulating salivary glands, gastric secretion, and motility.[78] Nutritionally, wine also supplies significant amounts of vitamins B_2 and B_6 and the essential elements manganese and iron, to a meal.[79] Ancient farmers might have shared, too, in the modern vintner wisdom that wine gently curbs the appetite by satisfying it. According to Adamo, the Tuscany vintner: "When you drink wine you eat less bread. You need less. When you drink water with bread you are hungrier."[80] As a mild sedative and vasodilator,[81] wine also helps to ease the tensions of the body. Hence, physiologically, those sharing a meal with wine, especially after a hard day of work, would likely feel relaxed together. Emotionally, meals in this relaxed state became expressions of fellowship.[82] Odysseus considered feasting to be the most pleasant occasion there was.[83] And this was probably true for Israelites as well. They would have enjoyed themselves by relaxing and feasting together. Fellowship is created or strengthened on the family level via shared drinking. Alcohol, as we shall see, marked the social sphere of the participants in significant ways.

Second, wine acts as a lubricant for social discourse. It eases communication by lowering inhibitions and, just as important, by marking boundaries of inclusion among the participants. Drinking, as Mary Douglas has argued, shapes a social context, not only by its impact on the central nervous system of those participating, but by the very fact of shared participation, which establishes a group identity.[84] The present analysis of the biblical texts concentrates on drinking as a social act. Biblical scenes of wine consumption, for sustenance or celebration, I argue, signify the esteem with which an interaction is held by its participants. Drinking, in other words, indicates the willingness of the participants to take the social risk. The vintner's trust dictates who will help in the vineyard (Judg 9:26) and share in its produce. The vintner drinks, in short, with those he trusts or wants to trust.

---

[77]Ross, "Meals," 316. See Gen 19:1-3; Judg 19:22; also, Mark 14:17.
[78]L. Grivetti, "Wine: The Food with Two Faces," in *OAHW*, 14.
[79]Grivetti, "Wine: The Food with Two Faces," 17. See table 1.1-1.2.
[80]Calabresi, "*Vin Santo* and Wine," 131.
[81]Grivetti, "Wine: The Food with Two Faces," 14.
[82]Ross, "Meals," in *IDB* 3:315.
[83]Homer *Odyssey* 9.5-6, 11.
[84]M. Douglas, introduction to *Constructive Drinking: Perspectives on Drink from Anthropology* (ed. M. Douglas; Cambridge: Cambridge University Press, 1987), 8.

The family was the primary locus of the vintner's trust, and we have to assume that the vintner himself drank the wine he produced. It becomes, then, a matter of discerning who, around the table, joined him. Women are depicted drinking in the Hebrew Bible (e.g., Judg 16:27; 19:19; Ruth 2:14; 1 Sam 1:9, 13-15, 18; Hos 2:7, 10 [ET 2:5, 8]; Amos 4:1; Jer 35:8; also Judg 12:13; Esth 1:9; 5:6; 7:2; Job 1:13, 18; Zech 9:17). These scenes sometimes carry negative valences. In Amos 4:1, for example, the women who call their husbands for drink are "cows of Bashan." In Hos 2:7 [ET 2:5] Hosea's wife, Gomer, has relied on lovers for her foodstuffs, including "my drink" (שִׁקֻּיָי). Her prostitution, not her drinking, however, is the cause of Hosea's/Yahweh's distress. For in her running around, she has failed to see that the real supplier of her "grain, wine, and oil" is and always has been Yahweh (Hos 2:10 [ET 2:8]). Hannah is mistaken by Eli for being drunk, though she is only praying (1 Sam 1:9, 13-15, 18). From these scenes, I surmise that wives did join their husbands in drinking wine, and that other issues besides their drinking make the wives in these three passages a target of prophetic taunting. At other times, drinking women are simply present in biblical narrative with no negative connotation. In later material, they even attend or host banquets with no editorial backlash (Judg 12:13; Esth 1:9; 5:6; 7:2; Job 1:13, 18).[85]

In this inclusion of women for social drinking, Israelite tradition differed from Greek and Roman traditions that speak of barring women from wine. Greek public drinking, as in the symposium at least, was strictly a male affair,[86] though private drinking customs may well have differed. Pliny states that in Rome, "women were not allowed to drink." He cites an instance where a man was acquitted of murder after he beat his wife to death for drinking wine from the vat.[87] While the biblical instances of women drinking are rare and not always positive, they surely lack the virulence of Pliny's tale. The examples of women drinkers are frequent enough to suggest that Israelite women were not forbidden to drink. In two cases, women are restricted from wine consumption:

---

[85]The Egyptian painting from the tomb of Khety at Beni Hasan depicts women at a banquet drinking: Darby et al., *Food: The Gift of Osiris*, 2: 585, fig. 14.13. In fact, one woman is apparently vomiting a dark liquid, while a friend or servant wipes her brow and extends a vessel towards her. Mesopotamian drinking customs too apparently included women, e.g., Assurbanipal's queen is drinking with him in the Nineveh relief: R. D. Barnett, "Assurbanipal's Feast," *Eretz-Israel* 18 (1985): 1*-6*. The wine lists of Nineveh mention women serving wine rations: Kinnier Wilson, *The Nimrud Wine Lists*. See Michalowski, "The Drinking Gods," 29-44, for numerous texts in which women drink with men.

[86]Burkert, "Oriental Symposia: Contrasts and Parallels," 7; Athenaeus notes that neither slave nor freeborn women drink wine: *Deipnosophistae* 10.429b; C. Robinson, *Everyday Life in Ancient Greece* (Oxford: Clarendon, 1933), 76.

[87]Pliny *Natural History* 14. 14.89; Darby et al., *Food: The Gift of Osiris*, 2:589.

Samson's mother, while pregnant with him (Judg 13:4), and the Rechabite wives and daughters (Jer 35:8). Their restriction, however, is something that is explicitly noted in the text, suggesting that abstention was unusual rather than customary. At least with the example of Samson's mother, a general abstention for pregnant and nursing mothers does not persist in later Jewish tradition. The Talmud, in fact, allocates an increased amount of wine consumption for nursing women on the grounds that it was beneficial for lactation.[88]

Since wine accompanied family meals and festivities, we would expect the woman and perhaps the children to be included in the drinking. Both women and children drink at meals in other patriarchally structured wine cultures such as modern France, Italy, and Greece. There exist only three biblical instances of children being associated with drinking, though their ages are not specified: the starving infants and babes ask their mothers for bread and wine in Lam 2:12; the Rechabite abstention from wine includes the children of the group (Jer 35:8); and Job's sons and daughters regularly feast together, but they are old enough to have their own houses (Job 1:4).

Deuteronomy 21:18-21 addresses the problem of a rebellious son who is drunken and gluttonous, but again age is not specified. Sons typically stayed under the authority of the father until the latter died.[89] If they lived and worked on the farm until their inheritance, then it is easy to surmise that a gluttonous, drunken son would be a problem, regardless of his age. Quite simply, he ingests too much of the farm's produce in a reckless manner and demonstrates his incompetence to one day manage the farm. His behavior, in other words, is much more than a social embarrassment. It threatens the livelihood of the family, and so the punishment recommended by this passage is severe, namely, death by stoning. Obviously, we cannot infer from this law that moderate or occasional drinking by children was sanctioned in ancient Israel. There simply are too few biblical texts to decide the issue. At the same time, however, there is no prohibition against children drinking[90] such as there was for Plato, who argued that boys under eighteen should not even taste wine.[91] Since wine was a basic element of the Israelite meal, there is no reason to assume children were forbidden from its use. The determining factor might have been practical, rather than ethical, dictated simply by how much wine was available.

---

[88]*b. Keth.* 65b. An Irish custom today is quite similar: pregnant women and nursing mothers are encouraged to drink a glass of Guinness stout daily to increase their milk.

[89]See chapter 2, pp. 63-68.

[90]Except for the Rechabite children (Jer 35:8) and Nazirite youths (1 Sam 1:11), in which the prohibition, of course, rests on grounds other than that of age.

[91]Plato *Laws* 2.666 a-b.

Unfortunately, the social quality of the Israelite daily meal remains obscure to the historian. We are reminded of Christopher Eyre's remarks that no one in antiquity bothered to write down poor daily life,[92] and so, we are forced to imagine its social dimensions. The cooperation between family members in agrarian households undoubtedly engendered a familial trust, through necessity and affection. And self-sufficiency remained an ideal for these families, evident in the field and vineyard, and by extension, at the table.[93] Families, in other words, ate what they spent the year growing. Sharing in those meals, then, was an expression of their cooperative livelihood. Trust is implicit, almost banal, even in the tedium of daily labors and meals among family members. The ideology of the family in the Hebrew Bible was in part formed from this basic economic reflex, that members relied on cooperation for their livelihood. Harmony, preferable no doubt for comfort in the home, also brought in the harvests most efficiently. We can assume that familial trust was a given for ancient Israel, though, as we shall see, it did not go undisturbed. A surprising number of stories with wine at their center involve a flaunting and betrayal of family trust.

Enjoyment through drinking was no doubt a fairly commonplace activity for families who had cultivated and harvested a vineyard and produced their own wine. Drinking that wine would likely have begotten a kind of shared pride and intimacy. A family unity was affirmed each time that people drank together from the fruit of their vines. In a sense, sipping wine was an implicit ritual reminder of their toil (Ps 128:2). Memories no doubt returned on occasion as they drank, memories of—whose feet had been in the press; who danced; what sorts of pests and climatic uncertainties had loomed; near misses with newly filled jars; a new participant, or a missing neighbor; maybe even a marriage engagement. There would quite naturally have been discussions comparing the taste and quantity of wine with last year's crop. The incidental chatter and habits are what comprised the culture, the everyday reminders of a shared livelihood, and marked a family as vintner farmers. These are the kinds of intimacies we imagine at a vintner's tables, though they go unrecorded in biblical texts. Nevertheless, we do have biblical scenes of family feasts, shorn of these details, but still reflecting intimacies. The family is shown celebrating with drink, and here the intimacies become more pronounced.

---

[92]Eyre, "The Agricultural Cycle," 186; see chapter 1, p. 22.
[93]Antoun, *Arab Village*, 9; Garnsey, *Famine and Food Supply*, 44, 56; S. B. Brush, *Mountain, Field, and Family: The Economy and Human Ecology of an Andean Valley* (Philadelphia: University of Pennsylvania Press, 1977), 69.

III. Banquets

Two types of banquets are preserved in biblical writings: one is termed a *mišteh* and the other a *marzēaḥ*. Small family farmers are portrayed drinking at the first kind, the *mišteh*. They might have participated at the *marzēaḥ* too, though the biblical evidence is meager for this type of banquet. Since the *mišteh* banquet is more frequent in biblical texts than the *marzēaḥ*, the following four parts (a, b, c, d) will concentrate on the biblical *mišteh*. The *marzēaḥ* banquet is discussed in section IV.

Banquets were enjoyed on the domestic level by the average family, as they were also by kings and the wealthy. These were occasions for festivity involving food, sumptuous at times, and wine, plenty of it. They resembled daily meals in that food and drink from the farm were consumed. Mary Douglas, in her study of the anthropology of dining, provides a useful perspective for analyzing Israelite family banquets. She has suggested that "the smallest, meanest meal metonymically figures the structure of the grandest."[94] To some extent, then, a banquet was a meal on a grander scale. It would likely have had the basic elements of a meal—bread, wine, and oil—but in greater quantities and with the addition of items not typically part of the daily meal, such as meat (Prov 23:20; Isa 22:13; Dan 10:3),[95] music (Isa 5:12; 24:8-9), and play (Gen 21:9; Exod 32:6; Judg 16:25).[96] Also, feasting would often include the same base group of participants, with the possible addition of others, and the banquet meal no doubt would have lasted longer than a daily meal.

A banquet also differed from daily dining and not simply by the quantities consumed. While enjoyment was a facet of the "smallest, meanest meal," it became a *central* activity of the banquet. In this sense, then, a banquet's festivity is not merely a daily, even incidental, pleasure writ large. Merrymaking was instead the intent, and wine's agency in this is obvious (2 Sam 13:28; 1 Kgs 4:20; Esth 1:10; Prov 15:15; Eccl 9:7; Isa 24:7). Indeed, the Hebrew term for banquet, מִשְׁתֶּה, is from the verb שָׁתָה, to "drink." Wine at a banquet was not only a detail of Mediterranean dining, though it was that. It became as well a *constitutive* element of the social gathering. Quite simply, drinking was done at the biblical *mišteh*, and probably a fair amount of it.

---

[94]M. Douglas, "Deciphering a Meal," in *Rules and Meanings: The Anthropology of Everyday Knowledge* (Harmondsworth, England: Penguin, 1973), 257.

[95]Homer *Odyssey* 9.159-163, Odysseus and his companions enjoy a feast on the island off the coast of the Cyclops' land, where they dine on unlimited wine and meat. Both elements in their abundance contribute to their pleasure, as does their leisure time. They feast all day until sunset.

[96]See below, pp. 233-36.

Festive drinking carries many symbolic and social meanings in any culture. My own premise for understanding biblical banquets is expressed by Mary Douglas: "the general tenor of the anthropological perspective is that celebration is normal and that in most cultures alcohol is a normal adjunct to celebration."[97] An increased intake of wine is, then, expected at the Israelite מִשְׁתֶּה. Wine's value as a social lubricant was sought, and so drinking was conspicuous. The increased drinking, in essence, turned a meal into a feast.[98] While a daily meal on occasion may have been transformed into a banquet by the levels of its wine consumption and festivity, if a banquet with these levels turned into a daily meal, presumably it would have been deemed a failure.

Wine consumption levels, then, at a banquet typically differed from those at the daily meal. While there are biblical texts detailing drunkenness (e.g., Gen 9:21; 1 Sam 1:13; 1 Kgs 16:9; 29:9; Jer 25:27; Nah 1:10), moderation (e.g., Ruth 3:7; 2 Sam 13:28; 1 Kgs 4:20; Esth 1:10; Prov 23:30; Eccl 9:7; Isa 24:7), as well as abstention (Lev 10:9; Num 6:3; Deut 29:6; Judg 13:4, 7; 1 Sam 1:15; Jer 35:6, 8; Ezek 44:21), none expressly deals with banquet protocol. The protocol Athenaeus offers for the Greek banquet can provide at least some idea of intake levels for merrymaking:

> Three bowls for the temperate; one to health,
> which they empty first, the second to
> love and pleasure, the third to sleep.
> (*Deipnosophistae* 2.36 b-c)[99]

And, in the *Odyssey*, it only takes three bowls of wine to cause the Cyclops to fall asleep.[100] The amounts of wine consumed, it would seem, varied by individual, occasion, availability, and, of course, bowl size.

The effects of wine on a drinker are most often described in the Bible as a "merry heart" (לֵב + טוֹב: Judg 16:25; 19:6, 9; 1 Sam 25:36; 2 Sam 13:28; Esth 1:10; Eccl 9:7) or "drunkenness" (שׁכר: e.g.,

---

[97]Douglas, *Constructive Drinking*, 4.

[98]Johnson, *Vintage*, 11; Burkert, "Oriental Symposia: Contrasts and Parallels," 13.

[99]He goes on to cover the next seven bowls. Amerine and Singleton provide a modern chart based on blood sugar levels: .03g, the drinker is "dull and dignified"; .05g, "dashing and debonair"; .15g,"daring and devilish" (and "under the influence" for driving standards); .30g, "delirious and disgusting"; .40g, coma; .45g, death: *Wine*, 326.

[100]Homer *Odyssey* 9.360, 371-74. Wine rations recorded for eighth-century BCE Calah, for those in service to the king of Assyria, seem to have been much lighter. They average 1.84 liters per ten people: Kinnier Wilson, *The Nimrud Wine Lists*, 5-6.

Gen 9:21; 43:34; 1 Sam 1:13; 1 Kgs 16:9; Isa 29:9; Jer 25:27; Nah 1:10). "Merry heart" seems to connote a general state of enjoyment and conviviality, though drunkenness may be included as well. For drunkenness, as a state in which the effects of alcohol are manifest in behavior, can still range from loudness, aggression, silliness, or charm, to a coma.[101] The Bible depicts certain specific behaviors caused by drunkenness, such as staggering (Isa 29:9), exposure (Gen 9:21, 23; Lam 4:21), nausea (Prov 23:34), vomiting (Jer 25:27; 48:26), sleep, and coma (Gen 9:21; 1 Sam 25:37; Jer 51:39, 57). That there are varying levels of intoxication is worth stating, for preferences and moods were likely to have played a role in determining the extent of festivity. In Plato's *Symposium*, for example, the participants discuss their drinking strategy for the evening and agree that they will drink only until it gives pleasure and not to the point of drunkenness.[102] Perhaps the biblical descriptions of a "merry heart" and drunkenness indicate a similar range of intoxication for ancient Israel.

The use of an intoxicant in social encounters, with its ability to loosen inhibitions and depress the central nervous system, can both solidify and threaten social relations.[103] The potential social benefit to using alcohol was an increased intimacy or the establishment of intimacy through a bond or agreement. The potential danger came in the form of trickery or betrayal in the relaxed atmosphere and feelings of goodwill that drinking engendered. Biblical traditions portray both the payoff and risk of drinking for the family.

Banquets, then, were occasions of increased social drinking, and they were a part of daily life. In the drinking and conviviality, they also provided a context for examining that life and talking about it. Hence, banquets were both features of Israelite daily life and the occasions for reflection upon it. Piotr Michalowski's insight into this function is instructive for analysis of the social effects of drinking in biblical banquets. A banquet, he suggests, is a "temporary structured bracketing of real world experience, a formalized lens that brings into focus social factors that cannot be easily expressed in everyday life."[104] Michalowski's view is more useful than theories that stress alcohol's function to blunt or escape reality. Magen Broshi, for example, interprets Israelite drinking to be a "universal need to escape—if only temporarily from the harsh grip of reality."[105] The participants depicted in biblical banquets, I shall argue, are not getting

---

[101] Amerine and Singleton, *Wine*, 326.

[102] They decide on this moderation in part because they were still suffering from the previous evening's drinking: Plato *Symposium* 176a, e.

[103] Grivetti, "Wine: The Food with Two Faces," 14. Wine's two faces are that it holds the potential for joy and disaster.

[104] Michalowski, "The Drinking Gods," 33.

[105] Broshi, "Wine in Ancient Palestine," 21.

together to escape their reality. Noah does drink to the point of
unconsciousness, and so escapes his newly postdiluvian world (Gen
9:21). Lot is intoxicated to the point that he is unaware of his
daughters' incestuous actions (Gen 19:31-35). These are stories of
drunken escapes, but they occur under cosmic duress. Noah, alone,
drinks after the world he knows has drowned; Lot's daughters believe
it is the end of the world, so they anesthetize their father in order to
propagate, i.e., to reconstitute their social world (Gen 19:32, 34).
Escape with alcohol is elsewhere reserved for times of duress (e.g.,
Prov 31:6) or admonished (Amos 6:6).

Banquet drinking differs from such crisis drinking.
Participants at a banquet drink together to forge or renew their
understanding of their social world. Hence, drinking in biblical
banquets represents, not an escape from, but an increased stake in that
social world. Social tensions come to a head and resolution with a
narrative efficiency that daily life undoubtedly did not share. The
following exegeses of biblical scenes of drinking focus on the social
effect—the celebration and change—for the family. There are several
occasions when banquets seemed to have played an important role in
the life of the Israelite family. They are: the vintage, the weaning of a
child, and the marriage of a family member.

a. The Vintage Feast

The vintage, as we saw in chapter 5, was an occasion for
celebration as well as labor. Music, dancing, and shouting are the
biblically attested expressions of joy for the grape harvest. Additional
expressions such as singing, giggling, and flirtations likely also added
to the overall pleasure of the vintage. The emotional energy of a
grape harvest remains high today as well. This is, in part, due to the
release of anxieties in getting this delicate crop to the press without
loss. The excitement and fun that a new year's batch of wine promises
add, as well, to the heightened emotions of the participants.

Festivity offered a communal outlet for this emotional energy
and so the grape harvest, as also the barley and wheat harvests, early
on appeared in the festival calendars of ancient Israel (Exod 23:14-19;
34:18-26; Lev 23:1-44; Deut 16:1-17). Since the harvests were
essential for survival, their successful completion no doubt elicited
relief. For this reason, they were not somber occasions marked only
by offerings to the deity. Instead, all three harvests were probably
marked with celebration feasts. In the Arab village of Baytin, for
example, after the wheat harvest and threshing, the owner of the crop
would slaughter a goat and feast with his family and anyone else who

helped in the labor. The feast often took place right on the threshing floor itself.[106]

Other peak periods of labor in ancient Israel, such as sheep shearing, concluded with wine celebrations (1 Sam 25:2-36; 2 Sam 13:23-28). Hence, it is reasonable to assume that the vintage, above all other harvests, would have included festive drinking. The grape harvest had the greatest potential for celebration because it brought in the festive yield, the juice that would "cheer God and mortals" (Judg 9:13). Exhausted vintners then were unlikely to miss the opportunity to cheer themselves and their coworkers with the product. Indeed, this may be understating the matter. Throughout France's history, Braudel notes, "the countryside resounded with merrymaking and near continuous feasting" during the vintage season.[107]

Judges 9:27, discussed in chapter 5, contains the one biblical description of a vintage feast, and so is worth citing again:

וַיֵּצְאוּ הַשָּׂדֶה וַיִּבְצְרוּ אֶת־כַּרְמֵיהֶם
וַיִּדְרְכוּ וַיַּעֲשׂוּ הִלּוּלִים
וַיָּבֹאוּ בֵּית אֱלֹהֵיהֶם וַיֹּאכְלוּ וַיִּשְׁתּוּ
וַיְקַלְלוּ אֶת־אֲבִימֶלֶךְ

They went out into the field and gathered
the grapes from their vineyards, trod them,
and praised. Then they went into the temple
of their god, ate and drank, and ridiculed
Abimelech.

(Judg 9:27)

This is the harvest and feast of the "lords of Shechem" and Gaal, whom they trust (v 26). Their "eating and drinking" is a feast that includes alcohol.[108] The feasting takes place after the grape treading and in a temple.[109] Virgil mentions restricting the treaders from eating and drinking while they were in the press, thereby suggesting that those around the press were already feasting.[110] No biblical tradition directly suggests that feasting began during the grape treading. Drinking during the Israelite vintage either went

---

[106]Antoun, *Arab Village*, 22. The Judean hill farmers studied by Turkowski also held a festival after the wheat harvest: "Peasant Agriculture in the Judaean Hills," 27.
[107]Braudel, *Identity of France*, 2:244.
[108]Dommershausen, "יַיִן," 62. Forbes, *Studies in Ancient Technology*, 3:110.
[109]See pp. 244-46.
[110]Virgil *Georgics* 2.6.

unrecorded or was not practiced. There were practical benefits, of course, to not drinking during this time. Alcohol would add an unpredictable element to wine production and detract from labor efficiency. Also, wine is a diuretic, so after a hot day in the sun, its consumption would further dehydrate workers and contribute to lethargy. Slipping, slumber, breaking jars all become the possibilities of drinking during work, yet they may have been tolerated. Aristophanes' vintner, Trygaeus, for example, remembers the vintage and its disarray with some fondness:

> wine-strainers, bleating
> flocks, the bosoms of women
> running errands to the fields,
> a drunken slave girl, an overturned jug,
> and many other good things.[111]

Drinking after grape treading is evident in the Judg 9:27 passage. Here relaxation is suggested by the note that the feasters made light of Abimelech. While gossip could, of course, occur in any social context, feasting provided an atmosphere especially conducive to it. Again, this is because part of the function of feasting lay in the catharsis of emotions—frustration, envy, hatred, no less than relief and joy. Shared food, wine, and relaxation in general often prompt the sharing of opinions. Alcohol facilitates that intimacy, to a point. The verse may simply describe a group of revelers mocking Abimelech, much as national leaders are today ridiculed in the bars and pubs of their countries.

Though Gaal is already trusted enough by the Shechem men to harvest and drink with them, his bond with them is solidified at this banquet. The participants bond first by harvesting, and then by drinking and mocking a common target. Their shared anti-Abimelech sentiment was strengthened and honed from mockery to a plot of insurrection. In this sense, a banquet and the preceding harvest have solidified relations between the Shechem lords and Gaal; Gaal is no longer an outsider in Shechem. However, the unity forged at the banquet costs the men dearly. They rebel against Abimelech, lose, and have to leave Shechem and their vineyards (vv 34-49). The vintage feast in Judg 9:27, then, first marked a social solidarity of having brought in the grape harvest, and then forged another, political solidarity that would eventually dissipate the group and have consequences to those vineyards. This vintage feast, in other words, was the farmers' last.

---

[111] Aristophanes *Peace* 535-37.

The celebrations of the vintage, such as singing, dancing, and shouting,[112] would quite naturally have extended to the evening meal, as Judg 9:27 perhaps suggests. Modern family vintages enjoy a night of feasting at the end of the harvest, complete with dancing and drinking.[113] Special dishes, clean clothes, and familial joy mark Adamo's vintage meal. The vintage is, for vintners such as Adamo, the occasion of the year. It marks a successful past year with its harvest, and has assured the family of wine for the following year. If the main Israelite feast was celebrated on the night of the vintage, then the wine in the vats or jars would have still been fermenting. The workers may have drunk some wine right from the vats to celebrate new wine and its fruity freshness. In addition, they may have reserved some wine from last year's harvest for the vintage feast. Continuity would be stressed by marking a successful harvest with the taste from the previous year's harvest. If the celebration was delayed or a matter of continuous days, then the fermenting wine would soon become available for drinking. The "new wine" of the year, drunk in fall banquets, likely did acquire a special term that is then reflected in one or more of the biblical terms for wine (e.g., תִּירוֹשׁ, עָסִיס).

Several biblical texts hint that feasting was a part of the overall celebration during the vintage. Isaiah 62:9 and Joel 2:24-26 associate the grape harvest with feasting in a general way as descriptions of the pleasures of farm life. In Isa 62:9, the promise is simply that those who gather the grapes will drink its wine. In Joel 2:24-26, with threshing floors full of grain and vats overflowing with wine and oil, one shall eat in plenty and be satisfied. Feasting is one of the general benefits of a grape harvest as these texts indicate; it may also be a facet of that harvest itself.

The passage in Isaiah recites an end to the joy that viticulture brought. It too is a general description of the pleasures of the farm, but it hints at merrymaking near the vines:

7 אָבַל תִּירוֹשׁ אֻמְלְלָה־גֶפֶן נֶאֶנְחוּ כָּל־שִׂמְחֵי־לֵב
8 שָׁבַת מְשׂוֹשׂ תֻּפִּים חָדַל שְׁאוֹן עַלִּיזִים
שָׁבַת מְשׂוֹשׂ כִּנּוֹר
9 בַּשִּׁיר לֹא יִשְׁתּוּ־יָיִן יֵמַר שֵׁכָר לְשֹׁתָיו

---

[112]Chapter 5, pp. 179-86.
[113]Friedl, *Vasilika: A Village in Modern Greece,* 32; Calabresi, *"Vin Santo* and Wine,"131. Feasts both private and public, for example, mark the vintage weekend in Napa Valley, California, in each town.

7. The wine dries up, the vine languishes,
all the merry-hearted sigh.
8. The mirth of the timbrels is stilled,
the noise of the jubilant has ceased,
the mirth of the lyre is stilled.
9. They will no longer drink wine with song,
date wine will be bitter to those drinking it.

(Isa 24:7-9)

The vintage gave the wine for the celebrations and meals of the coming year, and so it is probable that the grape harvest was the first occasion celebrated with drinking.

b. Weaning of a Child

While the vintage banquet was celebrated once a year, the Israelite family probably feasted more often, even for reasons unrelated to agricultural harvests. Family banquets in the Bible at times celebrate an event that altered family life. These banquets, then, differ from the seasonal regularity of harvest banquets. They mark rites of passage for the individual and family unit. One such event is portrayed in the story of Isaac's weaning in Gen 21:8-9. Abraham hosts the event:

וַיַּעַשׂ אַבְרָהָם מִשְׁתֶּה גָדֹול
בְּיֹום הִגָּמֵל אֶת־יִצְחָק
וַתֵּרֶא שָׂרָה אֶת־בֶּן־הָגָר הַמִּצְרִית
אֲשֶׁר־יָלְדָה לְאַבְרָהָם מְצַחֵק

Abraham made a great banquet
on the day Isaac was weaned.
And Sarah saw the son of Hagar
the Egyptian, Abraham's boy, playing.[114]

(Gen 21:8-9)

---

[114]J. A. Hackett, "Rehabilitating Hagar," in *Gender and Difference in Ancient Israel* (ed. P. Day: Minneapolis: Fortress, 1989), 20-21. On מְצַחֵק, see Joüon, *A Grammar of Biblical Hebrew*, § 52d: Piel as factitive, with the sense here then to laugh, make laugh, i.e., entertain, rather than mock.

Isaac is weaned, and this is an occasion for his father Abraham to make a "great banquet." Scholars note that weaning would likely have been in ancient Israel a "customary family feast."[115]   Although this passage is the only biblical example, weaning was doubtless marked as a rite of passage for the Israelite male child.

Unfortunately, no more details are provided for what distinguished Abraham's "great" banquet from an ordinary one.[116] Nor can we discern whether a great banquet would have been customary for a weaning or is rather a sign of Abraham's magnanimity, wealth, and prominence.   Still, several features are evident. First, and not surprisingly, the father initiates the banquet. Here, he apparently "made" whatever preparations were necessary for a banquet. As *paterfamilias*, he, rather than a wife or child, was most likely to determine the time and occasion for a banquet. He may have also had the primary responsibility for deciding what to serve (e.g., Gen 18:5-8; 19:3), depending on the availability of provisions and dining custom.[117]

That a son's weaning is cause for celebration is the next feature. Ishmael was the older son, and no such banquet is mentioned for him. The narrative omission is not all that surprising, as Ishmael's secondary status is pronounced in Gen 16:6-12 and Gen 21:9.[118]   A child's weaning occurred when he stopped feeding from his mother's breast, though an exact age is not given. 1 Samuel 1:23-24 mentions for Samuel's weaning only that he was young. 2 Maccabees 7:27 notes three years of nursing after nine months of pregnancy.   Various scholars estimate the weaning age to be between two and three years old, noting that this is typically the age for twentieth-century Palestinian women to wean their sons.[119] Weaning, then, was a rite of passage for the boy, an obvious shift toward independence from his mother.[120] It marked a son's survival from infancy and his separation from the mother into an eventual manhood. Another feature of this banquet for Isaac's weaning is that the entire family was involved. Indeed, the inclusivity is precisely the cause of the conflict; Sarah becomes angry when she sees Hagar's son "playing" (מְצַחֵק).

---

[115]J. Skinner, *A Critical and Exegetical Commentary on Genesis* (New York: Scribner's, 1910), 321; von Rad, *Genesis*, 231; C. Westermann, *Genesis II* (Neukirchen-Vluyn: Neukirchener, 1981), 2:414; Davidson, *Genesis 12-50*, 84.
[116]The rabbis of *Genesis Rabbah*, namely, R. Judah and R. Yudan, believe "great" to mean that God is present: *Genesis Rabbah*, 252.
[117]Queens Vashti and Esther make their own banquets (Esth 1:9; 5:4, 8).
[118]Though not in 17:18-20, 23-25.
[119]Skinner, *Genesis*, 321; Speiser, *Genesis*, 155; von Rad, *Genesis*, 232. See R. Patai, "Familism and Socialization," in *Readings in Arab Middle Eastern Societies and Cultures*, 579.
[120]Unwin notes that in France all the major rites of passage—first communion, wedding feast, then fishing and hunting, and maturity—are typically marked by libations: *Wine and the Vine*, 10.

מְצַחֵק, the Piel participle from the verb צחק, to "laugh," describes Ishmael's action (Gen 21:9). The boy apparently is both playing by himself and laughing or, in a causative sense, playing with Isaac and making him laugh. This play or entertainment is elsewhere an element of banqueting. This verb in the Piel occurs in three banquet scenes altogether. In Exod 32:6, the people hold a feast[121] after worshipping the golden calf:

וַיֵּשֶׁב הָעָם לֶאֱכֹל וְשָׁתוֹ וַיָּקֻמוּ לְצַחֵק

The people sat to eat and drink
and rose to play.

From this description, the "play" was an activity different from and subsequent to their eating and drinking. The participants, in other words, went away from the table to play. In Judg 16:25, as a Philistine party is underway, the prisoner Samson is brought out to entertain them:

וַיְצַחֵק לִפְנֵיהֶם

He played before them.

In this case, someone plays to entertain the revelers whose "hearts are merry" (v 25). These three biblical banquet scenes with צחק suggest that play, frivolity, was a facet of banqueting. Ishmael, then, would not seem to be guilty of behavior inappropriate to a banquet. Sarah's anger must have another cause, but to understand this, the importance of צחק in banquet scenes must first be addressed.

There are three other instances of a Piel form of צחק, viz., Gen 19:14; 26:8; and 39:17.[122] In Gen 19:14, Lot's sons-in-law think he is "joking" (מְצַחֵק) when he warns them to leave Sodom.[123] In 26:8, the grown and married Isaac is "playing" with Rebekah, and Abimelech understands from this action that she is Isaac's wife and not his sister. In 39:17, Potiphar's wife accuses Joseph of "playing with" or

---

[121]Early in the morning, perhaps to increase their sin with a daylong affair. Cf. Isa 5:11.

[122]All three passages are from the Yahwist, as Exod 32:6 is also thought to be.

[123]He was not.

"mocking" her, and some sort of conjugal intimacy is meant, since she has taken his garment to frame him. Thus, in 26:8 and 39:17, צחק suggests an intimacy between a man and a woman.[124]

The sense of צחק in Gen 21:9, though, does not seem to carry a sexual nuance. We expect laughter to be a facet of banqueting, and see that it was in two other banquets. Ishmael, then, is engaged in expected banquet festivities. Laughter and play were part of banqueting and would naturally be so for family occasions too. They were a desired part of banqueting, a part of the joy. Quite simply, צחק means to "make laugh" or "intensify laughter." As such, tickling might be what Ishmael is doing in Gen 26:8.[125]

Sarah's anger is motivated by something other than the demonstration of laughter at a family banquet, since this is expected. As these other occurrences of צחק demonstrate, the laughter implies a considerable intimacy between the characters experiencing it. It is not the laughter that angers Sarah, it is the presumption of intimacy that Ishmael's playing signals. The reason for Sarah's animosity toward Hagar is stated quite clearly in v 10: "this slave's son will not inherit along with my son, Isaac."[126] Ishmael would detract from Isaac's inheritance. Sarah uses the family celebration, a time of conviviality, to assert her displeasure. Ishmael's laughing or playing with her son is to her a reminder of a longstanding domestic concern. Ishmael, in other words, is enjoying himself at a family banquet *as if* he belonged there, with rights to family feasts and an inheritance. Minimally, in Sarah's eyes, Ishmael's enjoyment intrudes into Isaac's feast, his rite of passage toward manhood, toward his eventual inheritance of Abraham's estate.

Yet another allusion is evident in Ishmael's play. Isaac has gotten his name from צחק, laugh,[127] since, almost ludicrously, Abraham and Sarah were 100 and 90 years old respectively (17:17) when Isaac was born. In this particular banquet מְצַחֵק, then, something a little more important than festive playing or a wife's erratic rage is indicated. As Jo Ann Hackett has nicely argued, when Ishmael is מְצַחֵק, "he is not just laughing or playing—he is also 'Isaac-ing.'"[128] Sarah, in other words, saw Ishmael "doing something to indicate he was just like Isaac" and this angered her.[129] On the level of the narrative, of course, the pun with Isaac's name is evident. It

---

[124]The NRSV therefore renders the participle in 26:8 as "fondling," and the construct in 39:17 as "insult."

[125]Abimelech, in 26:8, might have observed not an overt sexual act between Isaac and Rebekah, but a quality of their tickling, which suggested that the two were married.

[126]Also in Gen 16:4-6.

[127]Gen 17:17, 19 (P); 18:12 (J).

[128]Hackett, "Rehabilitating Hagar," 20-21.

[129]Hackett, "Rehabilitating Hagar," 21.

matters little whether Ishmael was self-consciously posturing a status that was not his, or was simply a boy tickling his brother. What does finally matter is that there can be only one Isaac; Ishmael must not play at family banquets anymore. He must, in fact, go.

Ishmael's play at the banquet triggers the anger of Sarah, but it does not explain its intensity. In one verse, she demands the child and her mother leave the home altogether (v 10). Her harsh reaction has led interpreters to argue for additional nuances to Ishmael's מְצַחֵק. It becomes the child's hubris,[130] his "mocking,"[131] or even his seducing married women.[132] Furthest from the text is R. Ishmael's reading, viz., that Sarah saw "Ishmael building little altars, hunting locusts, and offering them up."[133] The problem in the text for these rabbinic interpreters, however, is not really Isaac's behavior; rather, it is the extent of Sarah's anger.

The severity of Sarah's response does have an explanation provided by the context. Since drinking wine was the focused activity of a banquet (מִשְׁתֶּה), and this was a "great banquet" (מִשְׁתֶּה גָדוֹל), we can surmise that the participants, Abraham, Sarah, and Hagar, if not the boys too, were drinking wine. Drinking, then, influenced both the confrontation and its resolution. Alcohol lowers inhibitions, and when it does, emotions that have lain dormant find expression. On the narrative level, Sarah's animosity toward Hagar began the moment Hagar conceived Ishmael (Gen 16:4-6). If Sarah had drunk even a moderate amount of wine, three bowls by Greek banquet standards, then her longstanding resentment towards Hagar could have boiled under wine's influence into open confrontation. Verse 10 may simply describe the quick flash of rage that alcohol can unleash.

To conclude: Laughter, otherwise desired in biblical banquets (Judg 16:26; Exod 32:6), becomes unwelcomed in this celebration of Isaac's weaning. Drinking at this banquet does not facilitate celebration and unity, only discord and expulsion. A web of familial inclusion has tightened for the primary couple and their son, but excluded forever the other half of the family. The celebration itself would certainly have come to a close after the confrontation. Sarah's mood had been the first to alter upon seeing Ishmael playing; Hagar and Ishmael themselves could no longer be festive, as they faced an expulsion in the morning; and Abraham too became distressed (v 11). Even Isaac would have been affected. His future inheritance is guaranteed, but his banquet is cut short, and he loses a brother with whom to play. Though Sarah's conflict is resolved, this is one family

---

[130]Hackett, "Rehabilitating Hagar," 21.
[131]S. Hirsch, *Genesis* (London: L. Honig & Sons, 1963), 354.
[132]*Genesis Rabbah*, R. Aqiba (citing Gen 39:17), 253.
[133]*Genesis Rabbah*, R. Ishmael, 253.

banquet that ended in discord and uneasy silence, rather than satisfaction.

c. Marriage Feasts

Acquiring women, through marriage or concubinage, is cause for celebration with banquets in several biblical stories (Judg 14:10-11; Gen 29:22; Judg 19). A marriage feast, when successful, brought two families together as it celebrated the union of a couple. Drinking facilitated the new bond and celebrated the rite of passage of two individuals. In Judg 14:10-11, Samson's marriage feast is slated to last seven days, and seems to include only men (v 10 בַּחוּרִים; v 11 מֵרֵעִים). However, his wife is present in some way, since she weeps for the duration of the feast before Samson (v 17) and is blackmailed by the men to betray him (v 15).[134] And, that she is his wife during the feast indicates that this wedding feast at least was not celebrated previous to consummation, as a kind of bachelor party, but was enjoyed as the couple came together.

At Jacob's wedding banquet also, only men are specified:

וַיֶּאֱסֹף לָבָן אֶת־כָּל־אַנְשֵׁי
הַמָּקוֹם וַיַּעַשׂ מִשְׁתֶּה

Laban gathered all the men of
the place and made a banquet.
(Gen 29:22)

For both Samson and Jacob, then, the woman's relatives seem to provide the feast, but how pervasive a custom this was cannot be said. King Ahasuerus, as groom and king, provides for the banquet[135] in which Esther becomes his queen (Esth 2:17-18). In the New Testament, in the Gospel of John, the groom himself provides the wine for the wedding of Cana (John 2:9-10). In Samson's case, the Timnite vineyards (v 5)[136] presumably were the source of the wine.

Drinking was of course a main activity of the wedding banquet. In Samson's feast, riddling is mentioned and it is likely that other such

---

[134]Either she gets the riddle's answer out of Samson, or they burn her and her father's house.

[135]Like Gen 21:8, the largesse of this banquet apparently went beyond regular banqueting.

[136]Also the source of Samson's riddle material.

drinking games were a part of general feasting traditions. The riddling goes too far, with frustration, enmity, and annulment of the marriage the result. Festive drinking results not in the unity of a new marriage, but in fractured relations between the parties.

The description of events at Jacob's wedding feast also shows the effects of drinking. In the evening, Laban gave Leah, the wrong daughter, to Jacob, and Jacob does not realize until morning, post coitus, that she is not Rachel (Gen 29:23, 25). Leah might have been veiled,[137] but it is hard to believe that drinking did not sufficiently impair Jacob's awareness to the point that he could make such a mistake. Additional identity markers, such as voice, smell, body shape, and gestures certainly could have helped him.[138] The mistake, after all, cost him another seven years of labor to acquire Rachel. The bond between families that a wedding banquet would characteristically forge was, in this story, forced. Jacob remains for another seven years, but not as a result of the goodwill celebrated at his wedding banquet. Laban used the occasion of the festivity to trick Jacob.[139] A social unity was established, but with the wrong bride. Drinking at wedding feasts apparently caused confusion on occasion in ancient Greece too, for Plato cautioned both bride and groom against excessive drinking during their wedding banquet.[140]

The goodwill and celebration enjoyed at a wedding feast could continue beyond it. In Judg 19, the Levite travels to his concubine's father to get back his woman after a dispute. The father and Levite spend first three days feasting (v 4), and then two more (v 8). The men drink together for five days (vv 4, 6-7, 9). That they were having fun is evident by the duration of their feasting, their "merry hearts" (vv 6, 9: וַיִּיטַב לְבָבֶךָ),[141] and that the reason for the visit, the concubine, goes unmentioned. It is an enjoyable and successful feasting period, for she does depart with the Levite when he eventually leaves (v 10). There is no formal resolution mentioned, simply that somewhere in the feasting the estrangement is resolved. Samson also went to his father-in-law after his wedding feast, a fiasco of betrayal and ill will. Subsequent feasting is precluded, as the father had given his daughter to another, Samson's friend, and so Samson is effectively cut out of the family.[142]

---

[137]Speiser considers this to be a sufficient explanation for mistaken identity in the marriage bed (cf. Gen 24:65): *Genesis*, 225.

[138]A narrative disinterest in the woman's characterization contributes to the ambiguity of the scene.

[139]Jacob, given his history tricking Esau, certainly had it coming.

[140]Plato *Laws* 6.775 b-c. Hence, Greek women do seem to have drunk at some public occasions.

[141]Verse 6 varies slightly: וַיִּטַב לִבְּךָ.

[142]The father offers his younger daughter, but Samson refuses.

d. Sibling Feasts

Several biblical scenes depict banquets among siblings without the parents present. These are: Gen 43:34; 2 Sam 13:28; and Job 1:4-5, 18. Drinking in these three instances provides the occasion for bonding and repositioning. I imagine as well that it would have been a chance to speak candidly about the parents though this is not captured in the texts.[143] In Gen 43:34, for example, Joseph makes a feast for his eleven brothers. An occasion is not specified, nor is a feast expected during a famine (v 1), even though Joseph is in a position of power. Neither Egyptian protocol nor family recognition explains this feast, since the brothers were not invited to one during their first trip to Egypt. In fact, Joseph, to say the least, was brusque: he

וַיִּכֵּר וַיִּתְנַכֵּר אֲלֵיהֶם וַיְדַבֵּר אִתָּם קָשׁוֹת

treated them like strangers
and spoke harshly with them
(Gen 42:7)

and then threw them into prison for three days (v 17).

In Gen 43:34, Joseph hosts a banquet and the twelve brothers all get drunk: "they drank and got drunk with him," וַיִּשְׁתּוּ וַיִּשְׁכְּרוּ עִמּוֹ. The banquet is unusual in several respects. It has all the signs of hospitality and inclusion: copious amounts of drink, even in a time of severe famine; an invitation; tables; even a seating arrangement. Yet, there are as many markers of distance. It *is* a family reunion, but only one participant knows that, and he is sitting at a separate table (v 32). In fact, Joseph is at one table, Egyptians are at a second, and the eleven brothers sit at a third and in the order of their birth (v 33). Their shared drinking, nevertheless, marks a certain bond among the participants. The brothers celebrate the securing of food and their relationship with the man who will provide it. Joseph celebrates a reunion, even needing to weep in another room at one point (v 30). He may also be celebrating his power over his brothers, as he delays

---

[143] Joseph does ask about their father before the banquet: Gen 43:27-28. Some discussion of the father might naturally have arisen during the meal in which Jacob steals Esau's birthright, Gen 25:30-34. Though Esau "ate and drank" (v 34), he did so under duress, not in celebration.

revealing himself and places them in jeopardy by planting "stolen" goods on them twice (42:25; 44:2).

From Joseph's perspective, the banquet is a family reunion, an occasion to reestablish intimate ties, yet its markers of exclusion are more telling. Joseph's deception approaches the sadistic as he forces the brothers to risk their father's favorite son, Benjamin (44:12). The father he asked about in 43:27 Joseph would now torment by seizing Benjamin. Judah states twice that Jacob would die if Benjamin does not return from Egypt with his brothers (44:22, 31) and poignantly concludes: "I fear to see the suffering that would come on my father," פֶּן אֶרְאֶה בָרָע אֲשֶׁר יִמְצָא אֶת־אָבִי (v 34). At this plea, Joseph finally reveals his identity (45:1).

Joseph frames his brothers by planting his silver cup into their sacks (44:2). His act further takes advantage of the trust implicit in banqueting, because he uses as his prop a vessel present during the shared drinking. With Joseph's own cup in their possession, then, the brothers are framed both for stealing and for violating the communal trust of a banquet. Presumably, this would be the same cup Joseph used in the previous night's banquet.[144] The brothers leave at dawn (v 3), and so they would have been hung over from the drunkenness of the previous evening. It is a grievous theft in Joseph's eyes because of the cup's function, not its value. By "stealing" his drinking cup from the night before, the brothers have insulted the host, and "returned evil for good" לָמָּה שִׁלַּמְתֶּם רָעָה תַּחַת טוֹבָה (v 4).

Joseph, in essence, accuses his brothers of theft at a banquet. Such an act signals a breach in the trust established the night before. His own cup, with which he presumably drank and perhaps toasted at the feast, is an emblem of his hospitality. Joseph's feigned indignation is over this breach of trust, and not the theft, since, as the brothers remark, they had earlier returned the (also planted) money (v 8). A similar, though more glaring, breach occurs in a banquet description from Mari, of a man who had drunk with the king:

> He drank from the cup and raised it . . . .
> But he went back on his word and he
> defecated into the cup from which he had drunk;
> He is hostile to his majesty.[145]

Indeed! Michalowski notes the insult to hospitality in this symbolic

---

[144]And routinely divines with, v 5.
[145]Cited in Michalowski, "The Drinking Gods," 35. Michalowski notes that the full text has not yet been published: 35, n. 32.

reversal of consumption and bodily expulsion.[146] Joseph stops short of such a colorful accusation, but not by much. And the breach in hospitality is of course on *his* end, by his framing and deceiving his guests, his own brothers. Reunion, relief, and forgiveness all occur for these brothers, and a bond is established anew, but despite a banquet not because of one.

Deception, separate tables, and even different languages (42:23) all severely restrict the intimacy potential of this banquet. The brothers get drunk together, but there can be no sharing of recollections or amends. After the banquet, a repositioning of family order does occur, as does a bond of perhaps bittersweet reunion. The brothers, who all sat according to the order of their birth at the banquet, are forced to profess that the youngest, Benjamin, but also Joseph, are the most valued members of the family. Joseph achieves, then, a reconciliation and reunion with his brothers, but not without asserting his supremacy over them. In this respect, Joseph is not that different from the boy who dreamed he was a sheaf towering over his brothers or the center of the celestial world (37:7, 9).

Deception occurs at another sibling banquet, that in 2 Sam 13:28. There, David's sons come together for a feast in the season of sheep shearing. David declines the invitation, as he does not want to burden Absalom, who is hosting: לֹא נִכְבַּד עָלֶיךָ (v 25). He, thus, indicates that hosting could be costly, even for a king's son. Festive drinking is evident at this sibling feast. Wine is the only element of the feast mentioned, and the murderers are instructed to wait until Amnon's "heart is merry with wine," רְאוּ נָא כְּטוֹב לֵב־אַמְנוֹן בַּיַּיִן (v 28). One may compare this drinking episode with another sheep-shearing banquet, in which the rich Nabal has a "merry heart" and is "very drunk": וְלֵב נָבָל טוֹב עָלָיו וְהוּא שִׁכֹּר עַד־מְאֹד (1 Sam 25:36). Amnon, then, is relaxed, tipsy, or even drunk. He has had enough wine, at any rate, for the effects to be noticeable as a cue to his assassins, and for him, thus, to be taken by surprise and murdered. Amnon's guard is down in part because of the drinking and in part because of the trust implied by a family banquet.

Actually, Amnon gets his comeuppance. For he too had taken advantage of the trust implicit in a family meal (without wine) to rape his sister Tamar (vv 8-19). Harm is not expected at family dining, as Tamar herself voiced: "such a thing is not done in Israel" (v 12). The expectations of trust and conviviality at a banquet make it a prime context for mischief and misconduct. Job, for example, always offered sacrifices by proxy for his children in case they had sinned while feasting (Job 1:4-5, 18). Harm comes to them anyway, and a messenger informs Job that all his children have died while feasting (v

---

[146]Michalowski, "The Drinking Gods," 35-36.

19). Oddly, David is also told that all his sons have been killed at
Absalom's banquet (2 Sam 13:30). Later, Jonadab can soften that Job-
like blow by reporting that only Amnon is dead (v 32-33). These two
post-banquet reports seem to reflect an Israelite understanding that
anything could happen during feasting. Nabal's banquet had also
proved fatal to him. In the morning, after the wine has left him:

וַיָּמׇת לִבּוֹ בְּקִרְבּוֹ וְהוּא הָיָה לְאָבֶן

his heart died within him
and he became like a stone.
(1 Sam 25:37)

This verse may well describe an alcohol-induced coma, even though
the text mentions that the wine "went out" from him, בְּצֵאת הַיַּיִן מִנָּבָל.
This odd phrase might signal part of Nabal's demise, from
intoxication, to morning vomiting, coma, and, ten days later, death (v
38).[147]

In sum, feasting, ceremonial drinking, was for the family,
probably enjoyable and relaxing. Family members could spend time
together, indulge in wine and food, exchange perhaps songs or
riddles, and talk in a way that everyday life did not always allow.
Banquets could celebrate and strengthen family relations: Sarah and
Isaac are made more secure; Jacob and Leah are wed; Esther becomes
Ahasuerus's queen; the Levite and his concubine's father remain close;
Joseph and his brothers are reunited; and Job's children enjoy
themselves by rotating their celebrations in the brothers' houses.
Festive drinking also brought emotions and tensions to the surface,
and so it entailed a risk: Hagar and Ishmael are expelled; Samson loses
his wife; Jacob is tricked; and Amnon is killed. These things
happened at a banquet precisely because of the apparent safety and
goodwill that drinking engendered. Drinking wine facilitated that
goodwill and relaxation, but could not guarantee them. In these
biblical banquet scenes, drinking remains an act of intimacy, because
it signals a trust that *should* be there, but may not be.

---

[147]With Yahweh's help.

IV. *Marzēaḥ*

The second kind of banquet mentioned in the Bible, besides the *mišteh*, is the *marzēaḥ*. It occurs far less frequently than does the *mišteh*, and so its social function is difficult to gauge.[148] The term *marzēaḥ* (מַרְזֵחַ) is used in two biblical texts, Amos 6:7[149] and Jer 16:5. In Amos 6, a lavish banquet is underway, complete with wine from unusual bowls, מִזְרְקֵי יַיִן,[150] oil, and meat (vv 4, 6), as well as music, singing, and "lounging on ivory beds" (vv 4-5; cf. Esth 1:6). In Jer 16, mourning rituals, such as gashing, consoling, and serving a meal to the bereaved, comprise the *marzēaḥ* (vv 6-7). Drinking vessels occur in both texts ("cup," כּוֹס: Jer 16:7; "bowl," מִזְרָק: Amos 6:6), and so drinking was a central activity of the *marzēaḥ*. In fact, conspicuous drinking is likely present in these texts. The Amos passage details the presence of wine and drinking vessels (v 6), the addition of food (v 4), and lounging (vv 4, 7). The Jeremiah text also suggests considerable drinking as it mentions a מִשְׁתֶּה (v 8), "sitting to eat and drink" (v 8) and a "sound of joy," וְקוֹל שִׂמְחָה (v 9), either in contrast or parallel with the מָרְזֵחַ of vv 5-7. Drinking, then, is evident for the *marzēaḥ* banquet, yet its social function seems considerably different from what it was for the *mišteh*.

Other ancient Near Eastern texts mention the *marzēaḥ* and provide valuable analogues for interpreting the Israelite *marzēaḥ*. They offer insight for delimiting how a *marzēaḥ* might have differed from other drinking feasts of ancient Israel. One Ugaritic text[151] has El enjoying a *marzēaḥ* in his house:

> *yṯb. b mrzḥ*
> *yš. [y]n. ῾d šb῾.*
> *trṯ. ῾d škr*

he sits in his *mrzḥ*
he drinks wine until sated,
fresh wine until drunken.

---

[148]J. C. Greenfield, "The *Marzēaḥ* as a Social Institution," *Acta Antiqua* 22 (1974): 451-55; M. H. Pope, "A Divine Banquet at Ugarit," in *The Use of the Old Testament in the New and Other Essays* (ed. J. Efird; Durham, N.C.: Duke University Press, 1972), 170-203; Porten, *Archives From Elephantine*, 179-86; P. J. King, "The *Marzēaḥ*: Textual and Archaeological Evidence," *EI* 20 (1989) (Yigael Yadin memorial volume) 98*-106*.

[149]It is vocalized מִרְזַח in the Masoretic Text.

[150]See above, p. 213.

[151]*KTU* 1.114. lines 3-4, 16.

El drinks in fact beyond plenty, into a drunkenness that lands him in his own excrement.[152] In his state of inebriation, he is "like a dead man,"[153] though it is not the severity of use that characterizes the *marzēaḥ*. Nabal too had been like a dead man (1 Sam 25:37) and then became one (v 38) the morning after his banquet, a *mišteh*. Drunkenness is clearly a facet of the *marzēaḥ* as depicted in this Ugaritic text.[154] Also, a cultic context is implied with the deity drinking.

A cultic context is not explicit in either of the biblical texts. The mention of meat in Amos 6:4 might allude to sacrifice, but it could as easily be only what the feasters were eating. The other hint of a cultic context for the Israelite *marzēaḥ* is Jer 16:5, which mentions a *marzēaḥ* house, בֵּית מַרְזֵחַ. The phrase בֵּית מַרְזֵחַ may reflect either a specific architectural structure for feasting,[155] or more generally any house in which such feasting occurred.[156] At Dan, a rectangular room was excavated above an altar room in the temple precinct. Nearby, a bronze-plated bowl dating to the eighth century BCE was discovered.[157] A later bowl, dating to the fourth century BCE,[158] has a Phoenician inscription with the term *mrzḥ*:

קבעם אנחן ‖ ערבת למרזח שמש

Two cups we offer to the
marzēaḥ of Shamash.[159]

---

[152]Like the Mari text cited above, excrement is the symbolic reminder of the somatic process and the end result of feasting.

[153]*KTU* 1.114, line 21: *il km mt*.

[154]Pope argues that since this is the only Ugaritic text explicitly stating that a god drank to the point of inebriation, it must be an extraordinary occasion "reflecting a human affair in which it was deemed proper or obligatory to drink to excess." For him, such an occasion was mourning: "A Divine Banquet at Ugarit," 178.

[155]*PRU* III, 88; Porten, *Archives From Elephantine*, 179; T. Lewis, "Banqueting Hall/House," in *ABD*, 1:581; King, "The *Marzēaḥ*: Textual and Archaeological Evidence," 98*.

[156]Holladay, *Jeremiah I*, 471.

[157]A. Biran, "Tel Dan, 1984," *IEJ* 35 (1985): 186-89; A. Biran, "The Dancer from Dan, the Empty Tomb and the Altar Room," *IEJ* 36 (1986): 168-87.

[158]N. Avigad and J. C. Greenfield, "A Bronze *Phialē* with a Phoenician Dedicatory Inscription," *IEJ* 32 (1982): 121.

[159]Avigad and Greenfield, "A Bronze *Phialē* with a Phoenician Dedicatory Inscription," 120-21.

Though its provenance is unknown, it would seem to be a vessel for cultic offering.[160]

Biran, the excavator of Dan, has suggested that the bronze bowl and rectangular room in the temple precinct may be evidence of an Iron Age *liškah*, a banquet hall.[161] A banquet hall for drinking is a possible context for the *marzēaḥ*.[162] Such a room is mentioned for the thirty feasters at Saul's anointing in 1 Sam 9:22: לִשְׁכָּה. Jeremiah 35:2 mentions a לִשְׁכָה of the temple in which wine is served.[163] Ezekiel mentions thirty such sacred precinct halls for the temple (40:17). Greek traditions of drinking halls apparently share this notion in the term *lesche*.[164]

Wine was included in offerings in the cultic legislation in the Hebrew Bible (Exod 29:40; Lev 23:13; Num 15:7, 10; 28:14; Deut 32:38). Wine, along with grain and meat, was offered to the deity as sacrifice. Some texts also show drinking and feasting in the temple. The Shechemite vintners had feasted in the temple of their god, (Judg 9:27: בֵּית אֱלֹהֵיהֶם). Elkanah and his family feasted at the temple of Shiloh on a yearly basis (1 Sam 1:7: בְּבֵית יְהוָה). Drinking wine was part of the feast (vv 9, 18), with drunkenness a possibility, and Eli wrongly suspects Hannah of it (v 12). Drinking wine in the temple is elsewhere mentioned in Amos 2:8; Deut 14:23, 26; and Jeremiah 35:2. In the Jeremiah text, the activity seems to involve only wine drinking and not food. Jeremiah is instructed to invite the Rechabites and

וַהֲבִאוֹתָם בֵּית יְהוָה אֶל־אַחַת הַלְּשָׁכוֹת וְהִשְׁקִיתָ אוֹתָם יָיִן

> bring them into the temple of Yahweh,
> into one of the banquet halls, and offer
> them wine to drink.[165]

(Jer 35:2)

If the *marzēaḥ* of Jer 16:5 had a cultic context, then the בֵּית מַרְזֵחַ could refer to the temple or banquet hall.

---

[160]Avigad and Greenfield, "A Bronze *Phialē* with a Phoenician Dedicatory Inscription," 118.

[161]Biran, "Tel Dan, 1984," 187.

[162]Holladay, *Jeremiah I, 1-25*, 470; Bright, *Jeremiah*, 111; Lewis, "Banqueting Hall/House," 581.

[163]See below.

[164]Burkert, "Oriental Symposia," 17.

[165]They refuse. In fact, they neither drink wine nor cultivate vines as their ancestor commanded them, and so the Rechabites become for Jeremiah an example of loyalty (vv 5-19).

The social group involved in a *marzēaḥ* is hard to assess. Wealth is evident in the lavishness of the feast in Amos 6:1-7, and in Jer 16:5-9 mourners are mentioned. The four other *mrzḥ* Ugaritic texts mention *mrzḥ* men, indicating that a defined social group was involved in this type of feasting.[166] One mentions a *mrzḥ* house for the men. In two of the texts,[167] vineyards are associated with the *mrzḥ*. Various scholars have suggested that these vineyards either provided the wine for the feasting or were jointly owned by the members of a *mrzḥ*.[168] An Aramaic ostracon from Elephantine (fifth century BCE) details the discussion of money for a *mrzḥ*.[169] A later text, from Palmyra, of the first to third centuries CE, lists nine members of a *mrzḥ*.[170] Three additional *mrzḥ* texts from Palmyra list the leader of such a group.[171] These ancient Near Eastern parallels attest to the existence of a *mrzḥ* tradition involving wine consumption, and suggest a level of social cohesion and organization for the function of this feast.

If a similar level of social organization lay behind the lavishness of the feast described in Amos, the text does not make this explicit. There is no mention of who provides the wine, oil, meat, or furniture and bowls. Amos 6:6 mentions bowls for wine. These were standard drinking vessels in antiquity, as stated earlier in this chapter.[172] However, Amos's term is unique: מִזְרְקֵי יַיִן. The root, זרק, meaning to "throw," to "sprinkle," is usually found in cultic contexts (e.g., Exod 27:3; 38:3; 1 Kgs 7:40; Jer 52:18), where presumably water, blood, incense, and wine are scattered as part of the offering. A bowl for sprinkling could then have been larger than an average domestic drinking bowl. Amos's admonition against the banqueting people may assume a cultic context in which the participants are misusing temple vessels (cf. Dan 5:23). Alternately, their feasting practices may not be the issue, but rather their timing or apparent insensitivity in a time of ruin (v 7).

Jeremiah 16:5-9 mentions a house for a *marzēaḥ* and one for a *mišteh*. Feasting, clearly, is one of the activities in which Jeremiah is not to participate. In this passage, the first banquet, the *marzēaḥ*, is associated with mourning (vv 5-7). The second, the *mišteh*, is associated with the "sound of joy": וְקוֹל שִׂמְחָה. In contrast to the Amos

---

[166] *PRU* III, 88, 130; IV, 230; *UT* 2032.

[167] *PRU* IV, 230; *UT* 2032.

[168] King, "The *Marzēaḥ*: Textual and Archaeological Evidence," 98*. One text from Palmyra does credit the leader individually with providing wine for the year. See Porten, *Archives from Elephantine*, 180, 182.

[169] Porten, *Archives from Elephantine*, 179.

[170] Porten, *Archives from Elephantine*, 182. The Palmyra inscriptions all mention priests, so here too a cultic context is indicated.

[171] See Porten, *Archives from Elephantine*, 182. Tesserae found at Palmyra mention a leader of the *mrzḥ* as well.

[172] See above, pp. 212-14.

passage, Jer 16:5-9 does note something of the social nature of the participants. In the *marzēaḥ,* a feast of bread and wine for mourning one's father and mother is mentioned (v 7). And the *mišteh* feast with sounds of joy includes a bride and groom (v 9). The Jeremiah passage suggests, then, the possibility that the *marzēaḥ* no less than the *mišteh* could have been familial occasions, providing a context for the full range of emotions brought on by life events. Death was perhaps an occasion for drinking together as much as a marriage was. It certainly marked a rite of passage. In this case, a banquet would have brought comfort and support for the bereaved (Jer 16:7), much as a wake does today. It too would have served a vital cohesive function, precisely by uniting a group shaken by the loss of one of its members. Social cohesion is always a desired function of banqueting, but in this circumstance, that function is heightened, and it provides a coping strategy for death. Drinking together would have facilitated emotions of grief no less than joy elsewhere, and have given the participants, in this case, the surviving family members, a chance to recollect, make plans, and simply spend time together in a relaxed fashion.

At any rate, these feasts, the *marzēaḥ* and the *mišteh,* are the details of daily life that Jeremiah is to relinquish. Marriage, fatherhood, grief rituals, and feasts are apparently all the treasured parts of a life with hope in the future. Jeremiah's abstention from all these plainly demonstrates his hopelessness in life as usual for Jerusalem. Banquets, of mourning and of joy, were valued parts of the daily life of even the doomed people of Jerusalem.

Conclusion

The grape vine was widely cultivated in ancient Israel by the Iron Age, and so viticulture became an essential component of the culture reflected in the Bible. Though not native to the Levant, grapes were adapted into the farmer's agricultural calendar, complementing, rather than competing with, the other main harvests of barley, wheat, summer fruits, such as figs and pomegranates, and olives. The vine offered the farmer a plant that would grow during the long summer period of annual drought. It, like horticulture in general, enabled the farmer to diversify his crops. At first, grapes (and olives) minimized a farmer's risks from drought and famine. But quickly, they added much more. They eased his existence. And, he could increase his vineyard, his livelihood from within; simply by cutting branches off his vines, he planted more.

Vines gave a wine that he could exchange for more or different foods and for tools, or that he could enjoy at the vintage, at the table, at his wedding, and up to his death. They gave a tasty addition to his diet in the form of grapes in the fall and raisins perhaps throughout the year. Most simply, when the soil cracked with dryness in the summer drought, that same semi-arid land gave in the vine something truly miraculous, a juice from sun-cracked soil. The vine was an object of esteem and even wonder in ancient Israel, even if the farmers had lost sight of its newness, worn down by fatigue and heat. Its value still emerges in the biblical appropriation of this plant. It is evident in Judg 9:13, where the production of wine is more noble than kingship, simply for the joy it brings. The vine, when asked to rule, asks only "what, would you have me cease making wine that cheers both God and mortals?" The question was rhetorical for ancient Israel, so welded into its psyche was viticulture.

Vineyards demanded labor, care, even love, and brought a measure of economic stability, along with joy, relaxation, and celebration. With this much human investment at stake, the vineyard offered an additional yield beyond that of grapes and wine. It offered too a rich source of imagery for the biblical writers to exploit for their theological and other ideological agendas. In such a way, vineyards truly did shape the biblical landscape, literally and figuratively—literally, since they comprised part of the economic regimen of daily life, and figuratively, since the writers so often drew on their metaphoric potential.

In Isa 5, for instance, vine tending was used to demonstrate the long-term, but not everlasting, care of Yahweh for his people. The poem provides details of viticultural tasks for historical reconstruction and sheds light on the esteem with which vineyards were held. Because the vintner in Isa 5 was the deity himself, Isaiah reflects the value of vine tending to ancient Israel. His vintner's care would resonate with the audience's own practices. Establishing that identification deftly enabled Isaiah to make his theological case, viz., that disobedience is literally "rotten" (as worthless grapes) and has its dire consequences. Isaiah's use of viticultural metaphor was, then, effective rhetorically, but also revealing. For the Song reflects the care, bordering on passion, of the vintner. Yahweh's love, according to Isaiah, was best expressed in the love a vintner felt for his vineyard.

This Song is not the lone example of enlisting viticulture in the service of theology. In a number of other biblical texts, Yahweh acts as vintner in relationship to people. He transplants the initial vine stock (the people Israel) into the ground Canaan (Ps 80:8-16; Jer 2:21); he harvests grapes (his people) by separating the good from the bad (Isa 65:8); and he stomps grapes (various peoples) in a winepress for gruesome punishment (Isa 63:2-3; Lam 1:15). This image of Yahweh gleeful and spattered from treading his human crop in a winepress is every bit as horrifying a divine image as that of the Canaanite goddess Anat "plunged knee-deep into the soldiers' blood."[1] Grape treading shared with human slaughter the copious splattering of red liquid, and so became a natural, if horrifying metaphor. In all of these instances of Yahweh's behavior, the divine tasks of love and destruction are conveyed through metaphors of vineyard maintenance. As we shall see below, serving and drinking wine also functioned as metaphors for Yahweh's actions. But these metaphoric meanings can only come into sharp focus through study of the actual processes

---

[1]*KTU* 1.3, lines 14-15; M. D. Coogan, ed. and trans., *Stories from Ancient Canaan* (Philadelphia: Westminster), 90.

involved in ancient vine tending and wine production, as the present
book has sought to do.

Most farmers likely did not begin their own vineyards as
Yahweh does in Isa 5. They, like Naboth, were more likely to have
inherited the farm and its vineyard from their fathers before them (1
Kgs 21:3). The importance of patrimonial inheritance is underscored
in the Deuteronomist's tale of Naboth's vineyard. Its importance for
an aging farmer with a vineyard was perhaps heightened still. For the
farmer gave his house and the fields to his son(s) as an inheritance, to
be a livelihood and home. With a vineyard, he gave the very plants he
had planted, some of which gave fruit and some still too young to do
so. The vine to which Judah would tie his donkey in Gen 49:11 was,
in a sense, already there, planted and standing in his father's vineyard.
There is a concrete and poignant power to the image of a grape vine
for all it can summon in the imagination: the delicacy of the fruit, the
promise of wine, and the memory that one's own father had planted it.
In the story of Naboth's vineyard, we can glimpse the importance of
such an inheritance. And, as with Isaiah, so for the Deuteronomist
telling the Naboth story, the choice of location, viz., a vineyard, is not
incidental.   A vineyard best illustrates what is at stake, viz., the
importance of inheritance, and the pain and offense of its having been
thwarted.

Israelite farmers, taking over their fathers' vineyards, would
not likely have had to select the site, clear the soil of stones, or build
installations, as Yahweh does in Isa 5. They were trained on vine
cultivation from their youth, by their fathers, so that by the time they
came to manage their vineyard, they were intimately familiar with the
skills necessary to tend it.   While they did not have to start their
vineyard, they would have spent their lives renewing it. A vineyard is
in a state of constant renewal, as shoots are taken from older vines for
new plantings.   Indeed, this is one of the true assets of vine
tending—one of the features that made it successful—that expansion
came from within the vineyard. A farmer could always expand his
vineyard simply by cutting shoots from his current plants and planting
them and waiting three to four years. We saw in chapter 6 that 275
vines would yield on average 694 liters of wine a year. A farmer's
decision to expand would secure him more wine, and require of him
the patience to wait three to four years for the new vines to yield
fruit.   These decisions would have been part of his agrarian
routine—some arrived at while tending his vines, some perhaps settled
at the family table by discussion.

The present project has been an exercise in social history,
demonstrating how the practice of vine tending shaped an Israelite
farmer's life, from his work to his social relations. The life of the
Israelite farmer was marked in all manner of ways by his vines, many

of these ways being captured, if refracted, in biblical traditions. He undoubtedly pruned, planted new vines, and hoed the soil several times a year to keep it from reverting to "thorns and thistles" (Isa 5:6). In the spring, he would start to smell and see the vine blossoms that signaled already the fall grape clusters. As the fall approached, he would wait and test for the ripening, the *véraison*, tasting berries for their ripeness and shooing away pesky foxes who wanted to do the same. There must have been a seasonal excitement and anticipation of this harvest. Other harvests, such as the barley and wheat harvests, were more important for survival as they brought the food for the year. But the grape harvest brought juice, sweetness, alcohol, and the promise of celebrations and relaxation. It brought a harvest that could "cheer God and mortals" (Judg 9:13). It also brought dancing, celebrations, the giggling of children and of adults, and flirtations.

Preparations of jar cleaning, basket mending, and making sure that the winepress was ready and clean were no doubt a part of the increased activity before the vintage. The vintner had a whole range of viticultural accoutrements—baskets, pruning hooks, strainers—and a variety of vessels, such as bowls, decanters, dipper juglets, and wineskins for travel. These were used throughout the year, probably on a daily basis, but were readied for the vintage. And, once it arrived, once he knew by the taste in his mouth that the bursting grapes were ready, the vintage began, a shared labor and a shared joy. Music, dancing, feasting all accompanied wine production. There were the details—of treading, of staining his legs and robes blood red, of stirring the must, and fending off intruders—and there were the emotions—the joy, relief, exhaustion. In all these ways, then, the tasks involved in vine tending marked the Israelite's daily life. In yet another major way, viticulture marked that life and Israelite culture, and that, obviously, was through drinking the wine.

Beginning with the celebrations of the vintage, which might have lasted for several days, the Israelite could plan or look with anticipation at the coming year's celebrations. A successful harvest was a relief and a guarantee of joyous celebrations in the year ahead. Other occasions were likely celebrated with wine, such as other harvests, and rites of passage. Seasonal drinking—after peaks of intensified labor, like sheep shearing and the grape, barley, and wheat harvests—differed somewhat in function from that during rites of passage. Both, though, celebrated an occasion and renewed a social cohesion among participants.

Seasonal drinking marked the success of a work period and early on became a part of the festival calendar of ancient Israel.[2] It eased the physical and emotional tensions of the agrarian cycle at

---

[2]Exod 23:14-19; 34:18-26; Lev 23:1-44; Deut 16:1-17.

regular intervals. Drinkers could afford to linger awhile in wine's sedative properties,[3] content with their accomplishment and temporarily free from subsequent chores. Emotionally, festive drinking in such a relaxed state became an expression of fellowship.[4] The participants, then, would take a break from routine as they marked a seasonal pattern of their livelihood. Drinking, here, in Joseph Gusfield's term, "cues" the transition from labor to ritualized celebration.[5] In terms of function, not frequency, it is akin to the modern "happy hour." Intoxication at seasonal banquets, then, is not the breakdown of social order, but rather celebrates its basis, in livelihood. Relaxation "under the influence" affirms the past accomplishment and renews energy for upcoming labor periods. It temporarily punctures routine in order to rejuvenate it.

Seasonal festivals offered a communal outlet, a catharsis of relief and joy, for the emotional energy of agricultural labor. Anticipation, laced with anxiety and hope, at these peak periods gave way to relief and joy, and hence the cathartic aspect of intoxication no doubt played a pronounced role. The Israelite drank his wine, at the end of exhausted days, with his family, and at village banquets; sometimes to relax, sometimes for joy to gladden his heart, and sometimes to excess, to the point of needing some help home. His drinking, no less than his pruning or transplanting or jarring wine, were part of the everyday tasks that made up his world.

Wine is an effective social lubricant for at least two reasons. It eases communication by lowering inhibitions and, just as important, by marking boundaries of inclusion among the participants. Drinking, as Mary Douglas has argued, shapes a social context, not only by its impact on the central nervous system of those participating, but by the very fact of shared participation—a group identity is established.[6] Drinking participants, now as then, demonstrate their trust of each other to lower their inhibitions and relax in shared drinking. One drinks in short with those one trusts or wants to trust.

Wine facilitated social interaction. Drinkers relaxed, laughed, played, and sang together. Alcohol lowered inhibitions, and so shared drinking became a context for expressing emotions that otherwise might have gone unexpressed, causing Pliny later to remark *in vino veritas*, "in wine there is truth."[7] This was probably the experience of the Israelites on occasion as well. Alcohol does not create

---

[3]Grivetti, "Wine: The Food with Two Faces," 14.
[4]It was, for these reasons, a pleasant, highly anticipated occasion. See Ross, "Meals," 315; Homer *Odyssey* 9.5-6, 11.
[5]J. Gusfield, *The Culture of Public Problems: Drinking, Driving and the Symbolic Order* (Chicago: University of Chicago Press, 1981).
[6]Douglas, Introduction, *Constructive Drinking*, 8.
[7]Pliny *Natural History* 14.28.141.

emotions; it can, however, lay them bare and distort them.   Hence, drinkers took a risk each time they drank together, a risk for greater cohesion and intimacy, and also for the increased potential of betrayal or trickery. With heavy drinking, that risk only increased.[8] And the dangers implicit with heavy drinking are evident in biblical scenes of festive drinking: Hagar and Ishmael were expelled (Gen 21:14); Jacob got the wrong wife (Gen 29:23, 25); Samson lost his bride as soon as he had acquired her (Judg 14:20); and Amnon was killed on orders by his brother Absalom (2 Sam 13:28).   Drinking facilitated goodwill and relaxation, but could never guarantee them.

In the stories of Noah, Lot, Elah, and Ben-hadad, alcohol drunk to excess serves to incapacitate these men so that consequences befall them: Noah is (somehow) dishonored by Ham (Gen 9:21-22); Lot is bedded by his daughters to secure the progeny of Moab and Ammon (Gen 19:31-35); Elah is murdered in a military coup d'état (1 Kgs 16:9), and Ben-hadad is caught by surprise by the Israelite army (1 Kgs 20:12, 16).   Yet not all biblical drunks are incapacitated. Uriah, for example, though plied with wine by King David, still refuses to sleep with his wife Bathsheba, causing David to hatch his more desperate, murderous plot (2 Sam 11:13).

In the cases of King Elah and Ben-hadad, some sort of character critique is implied by their intoxication.   Elah is caught off guard through his drunkenness and killed.   And since this is virtually the only detail we are given on his reign, Dtr's moral disapproval is evident. Still, it is Elah's "idolatry and leading Israel into sin," and not his intoxication, that are the stated reasons for Yahweh's anger (1 Kgs 16:13).   In the story of Ben-hadad's demise, licentiousness is more apparent, with the detail that Ben-hadad and his men were getting drunk in their tents at noon (1 Kgs 20:16).   Drunkenness during military exploits (cf. 2 Sam 11:11) and at midday (cf. Exod 32:6; Isa 5:11) was certainly viewed as folly or wicked.

Wine could leave one open to disaster and so became associated with licentiousness in some passages.   It could "take away understanding" (Hos 4:11, 18); make a fool of one (Prov 20:1); lead to an addictive disregard of livelihood (Prov 21:17; 23:21); and bring on hallucinations, nausea, and just plain sorrow (Prov 23:29, 32-34). The deleterious effects of intoxication were known to the biblical writers, who warned against them: staggering (Isa 29:9), exposure (Gen 9:21; Lam 4:21) nausea (Prov 23:34), vomiting (Jer 25:27; 48:26), and coma (1 Sam 25:37; Jer 51:39, 57).   This cautionary, negative wisdom, then, is a natural sort of common wisdom in drinking cultures.

---

[8]Grivetti, "Wine: The Food with Two Faces," 14. As seen earlier, wine's two faces are precisely that it holds the potential both for joy and for disaster.

Too much wine could lead to moral blindness, and the prophets especially warned of this risk (e.g., Isa 5:12; 28:7-8; 56:11-12; Amos 6:6; cf. Prov 31:4). Intoxication at that point could then serve as a moral symbol to impugn the character of the disobedient people. It also became a metaphor for divine punishment itself, where the people would stagger and pass out, but not from wine, as they might have preferred (Isa 29:9-10).

The tradition of Yahweh's "cup of wrath," from which he would force people to drink, is widespread in the Bible (e.g., Ps 75:9; Isa 51:17, 21; Jer 13:13; 25:15-28; 49:12; Lam 4:21; Ezek 23:31-33; Obad 16; Hab 2:15-16; Zech 12:2). These are harsh, bitter images of forced drinking, which leaves its victims retching, helpless, and guilty before an avenging God. Yahweh is host, but at a banquet to which no one wants to come. These images of drunkenness as punishment and the cautionary traditions against intemperance have often been used to argue that the Bible has a generally negative stance on drinking, but this is not supported by study of biblical materials or analysis of the social world of ancient Israel.[9]

Finally, the drunkenness of Noah is a special case. Since Noah is the Bible's first vintner, his scene of drunkenness merits some attention. After disembarking from the ark:

וַיָּחֶל נֹחַ אִישׁ הָאֲדָמָה וַיִּטַּע כָּרֶם
וַיֵּשְׁתְּ מִן־הַיַּיִן וַיִּשְׁכָּר וַיִּתְגַּל בְּתוֹךְ אָהֳלֹה

Noah, a farmer, was the first to plant a vineyard.
He drank from the wine and became drunk, and
lay exposed in his tent.

(Gen 9:21-22)

Noah definitely got drunk, and the Yahwist here is unflinching in his description of a passed out, exposed father. This is the Yahwist's story of the origins of viticulture. In this tale, then, viticulture was

---

[9]For a nice example of how biblical texts have been marshaled in arguments against drinking, see E. A. Watson, *Religion and Drink* (New York: Burr Printing House, 1914). A recent tractate form of argument against "the drinking problem" can be seen in S. Bacchiocchi, *Wine in the Bible: A Biblical Study on the Use of Alcoholic Beverages* (Berren Springs, Mich.: Biblical Perspective, 1989). For a thorough study on the history of how alcohol has been viewed as a threat to American religion, see N. H. Clarke, *Deliver Us From Evil: An Interpretation of American Prohibition* (New York: Norton, 1976). Also useful is W. J. Rorabaugh's *The Alcoholic Republic* (New York: Oxford University Press, 1979), which provides an historical essay on America's ambivalent relationship to alcohol.

brought about by the only human worth saving in Yahweh's eyes (Gen 7:1). Noah, the "righteous" (צדיק‎,Gen 6:9), cultivated a vineyard and then drank its wine. The story reflects the miracle of juice from the dry, harsh land. In Gen 3:17-19 and 4:12, farmers stumble and toil with soil that is unwieldy and dry. Adam would toil forever in that soil to get it to yield food; Cain would be cut off completely from it. As expressions of the semi-arid geological conditions of ancient Israel, these etiologies are apt. After banishment from the garden of Eden, it was not until Noah that things start to change for this dry farming. They change first by divine overkill, in a flood that takes 150 (Gen 8:3) or 40 (Gen 8:6) days to abate. As a primeval story, the writers apparently thought in extremes—from a parched and punished ground to one submerged in a flood in which no bird can find a twig to perch on.

With Noah, some of the extremes themselves cease. God even repents and vows not to send such a flood of destruction again. Noah no longer has to be the only righteous human in order to buy his right to survive. His drunkenness goes unpunished by the deity. After the harrowing ark trip, he settles by planting a vineyard. The act of vine cultivation, as I argued in chapter 3, demonstrates sedentary life, peace, and enough stability to commit the next four years and more to the land. Noah indeed remains righteous for this act of faith, a trust in a God who had earlier cursed and then drowned that land.

Noah is the first to plant a vineyard and so is also the first to become intoxicated. He drank alone, and this is nowhere else described in the biblical traditions. There is social fallout, for a son, Ham, sees him naked (9:22). Two other sons, Shem and Japheth, cover Noah (v 23). Wine brought with it social consequences. Here, it affected the family in the demonstration of filial loyalty, and in betrayal by Ham. The risks of viticulture are evident even in this story of origins. Noah is exposed to Ham's betrayal, but he suffers no long-term damage. His drunkenness elicits the sons' reactions, thereby ushering in the preferred order of sons, Shem and Japheth over Ham/Canaan. This scene is reminiscent of the Ugaritic tale of Aqhat, where the dutiful son who will become the father's heir will help his father when the need arises and "hold his hand when he is drunk, support him when he is full of wine."[10] Ham/Canaan failed to respond to the father's drunkenness and so is punished, while Shem and Japheth perform the filial duty of sons to an inebriated father and so gain dominance over their brother. They are the righteous sons of a uniquely righteous father. The story functions to legitimate Canaan's lesser status, rather than to mock intoxication.

A similar etiological function is at stake in Lot's story. The story of Lot's drunkenness functions as a political satire on the origins

---

[10]*KTU* 1.17, lines 30-31; Coogan, trans., *Stories from Ancient Canaan*, 33.

of the Moabites and Ammonites, by incest. Lot is intoxicated to the point that he is unaware of his daughters' incestuous actions (Gen 19:31-35). Escape with alcohol is elsewhere reserved for times of duress (e.g., Prov 31:6; Jer 25:27) or is admonished (Amos 6:6). Lot's daughters, like Noah, were acting under cosmic distress. They believe it to be the end of the world, so they anesthetize their father in order to propagate the species (Gen 19:32, 34). While these stories are certainly not showcases of Israelite ethical wisdom, neither are they stern fables against drinking, as scholars sometimes assume.[11] Instead, intoxication is put to etiological purpose to explain the origins of nations considered inferior to Israel.

The story of Noah's vineyard is short, yet it manages to capture some of the realities of viticultural life: "planting" vine stock rather than seed, drunkenness and its symptomatic loss of control, a family context, and, implicitly, patrimonial inheritance. It also manages to capture the pleasure viticulture brought to ancient Israel. In an earlier verse of J, Noah's name is explained:

וַיִּקְרָא אֶת־שְׁמוֹ נֹחַ לֵאמֹר זֶה יְנַחֲמֵנוּ מִמַּעֲשֵׂנוּ
וּמֵעִצְּבוֹן יָדֵינוּ מִן־הָאֲדָמָה אֲשֶׁר אֵרְרָהּ יְהוָה

He named him Noah, saying, "from the ground
that Yahweh cursed, this one will bring us relief
from our work and from the toil of our hands."
(Gen 5:29)

Viticulture did bring relief to Israelite farmers: relief from the difficulties of farming a semi-arid land with long summer droughts, relief in the form of liquid and variety in the diet, and relief in the merrymaking of wine consumption. Hence, even though the story of Noah's vineyard is short and anticlimactic, coming as it does after the

---

[11]von Rad, *Genesis*, 224; Broshi, "Wine in Ancient Palestine," 21; Ross, "Wine," 851; Dommershausen, "יין" 63; M. Greenberg, "Drunkenness," in *IDB* 1:872. Reactions against this position in commentaries somewhat belie this tendency: C. Westermann, in his commentary on Noah and Lot, is at pains to stress that the intoxication was not reprehensible: *Genesis 1-11: A Commentary* (Minneapolis: Augsburg, 1984), 487; *Genesis 12-36*, 313. W. Brueggeman asserts that Noah's plight is "not a negative comment on drinking, alcohol, or drunkenness": *Genesis: A Bible Commentary for Teaching and Preaching* (Atlanta: John Knox, 1982), 89. Von Rad argues that since Noah is the first to drink wine, the reader "must on no account morally condemn his drunkenness": *Genesis*, 136. Others ignore it completely: so on Noah, Speiser, *Genesis*, 62-63; so on Lot, merely noted by Speiser, *Genesis*, 145; Davidson adds only the cryptic "there are those for whom the pilgrimage of faith ends tragically": *Genesis 12-50*, 78.

flood, it does manage to reflect a good bit about a vintner's world. And, since viticulture was such an essential facet of Israelite culture, this story truly does belong in the primeval history, as one of the etiological stories of Israelite culture.

Bibliography

Abd Er-Raziq, M. "Die altägyptischen Weingärten (k3nw/k3mw) bis zum Ende des Neuen Reiches." *Mitteilungen des Deutschen archäologischen Instituts. Abteilung Kairo* 35 (1970): 227-47.

_____. "Bemerkungen zum Verhältnis des k3mw- und 't-nt-ht Gartens." Pages 38-21 in *Karl-Richard Lepsius (1810-1884): Akten des Tagung anlässlich seines 100. Todestages, 10-12.7.1984 in Halle.* Berlin: Akademie-Verlag, 1988.

Aesop. *Aesop's Fables.* Translated by T. James. London: John Murray, 1848.

Aesop. *Aesop's Fables.* Translated by J. Keller and L. C. Keating. Lexington, Ky.: University Press of Kentucky, 1993.

Aharoni, Y. "Hebrew Ostraca from Arad." *IEJ* 16 (1966): 1-7.

_____. "The Samaria Ostraca: An Additional Note." *IEJ* 12 (1962): 67-69.

_____. *The Land of the Bible.* Translated by A. F. Rainey. 2d ed. London: Burns & Oates, 1979.

_____. "Arad: Its Inscriptions and Temple." *BA* 31 (1968): 2-32.

_____. "Three Hebrew Ostraca from Arad." *BASOR* 197 (1970): 16-42.

Aharoni, Y., M. Evenari, L. Shanan, and N. H. Tadmor. "The Ancient Desert Agriculture of the Negev: An Israelite Agricultural Settlement at Ramat Matred." *IEJ* 10 (1960): 23-36, 97-111.

Ahlström, G. W. "Wine Presses and Cup-Marks of the Jenin-Megiddo Survey." *BASOR* 231 (1978): 19-49.

Albenda, P. "Grapevines in Ashurbanipal's Garden." *BASOR* 215 (1974): 5-17.

Albright, W. F. "The Site of Tirzah and the Topography of Western Manasseh." *JPOS* 11 (1931): 241-51.

_____. "The Gezer Calendar." *BASOR* 92 (1943): 16-26.

_____. "The Excavation of Tell Beit Mirsim." Vol. 3. *Annual of the American Schools of Oriental Research.* Nos. 21-22. New Haven, Conn.: American Schools of Oriental Research, 1943.

_____. "The Phoenician Inscriptions of the Tenth Century B.C. from Byblus." *JAOS* 67 (1947): 153-60.

_____. *The Archaeology of Palestine.* Baltimore: Penguin, 1960.

_____. *Yahweh and the Gods of Canaan.* London: Athlone, 1968.

_____. *Archaeology and the Religion of Israel.* 5th ed. Garden City, N.Y.: Anchor Books, 1969.

Albright, W. F., and J. L. Kelso. *The Excavation of Bethel: 1934-1960.* Cambridge, Mass.: American Schools of Oriental Research, 1968.

Amerine, M. A., and V. L. Singleton. *Wine: An Introduction.* 2d ed. Berkeley: University of California Press, 1965.

Amiran, R. *Ancient Pottery of the Holy Land: From Its Beginnings in the Neolithic Period to the End of the Iron Age.* Jerusalem: Massada, 1963.

_____. "A Note on the 'Gibeon Jar.'" *PEQ* 107 (1975): 131-32.

_____., ed. *Early Arad.* Jerusalem: Israel Exploration Society, 1978.

Ammar, H. "The Social Organization of the Community." Pages 109-34 in *Readings in Arab Middle Eastern Societies and Cultures.* Edited by A. Lutfiyya and C. Churchill. The Hague: Mouton, 1970.

Amouretti, M.-C., and J.-P. Brun, eds. *Oil and Wine Production in the Mediterranean Area. Bulletin de Correspondance Hellénique: Supplement* 26. Paris: Boccard, 1993.

Anderson, F. I., and D. N. Freedman. *Amos.* Anchor Bible 24A. New York: Doubleday, 1989.

Antoun, R. T. *Arab Village: A Social Structural Study of a Transjordanian Peasant Community*. Bloomington: Indiana University Press, 1972.

Applebaum, S., S. Dar, and Z. Safrai. "The Towers of Samaria." *PEQ* 110 (1978): 91-100.

Aristophanes. *Clouds; Wasps; Peace*. Edited and translated by Jeffrey Henderson. Loeb Classical Library. Cambridge, Mass.: Harvard University Press, 1998.

Aschenbrenner, S. "A Contemporary Community." Pages 47-63 in the *Minnesota Messenia Expedition: Reconstructing a Bronze Age Regional Environment*. Edited by W. A. McDonald and G. R. Rapp, Jr. Minneapolis: University of Minnesota Press, 1972.

Athenaeus. *Deipnosophistae*. Translated by C. Gulick. 7 vols. Loeb Classical Library. London: William Heinemann, 1928.

Austin, G. *Alcohol in Western Society from Antiquity to 1800*. Santa Barbara: ABC-Clio, 1985.

Avigad, N. "Two Hebrew Inscriptions on Wine Jars." *IEJ* 22 (1972): 1-9.

Avigad, N., and J. C. Greenfield. "A Bronze *Phialē* with a Phoenician Dedicatory Inscription." *IEJ* 32 (1982): 118-28.

Avimelech, M. "Geological History of the Yarkon Valley and Its Influence on Ancient Settlement." *IEJ* 1 (1950-51): 77-83.

Ayalon, E. "Tel Qâsile." *Excavations and Surveys in Israel* 13 (1995): 51.

Bacchiocchi, S. *Wine in the Bible: A Biblical Study on the Use of Alcoholic Beverages*. Berren Springs, Mich.: Biblical Perspective, 1989.

Badler, V. R. "The Archaeological Evidence for Winemaking, Distribution and Consumption at Proto-Historic Godin Tepe, Iran." Pages 45-65 in *The Origins and Ancient History of Wine*. Edited by P. McGovern, S. Fleming, and S. Katz. Luxemburg: Gordon & Breach, 1995.

Badler, V. R., P. McGovern, and R. H. Michel. "Drink and Be Merry! Infrared Spectroscopy and Ancient Near Eastern Wine." *MASCA Research Papers in Science and Archaeology* 7 (1990): 25-36.

Baines, J., and J. Malék. *Atlas of Ancient Egypt*. Oxford: Phaidon, 1984.

Baly, D. *The Geography of the Bible*. 2d ed. New York: Harper & Row, 1979.

Barkay, G. "A Group of Stamped Handles from Judah." (Hebrew) *EI* 23 (1992) (Avraham Biran volume): 113-28.

―――――. "The Iron Age II-III." Pages 302-73 in *The Archaeology of Ancient Israel*. Edited by A. Ben-Tor. New Haven: Yale University Press, 1992.

Barnett, R. D. "Assurbanipal's Feast." *EI* 18 (1985): 1*-6*.

Barthélemy, D., and J. T. Milik. *Discoveries in the Judaean Desert I: Qumran Cave I*. Oxford: Clarendon, 1955.

Basset, F. W. "Noah's Nakedness and the Curse of Canaan: A Case of Incest?" *VT* 21 (1971): 232-37.

Ben-Tor, A., ed. *The Archaeology of Ancient Israel*. New Haven: Yale University Press, 1992.

Ben-Tor, A., and R. Rosenthal. "The First Season of Excavations at Tel Yoqneʿam, 1977." *IEJ* 28 (1978) 57-82.

Ben-Tor, A., and Y. Portugali. *Tell Qiri: A Village in the Jezreel Valley: Report of the Archaeological Excavations 1975-1977*. *Qedem* 24. Jerusalem: The Institute of Archaeology, The Hebrew University of Jerusalem, 1987.

Biran, A. "Tel Dan, 1984." *IEJ* 35 (1985): 186-89.

―――――. "The Dancer from Dan, the Empty Tomb and the Altar Room." *IEJ* 36 (1986): 168-87.

―――――. "Tel Dan, 1987, 1988." *IEJ* 39 (1989): 93-96.

Bleibtreu, E. *Die Flora der neuassyrischen Reliefs*. Vienna: Verlag des Instituts für Orientalistik der Universität Wien, 1980.

de Blij, H. J. *Wine: A Geographic Appreciation*. Totowa, N.J.: Rowman & Allanheld, 1983.

Bloch-Smith, E. "The Cult of the Dead in Judah: Interpreting the Material Remains." *JBL* 111 (1992): 213-24.
_____. *Judahite Burial Practices and Beliefs about the Dead.* Sheffield, Eng.: JSOT Press, 1992.
Borowski, O. *Agriculture in Iron Age Israel.* Winona Lake, Ind.: Eisenbrauns, 1987.
Bottéro, J. "Boisson, banquet et vie sociale en Mésopotamie." Pages 3-13 in *Drinking in Ancient Societies.* Edited by L. Milano. History of the Ancient Near East/Studies-6; Padova: Sargon, 1994.
du Boulay, J. *Portrait of a Greek Mountain Village.* Oxford: Clarendon, 1974.
Bradford, J. "Fieldwork on Aerial Discoveries in Attica and Rhodes." *The Antiquaries Journal* 36 (1956): 172-80.
Braemer, F. *L'architecture domestique du Levant à l'âge du fer.* Paris: Éditions Recherche sur les civilisations, 1982.
Braudel, F. *The Identity of France.* Translated by Sian Reynolds. 2 vols. New York: Harper & Row, 1988-1990.
Bright, J. *Jeremiah.* Anchor Bible 21. Garden City, N.Y.: Doubleday, 1965.
_____. *A History of Israel.* 3d ed. Philadelphia/London: Westminster/SCM, 1981.
Brockelmann, C. *Lexicon Syriacum.* Halle: Niemeyer, 1928.
Brooke, G. J. "4Q500 1 and the Use of Scripture in the Parable of the Vineyard." *Dead Sea Discoveries* 2 (1995): 268-94.
Broshi, M. "Wine in Ancient Palestine—Introductory Notes." *IMJ* 3 (1984): 21-40.
_____. "The Diet of Palestine in the Roman Period—Introductory Notes." *IMJ* 5 (1986): 41-56.
Broshi, M., and I. Finkelstein. "The Population of Palestine in Iron Age II." *BASOR* 287 (1992): 47-60.
Brothwell, D., and P. *Food in Antiquity.* London: Thames & Hudson, 1969.
Brown, F., S. R. Driver, and C. A. Briggs. *A Hebrew and English Lexicon of the Old Testament.* Oxford: Clarendon, 1907.
Brown, J. "The Mediterranean Vocabulary of the Vine." *VT* 19 (1969): 146-70.
Brueggemann, W. *Genesis: A Bible Commentary for Teaching and Preaching.* Interpretation. Atlanta: John Knox, 1982.
Brush, S. B. *Mountain, Field, and Family: The Economy and Human Ecology of an Andean Valley.* Philadelphia: University of Pennsylvania Press, 1977.
Buhl, M. *Shiloh I.* Copenhagen: The National Museum of Denmark, 1969.
Burford, A. *Land and Labor in the Greek World.* Baltimore: Johns Hopkins University Press, 1993.
Burkert, W. "Oriental Symposia: Contrasts and Parallels." Pages 7-21 in *Dining in a Classical Context.* Edited by W. Slater. Ann Arbor, Mich.: University of Michigan Press, 1991.
Cahill, J. M., G. Lipton, and D. Tarler. "Tell el Hammah, 1988." *IEJ* 38 (1988): 191-94.
Calabresi, A. T. "*Vin Santo* and Wine in a Tuscan Farmhouse." Pages 122-34 in *Constructive Drinking: Perspectives on Drink from Anthropology.* Edited by M. Douglas. Cambridge, Eng.: Cambridge University Press, 1987.
Callaway, J. A. "The 1966 'Ai (et-Tell) Excavations." *BASOR* 196 (1969): 2-16.
_____. "Village Subsistence at Ai and Raddana in the Iron Age." Pages 51-66 in *The Answers Lie Below: Essays in Honor of Lawrence Edmund Toombs.* Edited by H. O. Thompson. Lanham, Md.: University Press of America, 1984.
Campbell, E. F. "The Shechem Area Survey." *BASOR* 190 (1968): 19-41.
Campbell, J. K. "Karpofora: Reluctant Farmers on a Fertile Land." Pages 207-21 in *Regional Variation in Modern Greece and Cyprus.* Edited by M. Dimen and E. Friedl. New York: New York Academy of Sciences, 1976.

Caneva, I. "Recipienti per liquidi nelle culture pastorali dell'alto Nilo." Pages 209-26 in *Drinking in Ancient Societies*. Edited by L. Milano. History of the Ancient Near East/Studies-6; Padova: Sargon, 1994.

Cassuto, U. "The Gezer Calendar and its Historical-Religious Value." Pages 211-28 in vol. 2 of *Biblical and Oriental Studies*. Translated by I. Abrahams. Jerusalem: Magnes, 1975.

Cato. *On Agriculture*. Translated by W. D. Hooper. Loeb Classical Library. Cambridge, Mass.: Harvard University Press, 1936.

Childe, V. G. "The Neolithic Revolution." Pages 67-72 in *Man Makes Himself*. London: Watts, 1951.

Childs, B. S. *The Book of Exodus: A Critical, Theological Commentary*. Old Testament Library. Philadelphia: Westminster, 1974.

Clark, C., and M. Haswell. *The Economics of Subsistence Agriculture*. London: Macmillan, 1967.

Clarke, N. H. *Deliver Us from Evil: An Interpretation of American Prohibition*. New York: Norton, 1976.

Cohen, A. *Arab Border-Villages in Israel: A Study of Continuity and Change in Social Organization*. Manchester: Manchester University Press, 1965.

Cohen, D. "On Viticulture and Wine—in Israel and the Ancient World." (Hebrew) *Beth Mikra* 37 (1991-92): 59-69.

Columella. *On Agriculture*. Translated by Harrison Boyd Ash. 3 vols. Loeb Classical Library. Cambridge, Mass.: Harvard University Press, 1941.

_____. *On Trees*. Translated by Harrison Boyd Ash. Loeb Classical Library. Cambridge, Mass.: Harvard University Press, 1941.

Coogan, M. D., ed. and trans. *Stories from Ancient Canaan*. Philadelphia: Westminster, 1978.

Conder, C. R., and Kitchener, H. H. *The Survey of Western Palestine*. Vol. 1. London: Palestine Exploration Fund, 1881.

Cox, J. *From Vines to Wines*. New York: Harper & Row, 1985.

Craigie, P. C., P. H. Kelly, and J. F. Drinkard. *Jeremiah 1-25*. Word Biblical Commentary. Dallas: Word Books, 1991.

Cross, F. M. "Epigraphic Notes on Hebrew Documents of the Eighth-Sixth Centuries B. C. I: A New Reading of a Place Name in the Samaria Ostraca." *BASOR* 163 (1961): 12-14.

_____. "Epigraphical Notes on Hebrew Documents of the Eighth-Sixth Centuries B.C. III: The Inscribed Jar Handles from Gibeon." *BASOR* 168 (1962): 18-23.

_____. "Jar Inscriptions from Shiqmona." *IEJ* 18 (1968): 226-33.

_____. "Judean Stamps." *EI* 9 (1969): 20-27.

_____. *Canaanite Myth and Hebrew Epic*. Cambridge: Harvard University Press, 1973.

Cross, F. M., and D. N. Freedman. *Early Hebrew Orthography*. New Haven, Conn.: American Oriental Society, 1952.

_____. *Studies in Ancient Yahwistic Poetry*. Missoula, Mont.: Scholars Press, 1975.

Crowfoot, G. M., and L. Baldensperger. *From Cedar to Hyssop: A Study in the Folklore of Plants in Palestine*. London: Sheldon, 1932.

Dagan, Y. "Bet Shemesh and Nes Harim Maps, Survey." *Excavations and Surveys in Israel* 13 (1995): 94-95.

Dalley, S. *Myths from Mesopotamia*. Oxford: Oxford University Press, 1991.

Dalman, G. *Arbeit und Sitte in Palästina*. 7 vols. Gütersloh: Bertelsmann, 1928-39.

_____. "Notes on the Old Hebrew Calendar-Inscription from Gezer." *Palestine Exploration Fund Quarterly Statement* 41 (1909): 118-19.

Dalton, G., ed. *Primitive, Archaic, and Modern Economies: Essays of Karl Polanyi*. Garden City, N.Y.: Doubleday, 1968.

Dar, S. *Landscape and Pattern: An Archaeological Survey of Samaria 800 BCE –636 CE.* 2 vols. Oxford: B.A.R. International Series, 1986.

Darby, W. J., P. Ghalioungui, and L. Grivetti. *Food: The Gift of Osiris.* 2 vols. London: Academic Press, 1977.

David, R. "La maison à piliers dans l'argumentation concernent l'émergence d'Israel en Palestine à l'époque du fer I." Pages 53-69 in *Où demeures-tu?* Edited by J.-C. Petit. Saint-Laurent, Québec, Canada: Éditions Fides, 1994.

Davidson, R. *Genesis 12-50.* Cambridge Bible Commentary. Cambridge, Eng.: Cambridge University Press, 1979.

_____. *Jeremiah.* 2 vols. Daily Study Bible. Philadelphia: Westminster, 1985.

Dayagi-Mendels, M. *Drink and Be Merry: Wine and Beer in Ancient Times.* Jerusalem: Israel Museum, 1999.

Detienne, M. *Dionysos at Large.* Cambridge, Mass.: Harvard University Press, 1989.

Detienne, M., and J.-P. Vernant. *The Cuisine of Sacrifice among the Greeks.* Chicago: University of Chicago Press, 1989.

Deutsch, R., and M. Heltzer. "A Wine Decanter with a Hebrew Inscription." Pages 23-46 in *Forty New Ancient West Semitic Inscriptions.* Tel Aviv-Jaffa: Archaeological Center, 1994.

Dietrich, M., O. Loretz, and J. Sanmartín. *Die Keilalphabetischen Texte aus Ugarit.* Neukirchen-Vluyn: Neukirchener, 1976.

Dimen, M., and E. Friedl, eds. *Regional Variation in Modern Greece and Cyprus: Toward a Perspective on the Ethnography of Greece.* New York: New York Academy of Sciences, 1976.

Dixon, S. *The Roman Family.* Baltimore: Johns Hopkins University Press, 1992.

Dolgin, J., D. Kemmitzer, and D. Schneider, eds. *Symbolic Anthropology: A Reader in the Study of Symbols and Meanings.* New York: Columbia University Press, 1977.

Dommershausen, W. "יין." Pages 59-64 in vol. 6 of *Theological Dictionary of the Old Testament.* Edited by G. J. Botterweck and H. Ringgren. Grand Rapids, Mich.: Eerdmans, 1985.

Donner, H., and W. Röllig. *Kanaanäische und aramäische Inschriften.* Wiesbaden: Harrassowitz, 1966.

Dothan, T. "In the Days When the Judges Ruled—Research on the Period of the Settlement and the Judges." Pages 35-41 in *Recent Archaeology in the Land of Israel.* Edited by H. Shanks and B. Mazar. Washington, D.C.: Biblical Archaeology Society, 1984.

Dothan, T., and S. Gitin. "Miqne, Tel (Ekron)." Pages 1051-59 in vol. 3 of *The New Encyclopedia of Archaeological Excavations in the Holy Land.* Edited by E. Stern. 4 vols. Jerusalem: Israel Exploration Society and Carta, 1993.

_____. "Ekron." Pages 415-22 of vol. 2 of *Anchor Bible Dictionary.* Edited by D. N. Freedman. 6 vols. New York: Doubleday, 1992.

Douglas, M. Introduction to *Constructive Drinking: Perspectives on Drink from Anthropology.* Cambridge, Eng.: Cambridge University Press, 1987.

_____., ed. *Constructive Drinking: Perspectives on Drink from Anthropology.* Cambridge, Eng.: Cambridge University Press, 1987.

_____., ed. *Rules and Meanings: The Anthropology of Everyday Knowledge.* Harmondsworth, Eng.: Penguin, 1973.

_____., ed. *Implicit Meanings: Essays in Anthropology.* London: Routledge & Kegan Paul, 1978.

Driver, S. R. *A Critical and Exegetical Commentary on Deuteronomy.* 3d ed. International Critical Commentary 3. Edinburgh: T. & T. Clark, 1902.

Driver, S. R., and G. B. Gray. *The Book of Job.* New York: Scribner's, 1921.

Edelstein, G., and S. Gibson. "Ancient Jerusalem's Rural Food Basket." *BAR* 8 (1982): 46-54.

Edelstein, G., and M. Kislev. "Mevasseret Yerushalayim: The Ancient Settlement and Its Agricultural Terraces." *BA* 44 (1981): 53-56.

Eickelman, D. F. *The Middle East: An Anthropological Approach.* Englewood Cliffs, N. J.: Prentice-Hall, 1981.

Einsett, J., and B. H. Barritt. "The Inheritance of Three Major Fruit Colors in Grapes." *Journal of the American Society of Horticultural Science* 94 (1969): 87-89.

Eitam, D. "Olive Presses of the Israelite Period." *Tel Aviv* 6 (1979): 146-55.

Euripides. *Bacchae.* Translated by E. R. Dodds. Oxford: Clarendon, 1960.

Eyre, C. "The Agricultural Cycle, Farming, and Water Management in the Ancient Near East." Pages 175-89 of vol. 1 in *Civilizations of the Ancient Near East.* Edited by J. M. Sasson. 4 vols. New York: Scribner's, 1995.

Fales, F. M. "A Fresh Look at the Nimrud Wine Lists." Pages 361-80 in *Drinking in Ancient Societies.* Edited by L. Milano. History of the Ancient Near East/Studies–6; Padova: Sargon, 1994.

Finkelstein, I. *'Izbet Sartah: An Early Iron Age Site near Rosh Ha'ayin, Israel.* Oxford: B.A.R. International Series, 1986.

_____. *The Archaeology of the Israelite Settlement.* Jerusalem: Israel Exploration Society, 1988.

Finkelstein, I., and R. Gophna. "Settlement, Demographic, and Economic Patterns in the Highlands of Palestine in the Chalcolithic and Early Bronze Periods and the Beginning of Urbanism." *BASOR* 289 (1993): 1-22.

Finkelstein, I., and N. Na'aman, eds. *From Nomadism to Monarchy: Archaeological and Historical Aspects of Early Israel.* Jerusalem: Israel Exploration Society and Yad Izhak Ben-Zvi, 1994.

Finley, M. I. *The Ancient Economy.* London: Chatto & Windus, 1973.

Fisher, H. S. E. "Wine: The Geographic Elements. Climate, Soil and Geology Are the Crucial Catalysts." *Geographical Magazine* 51 (1978): 86.

Forbes, R. J. *Studies in Ancient Technology.* Vol. 3. Leiden: Brill, 1955.

Foster, M. L., and S. Brandes, eds. *Symbol as Sense: New Approaches to the Analysis of Meaning.* New York: Academic Press, 1980.

Fox, M. *Qohelet and His Contradictions.* Sheffield, Eng.: Almond, 1989.

Frankel, R. *The History of the Processing of Wine and Oil in Galilee in the Period of the Bible, the Mishna and the Talmud* (Hebrew with English Summary). Tel Aviv: Tel Aviv University Press, 1984.

_____. "Screw Weights from Israel." Pages 35-42 in *Oil and Wine Production in the Mediterranean Area, Bulletin de Correspondance Hellénique*: Supplement 26. Edited by M.-C. Amouretti and J.-P. Brun. Paris: Boccard, 1993.

_____. *Wine and Oil Production in Antiquity in Israel and Other Mediterranean Countries.* JSOT/ASOR Monographs 10; Sheffield, Eng.: Sheffield University Press, 1999.

Frankel, R., S. Avitsur, and E. Ayalon. *History and Technology of Olive Oil in the Holy Land.* Translated by J. C. Jacobson. Tel Aviv: Eretz Israel Museum, 1994.

Freedman, D. N. *Pottery, Poetry, and Prophecy: Studies in Early Hebrew Poetry.* Winona Lake, Ind.: Eisenbrauns, 1980.

_____., ed. *Anchor Bible Dictionary.* 6 vols. New York: Doubleday, 1992.

Frick, F. S. *The City in the Old Testament.* Missoula, Mont.: Scholars Press, 1977.

_____. *The Formation of the State in Ancient Israel.* Sheffield, Eng.: Almond, 1985.

_____. "The Rechabites Reconsidered." *JBL* 90 (1971): 279-87.

Friedl, E. *Vasilika: A Village in Modern Greece.* New York: Holt, Rinehart & Winston, 1962.

Gal, Z. "Loom Weights or Jar Stoppers?" *IEJ* 39 (1989): 281-83.

Gardiner, A. H. *Ancient Egyptian Onomastica.* 2 vols. London: Oxford University Press, 1947.

Garnsey, P. *Famine and Food Supply in the Graeco-Roman World*. Cambridge, Eng.: Cambridge University Press, 1988.

Gaster, T. H. *Myth, Legend, and Custom in the Old Testament*. New York: Harper & Row, 1969.

Geertz, C. *Myth, Symbol, and Culture*. New York: Norton, 1971.

Gelb, I. J. "Household and Family in Early Mesopotamia." Pages 1-97 in *State and Temple Economy in the Ancient Near East*. Edited by E. Lipinski. Leuven: Departemant Oriëntaliek, 1979.

Genesis Rabbah. I. Translated by H. Freedman and M. Simon. London: Soncino. 1983.

Gesenius, W. *Gesenius' Hebrew Grammar*. Edited by E. Kautzsch. Translated by A. E. Cowley 2d ed. Oxford: Oxford University Press, 1910.

de Geus, C. H. J. "The Importance of Archaeological Research into the Palestinian Agricultural Terraces with an Excursus on the Hebrew word *gbi*." *PEQ* 107 (1975): 65-74.

Gitin, S. "Tel Miqne Ekron in the 7th c BCE: City Plan and Development." Pages 81-97 in *Olive Oil in Antiquity: Israel and Neighbouring Countries from Neolith to Early Arab Period*. Edited by M. Heltzer and D. Eitam. Haifa: University of Haifa Press, 1987.

Goody, J. *The Oriental, the Ancient and the Primitive*. Cambridge, Eng.: Cambridge University Press, 1990.

Goor, A. "The History of the Grape-vine." *Economic Botany* 20 (1966): 46-64.

Goor, A., and M. Nurock. *The Fruits of the Holy Land*. Jerusalem: Israel Universities, 1968.

Gordis, R. *Koheleth: The Man and His World*. New York: Jewish Theological Seminary, 1951.

_____. *The Song of Songs and Lamentations*. 2d ed. New York: Ktav, 1974.

Gordon, C. H. "Azitawadd's Phoenician Inscription." *JNES* 8 (1949): 108-15.

_____. *Ugaritic Literature*. Rome: Pontifical Biblical Institute, 1949.

_____. *Ugaritic Textbook*. Rome: Pontifical Biblical Institute, 1965.

Gorney, R. "Viniculture and Ancient Anatolia." Pages 133-74 in *The Origins and Ancient History of Wine*. Edited by P. McGovern, S. Fleming, and S. Katz. Luxemburg: Gordon & Breach, 1995.

Gove, P., ed. *Webster's Third New International Dictionary*. 3d ed. Springfield, Mass.: Merriam, 1965.

Grace, V. R. "The Canaanite Jar." Pages 80-109 in *The Aegean and the Near East*. Edited by S. Weinberg. Locust Valley, N.Y.: J. J. Augustin, 1956.

_____. *Amphoras and the Ancient Wine Trade*. Princeton: American School of Classical Studies at Athens, 1961.

Graham, J. N. "'Vinedressers and Plowmen': 2 Kings 25:12 and Jeremiah 52:16." *BA* 47 (1984): 55-58.

Grant, E. *Rumeileh, Being Ain Shems Excavations (Palestine)*. Part 3. Haverford: Haverford College, 1934.

Grant, E., and Wright, G. E. *Ain Shems Excavations (Palestine)*.Part 4 (Pottery). Haverford: Haverford College, 1938.

_____. *Ain Shems Excavations (Palestine)*. Part 5 (Text). Haverford: Haverford College, 1939.

Grant, M. *Cities of Vesuvius: Pompeii and Herculaneum*. London: Weidenfeld & Nicolson, 1971.

Gray, G. B. "An Old Hebrew Calendar-Inscription From Gezer, 2." *Palestine Exploration Fund Quarterly Statement* 41 (1909): 30-33.

_____. "The Gezer Inscription." *Palestine Exploration Fund Quarterly Statement* 41(1909): 189-93.

Gray, G. B., and A. S. Peake. *A Critical and Exegetical Commentary on the Book of Isaiah*. International Critical Commentary 18. Edinburgh: T. & T. Clark, 1912.

Greenberg, M. "Drunkenness." Page 872 in vol. 1 of *The Interpreter's Dictionary of the Bible*. Edited by G. Buttrick et al. Nashville: Abingdon, 1962.

Greenfield, J. C. "The *Marzēaḥ* as a Social Institution." *Acta Antiqua* 22 (1974): 451-55.

Grivetti, L. "Wine: The Food with Two Faces." Pages 9-22 in *The Origins and Ancient History of Wine*. Edited by P. McGovern, S. Fleming, and S. Katz. Luxemburg: Gordon & Breach, 1995.

Gusfield, J. *The Culture of Public Problems: Drinking, Driving and the Symbolic Order*. Chicago: University of Chicago, 1981.

Hackett, J. A. "Rehabilitating Hagar." Pages 12-27 in *Gender and Difference in Ancient Israel*. Edited by P. L. Day. Minneapolis: Fortress, 1989.

Haman, M. "The Iron Age II Sites of the Western Negev Highlands." *IEJ* 44 (1994): 36-61.

Hamel, G. *Poverty and Charity in Roman Palestine, First Three Centuries C.E.* Berkeley: University of California Press, 1990.

Hayes, W. C. "Daily Life in Ancient Egypt." *National Geographic* 80 (1941): 419-516.

_____. "Inscriptions from the Palace of Amenhotep III." *Journal of Near Eastern Studies* 10 (1951): 35-56, 82-112, 156-83.

Helbaek, H. "Les empreintes de céréals," Appendix I. Pages 205-7 in *Hama, Fouilles et recherches de la fondation Carlsberg II, 3*. Edited by P. J. Riis. Copenhagen: Gyldendalske Boghandel, 1948.

_____. "Plant Economy in Ancient Lachish," Appendix A. Pages 309-17 in *Lachish IV*. Edited by O. Tufnell. London: Oxford University Press, 1958.

_____. "Late Cypriot Vegetable Diet at Apliki." Pages 17-86 in vol. 4 of *Opuscula Atheniensia*. Lund: CWK Gleerup, 1962.

Heltzer, M. *The Rural Community in Ancient Ugarit*. Wiesbaden: Ludwig Reichert, 1976.

_____. "Vineyards and Wine in Ugarit (Property and Distribution)." *UF* 22 (1990): 119-35.

Heltzer, M., and D. Eitam, eds. *Olive Oil in Antiquity: Israel and Neighbouring Countries from Neolith to Early Arab Period*. Haifa: University of Haifa Press, 1987.

_____. *The History and Technology of Olive Oil in the Holy Land*. Olearius Editions: Arlington, Va. and Tel Aviv: Eretz Israel Museum, 1994.

Herdner, A. *Corpus des Tablettes en Cunéiformes Alphabétiques*, X. Paris: Librairie Orientaliste de Paul Geuthner, 1963.

Herodotus. *Herodotus*. Translated by A. D. Godley. 4 vols. Loeb Classical Library. Cambridge, Mass.: Harvard University Press, 1928-30.

Herzog, Z. "Persian Period Stratigraphy and Architecture (Strata XI-VI)." Pages 88-114 in *Excavations at Tel Michal, Israel*. Edited by Z. Herzog, G. Rapp, Jr., and O. Negbi. Minneapolis: University of Minnesota Press and Tel Aviv: Sonia and Marco Nadler Institute of Archaeology, Tel Aviv University Press, 1989.

_____. "Hellenistic Period Stratigraphy and Architecture (Strata V-III)." Pages 165-76 in *Excavations at Tel Michal, Israel*. Edited by Z. Herzog, G. Rapp, Jr., and O. Negbi. Minneapolis: University of Minnesota Press and Tel Aviv: Sonia and Marco Nadler Institute of Archaeology, Tel Aviv University Press, 1989.

_____. "A Complex of Iron Age Winepresses (Strata XIV-XIII)." Pages 73-75 in *Excavations at Tel Michal, Israel*. Edited by Z. Herzog, G. Rapp, Jr., and O. Negbi. Minneapolis: University of Minnesota Press and Tel Aviv: Sonia and Marco Nadler Institute of Archaeology, Tel Aviv University Press, 1989.

_____. "Fortifications: Bronze and Iron Ages." Pages 322-26 in vol. 2 of *The Oxford Encyclopedia of Archaeology in the Near East*. Edited by E. M. Meyers. 5 vols. New York: Oxford University Press, 1997.

Herzog, Z., G. Rapp, Jr., and O. Negbi. *Excavations at Tel Michal, Israel.*
Minneapolis: University of Minnesota Press and Tel Aviv: Sonia and Marco
Nadler Institute of Archaeology, Tel Aviv University Press, 1989.

Hesiod. *Works and Days.* Pages 3-65 in *The Homeric Hymns and Homerica.*
Translated by H. G. Evelyn-White. Loeb Classical Library. London: William
Heinemann, 1920.

Hillers, D. R. *Lamentations.* Anchor Bible 7A. Garden City, N.Y.: Doubleday,
1972.

Hirsch, S. *Genesis.* London: L. Honig and Sons, 1963.

Holladay, J. S., Jr. "House, Israelite." Pages 308-17 in vol. 3 of *Anchor Bible
Dictionary.* Edited by D. N. Freedman. 6 vols. New York: Doubleday, 1992.

_____. "The Kingdoms of Israel and Judah: Political and Economic
Centralization in the Iron IIA-B (CA. 1000-750 BCE)." Pages 368-98 in *The
Archaeology of Society in the Holy Land.* Edited by T. E. Levy. New York:
Facts on File, 1995.

_____. "Houses, Syro-Palestinian." Pages 94-114 in vol. 3 of *The Oxford
Encyclopedia of Archaeology in the Near East.* Edited by E. M. Meyers. 5
vols. New York: Oxford University Press, 1997.

Holladay, W. L. *A Concise Hebrew and Aramaic Lexicon of the Old Testament.*
Grand Rapids, Mich.: Eerdmans, 1978.

_____. *Jeremiah 1.* Hermeneia. Philadelphia: Fortress, 1986.

_____. *Jeremiah 2.* Hermeneia. Minneapolis: Fortress, 1991.

Homer. *Iliad.* Translated by A. T. Murray. 2 vols. Loeb Classical Library.
Cambridge, Mass.: Harvard University Press, 1976-78.

_____. *Odyssey.* Translated by A. T. Murray. Revised by G. Dimock. 2 vols.
Loeb Classical Library. Cambridge, Mass.: Harvard University Press, 1995.

Honeyman, A. M. "Epigraphic Discoveries at Karatepe." *PEQ* (1949): 21-39.

Hope, C. *Jar Sealings and Amphorae of the 1st Dynasty: A Technological Study.*
Warminster, Eng.: Aris & Phillips, 1977.

Hopf, M. "Jericho Plant Remains." Pages 576-621 in vol. 5 of *Excavations at
Jericho.* Edited by K. Kenyon and T. A. Holland. Jerusalem: British School
of Archaeology in Jerusalem, 1983.

Hopf, M. "Plant Remains, Strata V-I." Pages 64-87 in *Early Arad.* Edited by R.
Amiran. Jerusalem: Israel Exploration Society, 1978.

Hopkins, D. C. "The Dynamics of Agriculture in the Early Iron Age." SBL Seminar
Papers 22. Atlanta: Scholars Press, 1983.

_____. *The Highlands of Canaan: Agricultural Life in the Early Iron Age.* The
Social World of Biblical Antiquity Series 3. Sheffield, Eng.: Almond, 1985.

Hopkins, K. "On the Probable Age Structure of the Roman Population." *Population
Studies* 20 (1966): 245-64.

Hunt, M. "The Pottery." Pages 139-223 in *Tell Qiri: A Village in the Jezreel Valley:
Report of the Archaeological Excavations 1975-1977.* Edited by A. Ben-Tor
and Y. Portugali. Jerusalem: The Institute of Archaeology, The Hebrew
University of Jerusalem, 1987.

Hyams, Edward. *Dionysus: A Social History of the Wine Vine.* London: Thames &
Hudson, 1965.

Ibrahim, M. "The Collared-Rim Jar of the Early Iron Age." Pages 117-26 in
Archaeology in the Levant. Edited by R. Moorey and P. Parr. Warminster,
Eng.: Aris & Phillips, 1978.

Ingholt, H. "Un nouveau thiase à Palmyré." *Syria* 7 (1926): 128-41.

Israel, Y. "Ashqelon." *Excavations and Surveys in Israel* 13 (1995): 102-3.

Jacobsen, T. *Salinity and Irrigation Agriculture in Antiquity.* Malibu, Calif.:
Undena, 1982.

Japhet, S. *I and II Chronicles: A Commentary.* Old Testament Library. Louisville
Ky.: Westminster/John Knox, 1993.

Jastrow, M. *A Dictionary of the Targumim, The Talmud Babli and Yerushalmi, and the Midrashic Literature.* New York: Putnam, 1903.
_____. "Wine in the Pentateuchal Codes." *JAOS* 33 (1913): 180-92.
Johnson, H. *Vintage: The Story of Wine.* New York: Simon & Schuster, 1989.
Jones, R. N. "Paleopathology." Pages 60-69 in vol. 5 of *Anchor Bible Dictionary.* Edited by D. N. Freedman. 6 vols. New York: Doubleday, 1992.
Joüon, P. *A Grammar of Biblical Hebrew.* Translated and edited by T. Muraoka. Rome: Pontifical Biblical Institute, 1993.
Kaiser, O. *Isaiah 1-12: A Commentary*, Old Testament Library. Philadelphia: Westminster, 1972.
Kaufman, I. T. "The Samaria Ostraca: A Study in Ancient Hebrew Palaeography. Text and Plates." Th.D. diss., Harvard University, 1966.
_____. "The Samaria Ostraca: An Early Witness to Hebrew Writing." *BA* 45 (1982): 229-39.
_____. "Samaria (Ostraca)." Pages 921-26 in vol. 5 of *Anchor Bible Dictionary.* Edited by D. N. Freedman. 6 vols. New York: Doubleday, 1992.
Kees, H. *Ancient Egypt.* London: Faber & Faber, 1961.
Kelm, G. L., and A. Mazar. "Three Seasons of Excavations at Tel Batash—Biblical Timnah." *BASOR* 248 (1982): 1-36.
_____. "Tel Batash (Timnah) 1987-1988." *IEJ* 39 (1989): 108-10.
Kelso, J. L. "The Ceramic Vocabulary of the Old Testament." *BASOR Supplemental Studies* 5-6 (1948): 3-48.
Kempinski, A. "Shiloh." Pages 1364-66 in vol. 4 of *The New Encyclopedia of Archaeological Excavations in the Holy Land.* Edited by E. Stern. 4 vols. Jerusalem: Israel Exploration Society and Carta, 1993.
King, P. J. *Amos, Hosea, Micah: An Archaeological Commentary.* Philadelphia: Westminster, 1988.
_____. "The *Marzēaḥ*: Textual and Archaeological Evidence." *EI* 20 (1992) (Yigael Yadin memorial volume): 98*-106*.
Kislev, M. "Food Remains." Pages 354-61 in *Shiloh: The Archaeology of a Biblical Site.* Edited by I. Finkelstein, S. Bunimovitz, and Z. Lederman.. Tel Aviv: Tel Aviv University Press, 1993.
Kjaer, H. "The Excavation of Shiloh, Preliminary Report." *JPOS* 10 (1930): 87-174.
Klein, E. D. *A Comprehensive Etymological Dictionary of the Hebrew Language for Readers of English.* New York: Macmillan, 1987.
Kochavi, M., ed. *Yehudah, Shomron ve-Golan: Seker arke'ologi bi-shenat tav-shin -kaf-het.* (Hebrew). Jerusalem: Hotsaat ha-Agudah le seker arkhiologi shel Yisrael al yede Karta Yerushalayim, 1972.
_____. "The First Two Seasons of Excavations at Aphek-Antipatris." *Tel Aviv* 2 (1975): 17-42.
_____. "The History and Archaeology of Aphek-Antipatris." *BA* 44 (1981): 75 -86.
Koehler, L., and W. Baumgartner. *Hebräisches und Aramäisches Lexikon zum Alten Testament.* 5 vols. Leiden: Brill, 1990.
_____. *The Hebrew and Aramaic Lexicon of the Old Testament.* 4 vols. Leiden: Brill, 1994-95.
Kramer, C. "An Archaeological View of a Contemporary Kurdish Village: Domestic Architecture, Household Size, and Wealth." Pages 139-63 in *Ethnoarchaeology: Implications of Ethnography for Archaeology.* Edited by C. Kramer. New York: Columbia University Press, 1979.
Kramer, S. N. "The Deluge." In *ANET.* [See Pritchard, 1969] 42-44.
Kutscher, E. *The Language and Linguistic Background of the Isaiah Scroll.* Leiden: Brill, 1974.
Lance, H. D. "The Royal Stamps and the Kingdom of Josiah." *HTR* 64 (1971): 315-32.

Lapp, P. "The 1963 Excavation at Ta'annek." *BASOR* 173 (1964): 4-44.
Lapp, N. L. "Pottery Chronology of Palestine." Pages 433-44 in vol. 5 of *Anchor Bible Dictionary*. Edited by D. N. Freedman. 6 vols. New York: Doubleday, 1992.
Larsen, M. T. *The Old Assyrian City-State and Its Colonies*. Mesopotamia: Copenhagen Studies in Assyriology 4. Copenhagen: Akademisk, 1976.
Lehmann, M. R. "Biblical Oaths." *ZAW* 81 (1969): 74-92.
Lemaire, A. "*Zāmīr* dans la tablette de Gezer et le Cantique des Cantiques." *VT* 25 (1975): 15-26.
_____. *Inscriptions hébraïques*. Vol.1. Les Ostraca. Paris: Cerf, 1977.
Lesko, L. *King Tut's Wine Cellar*. Berkeley: B. C. Scribe, 1977.
Levenson, J. D. "On the Promise of the Rechabites." *CBQ* 38 (1976): 508-14.
Levy, T. E., ed. *The Archaeology of Society in the Holy Land*. New York: Facts on File, 1995.
Lewis, T. J. "Banqueting Hall/House." Pages 581-82 in vol. 1 of *Anchor Bible Dictionary*. Edited by D. N. Freedman. 6 vols. New York: Doubleday, 1992.
Lichtheim, M., ed. *Ancient Egyptian Literature*. Vol. 1. Berkeley: University of California Press, 1973.
Lidzbarski, M. "An Old Hebrew Calendar-Inscription from Gezer." *Palestine Exploration Fund Quarterly Statement* 41 (1909): 26-29.
_____. "The Calendar Inscription from Gezer." *Palestine Exploration Fund Quarterly Statement* 42 (1910): 238.
Lindenberger, J. M. *Ancient Aramaic and Hebrew Letters*. Atlanta: Scholars Press, 1994.
Lisitsina, G. N. "The Caucasus: A Centre of Ancient Farming in Eurasia." Pages 285-92 in *Plants and Ancient Man: Studies in Palaeoethnobotany*. Edited by W. van Zeist and W. A. Casparie. Rotterdam/Boston: A. A. Balkema, 1984.
Liverani, M. "Economy of Ugaritic Royal Farms." Pages 127-68 in *Production and Consumption in the Ancient Near East*. Edited by C. Zaccagnini. Budapest: University of Budapest Press, 1989.
Livy. *Livy*. 14 vols. Translated by B. O. Foster. Loeb Classical Library. Cambridge, Mass.: Harvard University Press, 1967-84.
Lloyd, S. *The Archaeology of Mesopotamia*. London: Thames & Hudson, 1978.
Lucas, A., and J. R. Harris. *Ancient Egyptian Materials and Industries*. 4th ed. London: Edward Arnold, 1962.
Luckenbill, D. D. *Ancient Records of Assyria and Babylonia*. 2 vols. Chicago: University of Chicago Press, 1927.
Lutfiyya, A. *Baytīn: A Jordanian Village: A Study of Social Institutions and Social Change in a Folk Community*. The Hague: Mouton, 1966.
_____. "The Family." Pages 505-25 in *Readings in Arab Middle Eastern Societies and Cultures*. Edited by A. Lutfiyya and C. Churchill. The Hague: Mouton, 1970].
Lutfiyya, A., and C. Churchill, eds. *Readings in Arab Middle Eastern Societies and Cultures*. The Hague: Mouton, 1970.
Lutz, H. F. *Viticulture and Brewing in the Ancient Orient*. Leipzig: J. C. Hinrichs, 1922.
Lyons, R. *Vine to Wine*. Napa, Calif.: Stonecrest, 1985.
MacDonald, D. B. "The Pre-Abrahamic Stories of Genesis as a Part of the Wisdom Literature." *Studia Semitica et Orientalia* (1920): 115-25.
Machinist, P. "Palestine, Administration of (Assyrian and Babylonian)." Pages 69-81 in vol. 5 of *Anchor BibleDictionary*. Edited by D. N. Freedman. 6 vols. New York: Doubleday, 1992.
MacRae, G. W. "The Meaning and Evolution of the Feast of Tabernacles." *CBQ* 22 (1960): 251-76.
Maisler, [Mazar], B. "The Historical Background of the Samaria Ostraca." *JPOS* 21 (1948): 117-33.

_____. "The Excavations at Tell Qâsile: Preliminary Report." *IEJ* 1 (1950-51): 61-76, 125-40, 194-218.

Marcus, R., and I. J. Gelb. "The Phoenician Stele Inscription from Cilicia." *JNES* 8 (1949): 116-20.

Mau, A. *Pompeii: Its Life and Art*. Washington, D.C.: McGrath, 1973.

Mazar, A. "Giloh: An Early Israelite Settlement Site near Jerusalem." *IEJ* 31 (1981): 1-36.

_____. "Iron Age Fortresses in the Judaean Hills." *PEQ* 114 (1982): 87-109.

_____. *Archaeology of the Land of the Bible*. New York: Doubleday, 1990.

McCown, C. C. *Excavations at Tell en-Nasbeh I: Archaeological and Historical Results*. Berkeley and New Haven: The Palestine Institute of the Pacific School of Religion and the American Schools of Oriental Research, 1971.

McDowell, A. "Agricultural Activity by the Workmen of Deir El-Medina." *JEA* 78 (1992): 195-206.

McGee, R. J., and R. L. Warms. "Symbolic and Interpretive Anthropology." Pages 430-79 in *Anthropological Theory: An Introductory History*. Mountain View, Calif.: Mayfield Publishing, 1996.

McGovern, P., S. Fleming, and S. Katz, eds. *The Origins and Ancient History of Wine*. Luxemburg: Gordon & Breach, 1995.

McLaughlin, J. L. "The Marzeah at Ugarit: A Textual and Contextual Study." *UF* 23 (1991): 265-81.

McNutt, P. M. *The Forging of Israel*. Sheffield, Eng.: Almond, 1990.

Meek, T. J. "The Code of Hammurabi." In *ANET* [See Pritchard, 1969] 163-80.

Meeks, D. "Oléiculture et viticulture dans l'Égypte pharaonique." Pages 3-38 in *Oil and Wine Production in the Mediterranean Area, Bulletin de Correspondance Hellénique*: Supplement 26. Edited by M.-C. Amouretti and J.-P. Brun. Paris: Boccard, 1993.

Meyers, E. M., ed. *The Oxford Encyclopedia of Archaeology in the Near East*. 5 vols. New York: Oxford University Press, 1997.

Michalowski, P. "Negation as Description: The Metaphor of Everyday Life in Early Mesopotamian Literature." *Aula Orientalis* 9 (1991): 131-36.

_____. "The Drinking Gods: Alcohol in Mesopotamian Ritual and Mythology." Pages 29-44 in *Drinking in Ancient Societies*. Edited by L. Milano. History of the Ancient Near East/Studies-6. Padova: Sargon, 1994.

Midrash Koheleth. Translated by A. Cohen. London: Soncino, 1951.

Milano, L., ed. *Drinking in Ancient Societies*. History of the Ancient Near East/Studies-6. Padova: Sargon, 1994.

_____. "Vino e birra in Oriente. Confini geografici e confini culturali." Pages 421-40 in *Drinking in Ancient Societies*. Edited by L. Milano. History of the Ancient Near East/Studies-6. Padova: Sargon, 1994.

Milgrom, J. *Leviticus 1-16*. Anchor Bible 3. New York: Doubleday, 1991.

_____. "Priestly ("P") Source." Pages 454-61 in vol. 5 of *Anchor Bible Dictionary*. Edited by D. N. Freedman. 6 vols. New York: Doubleday, 1992.

Miller, J. M., and J. H. Hayes. *A History of Ancient Israel and Judah*. Philadelphia: Westminster, 1986.

Miller, N., and K. Gleason, eds. *The Archaeology of Garden and Field*. Philadelphia: University of Pennsylvania Press, 1994.

Moldenke, H. N., and A. L. *Plants of the Bible*. New York: Dover Publications, 1952.

de Moor, J. C. *An Anthology of Religious Texts from Ugarit*. Nisaba. Leiden: Brill, 1987.

Na'aman, N. "Hezekiah's Fortified Cities." *BASOR* 261 (1986): 5-21.

_____. "The Kingdom of Judah under Josiah." *Tel Aviv* 18 (1991): 3-83.

Neumann, J. "On the Incidence of Dry and Wet Years." *IEJ* 6 (1956): 58-63.

Newberry, P. *Beni Hasan* I. London: Kegan Paul, Trench, Trübner, 1893.

Nicholson, E. W. *Deuteronomy and Tradition*. Philadelphia: Fortress, 1967.

Noth, M. "Der Beitrag der samarischen Ostraka zur Lösung topographischen Fragen." *Palästinajahrbuch* (1932): 54-57.

Nougayrol, J., ed. *Le Palais Royal d'Ugarit*. III. Paris: Imprimerie Nationale, 1955.

_____., ed. *Le Palais Royal d'Ugarit*. IV. Paris: Imprimerie Nationale, 1956.

_____., ed. *Le Palais Royal d'Ugarit*. VI. Paris: Imprimerie Nationale, 1970.

Nuñez, D. G., and M. J. Walker. "A Review of Palaeobotanical Findings of Early *Vitis* in the Mediterranean and of the Origins of Cultivated Grape-vines, with Special Reference to New Pointers to Prehistoric Exploitation in the Western Mediterranean." *Review of Palaeobotany and Palynology* 61 (1989): 205-37.

Oates, J. *Babylon*. 2d ed. New York: Thames & Hudson, 1986.

Ofer, A. "'All the Hill Country of Judah': From a Settlement Fringe to a Prosperous Monarchy." Pages 93-121 in *From Nomadism to Monarchy: Archaeological and Historical Aspects of Early Israel*. Edited by I. Finkelstein and N. Na'aman. Jerusalem: Israel Exploration Society, 1994.

Olmo, H. P. "Grapes." Pages 294-98 in *Evolution of Crop Plants*. Edited by N. W. Simmonds. Essex, Eng.: Longman Scientific & Technical, 1976.

_____. "The Origin and Domestication of the *Vinifera* Grape." Pages 31-44 in *The Origins and Ancient History of Wine*. Edited by P. McGovern, S. Fleming, and S. Katz. Luxemburg: Gordon & Breach, 1995.

Oppenheim, A. L. "Babylonian and Assyrian Historical Texts." In *ANET*. [See Pritchard, 1969] 265-317.

_____. *Ancient Mesopotamia: Portrait of a Dead Civilization*. Revised and edited by Erica Reiner. Chicago: University of Chicago Press, 1977.

Oppenheim, A. L., and L. Hartman. *On Beer and Brewing Techniques in Ancient Mesopotamia*. *JAOS* Supplement 10. Baltimore, Md.: American Oriental Society, 1950.

Orni, E., and Efrat, E. *Geography of Israel*. 2d ed. Jerusalem: Israel Program for Scientific Translations, 1966.

Oswalt, J. N. *The Book of Isaiah: Chapters 1-39*. Grand Rapids, Mich.: Eerdmans, 1986.

Pardee, D. *Handbook of Ancient Hebrew Letters*. Missoula, Mont.: Scholars Press, 1980.

Patai, R. "The Middle East as a Culture Area." Pages 187-204 in *Readings in Arab Middle Eastern Societies and Cultures*. Edited by A. Lutfiyya and C. Churchill. The Hague: Mouton, 1970.

_____. "The Dynamics of Westernization in the Middle East." Pages 235-51 in *Readings in Arab Middle Eastern Societies and Cultures*. Edited by A. Lutfiyya and C. Churchill. The Hague: Mouton, 1970.

_____. "Familism and Socialization." Pages 578-82 in *Readings in Arab Middle Eastern Societies and Cultures*. Edited by A. Lutfiyya and C. Churchill. The Hague: Mouton, 1970.

Paul, S. M. *Amos*. Hermeneia. Minneapolis: Fortress, 1991.

Paul, S. M., and W. G. Dever. *Biblical Archaeology*. New York: Quadrangle/New York Times Book Co., 1974.

Pedersen, J. *Israel: Its Life and Culture*. 2 vols. London: Oxford University Press, 1926.

Perevolotsky, A. "Orchard Agriculture in the High Mountain Region of Southern Sinai." *Human Ecology* 9 (1981): 331-58.

Petrie, W. M. F. *Tools and Weapons*. London: University College Press, 1917.

Pilcher, E. J. "The Handwriting of the Gezer Tablet." *Palestine Exploration Fund Quarterly Statement* 42 (1910): 32-39.

Pinnock, F. "Considerations on the 'Banquet Theme' in the Figurative Art of Mesopotamia and Syria." Pages 15-26 in *Drinking in Ancient Societies*. Edited by L. Milano. History of the Ancient Near East/Studies-6; Padova: Sargon, 1994.

Plato. *Lysis, Symposium, Gorgias*. Translated by W. R. Lamb. Loeb Classical Library. Cambridge, Mass.: Harvard University Press, 1975.

_____. *Plato*. Translated by H. N. Fowler. 10 vols. Loeb Classical Library. Cambridge: Harvard University Press, 1914-29.

Plattner, S., ed. *Economic Anthropology*. Stanford: Stanford University Press, 1989.

Pliny, *Natural History*. Translated by J. Bostock and H. T. Riley. Loeb Classical Library. 10 vols. London: Henry G. Bohn, 1855-1857.

Poo, M.-C. *Wine and Wine Offering in the Religion of Ancient Egypt*. London: Kegan Paul International, 1995.

Pope, M. H. "A Divine Banquet at Ugarit." Pages 170-203 in *The Use of the Old Testament in the New and Other Essays*. Edited by J. Efird. Durham, N.C.: Duke University, 1972.

Porten, B. *Archives From Elephantine*. Berkeley: University of California Press, 1968.

Postgate, J. N. *Early Mesopotamia: Society and Economy at the Dawn of History*. London: Routledge, 1992.

Powell, A. *Food Resources and Food Systems in Two West Bank Villages*. Jerusalem: Arab Thought Forum, 1987.

Powell, M. A. "Weights and Measures." Pages 897-908 in vol. 6 of *Anchor Bible Dictionary*. Edited by D. N. Freedman. 6 vols. New York: Doubleday, 1992.

_____. "Wine and the Vine in Ancient Mesopotamia: The Cuneiform Evidence." Pages 97-121 in *The Origins and Ancient History of Wine*. Edited by P. McGovern, S. Fleming, and S. Katz. Luxemburg: Gordon & Breach, 1995.

Pritchard, J. B., ed. *Hebrew Inscriptions and Stamps from Gibeon*. Philadelphia: University Museum, 1959.

_____. *Gibeon: Where the Sun Stood Still: The Discovery of the Biblical City*. Princeton: Princeton University Press, 1962.

_____. *Winery, Defenses, and Soundings at Gibeon*. Philadelphia: University Museum, 1964.

_____., ed. *Ancient Near Eastern Texts Relating to the Old Testament*. 3d ed. Princeton: Princeton University Press, 1969.

_____., ed. *The Ancient Near East in Pictures Relating to the Old Testament*. 2d ed. Princeton: Princeton University Press, 1969.

_____. "Gibeon." Pages 511-14 in vol. 2 of *The New Encyclopedia of Archaeological Excavations in the Holy Land*. Edited by E. Stern. 4 vols. Jerusalem: Israel Exploration Society and Carta, 1993.

Rad, G. von. *Genesis*. Old Testament Library. Philadelphia: Westminster, 1972.

Rainey, A. F. "Administration in Ugarit and the Samaria Ostraca." *IEJ* 12 (1962): 62-63.

_____. "The Samaria Ostraca in the Light of Fresh Evidence." *PEQ* 99 (1967): 32-41.

_____. "Semantic Parallels to the Samaria Ostraca." *PEQ* 102 (1970): 45-51.

_____. "Wine From the Royal Vineyards." *BASOR* 245 (1982) 57-62.

Rast, W. E. *Taanach I: Studies in The Iron Age Pottery*. Cambridge, Mass.: American Schools of Oriental Research, 1978.

_____. *Through the Ages in Palestinian Archaeology: An Introductory Handbook*. Philadelphia: Trinity, 1992.

Reifenberg, A. *The Soils of Palestine*. Translated by C. L. Whittles. London: Thomas Murby and Co., 1947.

Reisner, G. A., C. S. Fisher, and D. G. Lyon. *Harvard Excavations at Samaria 1908-1910*. Vol.1. Cambridge, Mass.: Harvard University Press, 1924.

Renfrew, J. M. *Palaeoethnobotany: The Prehistoric Food Plants of the Near East and Europe.* New York: Columbia University Press, 1973.

Reviv, H. "The History of Judah from Hezekiah to Josiah." Pages 192-221 in vol. 4 of *World History of the Jewish People.* Edited by H. Ben-Sasson. Jerusalem: Massada, 1979.

Rice, P. *Pottery Analysis: A Sourcebook.* Chicago: University of Chicago Press, 1987.

Richards, I. A. *The Philosophy of Rhetoric.* New York: Oxford University Press, 1936.

Roaf, M. *Cultural Atlas of Mesopotamia and the Near East.* New York: Facts on File, 1990.

Robertson, D. S. *A Handbook of Greek and Roman Architecture.* Cambridge: Cambridge University Press, 1929.

Robertson, N. "Myth, Ritual, and Livelihood in Early Greece." Pages 3-34 in *Ancient Economy in Mythology: East and West.* Edited by M. Silver. Savage, Md.: Rowman & Littlefield, 1991.

Robinson, C. *Everyday Life in Ancient Greece.* Oxford: Clarendon, 1933.

Roll, I., and E. Ayalon. "Two Large Wine Presses in the Red Soil Regions of Israel." *PEQ* 113 (1981): 111-25.

Röllig, W. *Das Bier im Alten Mesopotamien.* Berlin: Gesellschaft für die Geschichte und Bibliographie des Brauwesens, 1970.

Ronzevalle, S. "The Gezer Hebrew Inscription." *Palestine Exploration Fund Quarterly Statement* 41 (1909): 107-112.

Ron, Z. "Agricultural Terraces in the Judean Mountains." *IEJ* 16 (1966): 33-49, 111-22.

_____. "Stone Huts as an Expression of Terrace Agriculture in the Judean and Samarian Hills." Ph.D. diss., Tel Aviv University, 1977.

Rorabaugh, W. J. *The Alcoholic Republic.* New York: Oxford University Press, 1979.

Rosen, B. "Subsistence Economy in Iron Age I." Pages 339-51 in *From Nomadism to Monarchy: Archaeological and Historical Aspects of Early Israel.* Edited by I. Finkelstein and N. Na'aman. Jerusalem: Israel Exploration Society, 1994.

Ross, J. F. "Wine." Pages 849-52 in vol. 4 of *The Interpreter's Dictionary of the Bible.* Edited by G. Buttrick et al. Nashville: Abingdon, 1962.

_____. "Meals." Pages 315-18 in vol. 3 of *The Interpreter's Dictionary of the Bible.* Edited by G. Buttrick et al. Nashville: Abingdon, 1962.

_____. "Vine." Pages 784-86 in vol. 4 of *The Interpreter's Dictionary of the Bible.* Edited by G. Buttrick et al. Nashville: Abingdon, 1962.

Russell, J. M. *Sennacherib's Palace without Rival at Nineveh.* Chicago: University of Chicago Press, 1991.

Saller, S., and E. Testa. *The Archaeological Setting of the Shrine of Bethphage.* Jerusalem: Franciscan Press, 1961.

Sanmartin-Ascano, J. "דוד." Pages 143-56 in vol. 3 of *Theological Dictionary of the Old Testament.* ed. G. J. Botterweck and H. Ringgren,. Grand Rapids, Mich.: Eerdmans, 1978.

Segert, S. *A Grammar of Phoenician and Punic.* Munich: Beck, 1980.

Seitz, C. R. *Isaiah 1-39.* Interpretation. Louisville: John Knox, 1993.

Seligman, J. "Jerusalem, Pisgat Ze'ev (East A)." *Excavations and Surveys in Israel* 12 (1993): 52-54.

_____. "A Late Iron Age Farmhouse at Ras Abu Ma'aruf, Pisgat Ze'ev A." *'Atiqot* 25 (1994) 63-75.

Selms, A. van. "The Etymology of *Yayin,* 'Wine'." *JNES* 3 (1974): 76-84.

Seltman, C. T. *Wine in the Ancient World.* London: Routledge & Kegan Paul, 1957.

Semple, E. C. *The Geography of the Mediterranean: Its Relation to Ancient History.* New York: Henry Holt, 1931.

Shavit, E. "Rishon Leziyyon." *Excavations and Surveys in Israel* 13 (1995): 57.

Shea, W. H. "The Date and Significance of the Samaria Ostraca." *IEJ* 27 (1977): 16-27.

Shewell-Cooper, W. E. *Plants, Flowers and Herbs of the Bible*. New Canaan, Conn.: Keats, 1977.

Shiloh, Y. "The Four-Room House: Its Situation and Function in the Israelite City." *IEJ* 20 (1970): 180-90.

_____. "The City of David Archaeological Project: The Third Season—1980." *BA* 44 (1981): 161-70.

_____. "Jerusalem: The Period of the Monarchy (Strata 14-10)." Pages 702-12 in vol. 2 of *The New Encyclopedia of Archaeological Excavations in the Holy Land*. Edited by E. Stern. 4 vols. Jerusalem: Israel Exploration Society and Carta, 1993.

Silver, M., ed. *Ancient Economy in Mythology: East and West*. Savage, Md.: Rowman & Littlefield, 1991.

Skinner, J. *A Critical and Exegetical Commentary on Genesis*. International Critical Commentary. New York: Scribner's, 1910.

Smith, J. Z. *Map Is Not Territory*. Chicago: University of Chicago Press, 1978.

in der Smitten, W. T. "מהר." Pages 465-70 in vol. 4 of *Theological Dictionary of the Old Testament*. Edited by G. J. Botterweck and H. Ringgren. Grand Rapids, Mich.: Eerdmans, 1980.

Speiser, E. A. "Southern Kurdistan in the Annals of Ashurnasirpal and Today." *Annual of the American Schools of Oriental Research* 8 (1926-27): 1-42.

_____. *Genesis*. Anchor Bible 1. Garden City, N.Y.: Doubleday, 1964.

_____. "Atrahasis." In *ANET*. [See Pritchard, 1969] 104-6.

Stager, L. E. "Agriculture." Pages 11-13 in *The Interpreter's Dictionary of the Bible Supplementary Volume*. Edited by K. Crim. Nashville: Abingdon, 1976.

_____. "Farming in the Judean Desert during the Iron Age," *BASOR* 221 (1976): 145-58.

_____. "The Archaeology of the East Slope of Jerusalem and the Terraces of the Kidron." *Journal of Near Eastern Studies* 41 (1982): 111-21.

_____. "The Finest Oil in Samaria." *JSS* 28 (1983): 241-45.

_____. "Merenptah, Israel and Sea Peoples: New Light on an Old Relief." *EI* 18 (1985) (Nahman Avigad volume): 56*-65.*

_____. "The Archaeology of the Family in Ancient Israel." *BASOR* 260 (1985): 1-35.

_____. "The Firstfruits of Civilization." Pages 177-88 in *Palestine in the Bronze and Iron Ages*. Edited by J. M. Tubb. London: Institute of Archaeology, 1985.

_____. "Shemer's Estate." *BASOR* 277/278 (1990): 93-107.

_____. "The Fury of Babylon: Ashkelon and the Archaeology of Destruction." *BAR* 22 (1996): 56-69, 76-77.

_____. "Jerusalem and the Garden of Eden." *EI* 26 (1999) (Frank Moore Cross volume): 183*-94*.

Steigelmann, W. *Der Wein in der Bibel*. Neustadt an der Weinstrasse: D. Meininger, 1971.

Steinberg, N. *Kinship and Marriage in Genesis*. Minneapolis: Fortress, 1993.

Steindorff, G., and K. C. Steele. *When Egypt Ruled the East*. Chicago: University of Chicago Press, 1942.

Stern, E., ed. *The New Encyclopedia of Archaeological Excavations in the Holy Land*. 4 vols. Jerusalem: Israel Exploration Society and Carta, 1993.

Stern, P. D. "The Origin and Significance of 'The Land Flowing with Milk and Honey'." *VT* 42 (1992): 554-57.

Strabo. *The Geography of Strabo*. Translated by H. L. Jones. 8 vols. Loeb Classical Library. London: William Heinemann, 1949-54.

Stronach, D. "The Imagery of the Wine Bowl: Wine in Assyria in the Early First
    Millennium B.C." Pages 175-95 in *The Origins and Ancient History of
    Wine*. Edited by P. McGovern, S. Fleming, and S. Katz. Luxemburg:
    Gordon & Breach, 1995.
Sweeney, M. A. *Isaiah 1-39*. The Forms of Old Testament Literature 16. Grand
    Rapids, Mich.: Eerdmans, 1996.
Sweet, L. "A Day in a Peasant Household." Pages 218-23 in *Readings in Arab
    Middle Eastern Societies and Cultures*. Edited by A. Lutfiyya and C.
    Churchill. The Hague: Mouton, 1970.
Tadmor, M., ed. *Inscriptions Reveal*. Jerusalem: Israel Museum, 1973.
Talmon, S. "The Gezer Calendar and the Seasonal Cycle of Ancient Canaan." *JAOS*
    83 (1963): 177-87.
Theognis of Megara. *Elegy and Iambus*. Translated by J. M. Edmonds. Loeb
    Classical Library. Cambridge, Mass.: Harvard University Press, 1982.
Theophrastus, *Enquiry into Plants*. Translated by B. Einarson and G. Link. 2 vols.
    Loeb Classical Library. Cambridge, Mass.: Harvard University Press, 1916,
    1926.
Todorov, T. *Symbolism and Interpretation*. Ithaca, N.Y.: Cornell University Press,
    1982.
Trigger, B. G., B. J. Kemp, D. O'Connor, and A. B. Lloyd. *Ancient Egypt: A
    Social History*. Cambridge, Eng.: Cambridge University Press, 1983.
Turkowski, L. "Peasant Agriculture in the Judaean Hills." *PEQ* 101 (1969): 21-33,
    101-12.
Turner, V. *Blazing the Trail: Way Marks in the Exploration of Symbols*. Tucson:
    University of Arizona Press, 1992.
_____. *The Forest of Symbols: Aspects of Ndembu Ritual*. Ithaca, N.Y.: Cornell
    University Press, 1967.
Tushingham, A. D. "New Evidence Bearing on the Two-Winged LMLK Stamp."
    BASOR 287 (1992): 61-65.
_____. "A Royal Israelite Seal (?) and the Royal Jar Handle Stamps (Part One)."
    BASOR 200 (1970): 71-8.
_____. "A Royal Israelite Seal (?) and the Royal Jar Handle Stamps (Part Two)."
    BASOR 201 (1971): 23-35.
Unwin, T. *Wine and the Vine: An Historical Geography of Viticulture and the Wine
    Trade*. London: Routledge, 1991.
Ussishkin, D. *The Conquest of Lachish by Sennacherib*. Tel Aviv: Tel Aviv
    University Press, 1982.
_____. "Excavations at Tel Lachish 1978-1983: Second Preliminary Report." *Tel
    Aviv* 10 (1983): 97-177.
Varro. *On Agriculture*. Translated by L. Storr-Best. Loeb Classical Library. London:
    G. Bell, 1912.
de Vaux, R. *Ancient Israel: Its Life and Institutions*. New York: McGraw-Hill,
    1961.
Vermeule, E. *Greece in the Bronze Age*. Chicago: University of Chicago Press,
    1964.
Vernant, J.-P. *Myth and Society in Ancient Greece*. New York: Zone Books, 1990.
Vickery, K. F. *Food in Early Greece*. Urbana, Ill.: University of Illinois Press,
    1936.
Virgil. *Georgics*. Translated by H. A. Fairclough. Loeb Classical Library.
    Cambridge, Mass.: Harvard University Press, 1986.
Virolleaud, C., ed. *Le Palais Royal d'Ugarit*. II. Paris: Imprimerie nationale, 1957.
_____., ed. *Le Palais Royal d'Ugarit*. V. Paris: Imprimerie nationale, 1957.
Waltke, B. K., and M. O'Connor. *An Introduction to Biblical Hebrew Syntax*.
    Winona Lake, Ind.: Eisenbrauns, 1990.
Watkins, C., ed. *American Heritage Dictionary of Indo-European Roots*. Boston:
    Houghton Mifflin, 1985.

Watson, E. A. *Religion and Drink.* New York: Burr Printing House, 1914.

Watson, P. J. *Archaeological Ethnography in Western Iran.* Viking Fund Publications in Anthropology 57. Tucson: University of Arizona Press, 1979.

Watts, J. D. W. *Isaiah 1-33.* Word Biblical Commentary 24. Waco, Tex.: Word, 1985.

Weinfeld, M. *Deuteronomy 1-11.* Anchor Bible 5. New York: Doubleday, 1991.

Weinhold, R. *Vivat Bacchus: A History of the Vine and its Wine.* Watford, Eng.: Argus Books, 1978.

Weizman, Z. "Ethnology, Etiology, Genealogy, and Historiography in the Tale of Lot and His Daughters (Genesis 19:30-38)." Pages 43-52 in *Sha'arei Talmon.* Edited by M. Fishbane and E. Tov. Winona Lake, Ind.: Eisenbrauns, 1992.

Wellhausen, J. *Prolegomena to the History of Ancient Israel.* New York: Meridian, 1957.

Westbrook, R. *Property and the Family in Biblical Law.* JSOT Supplement 113. Sheffield, Eng.: JSOT, 1991.

Westermann, C. *Genesis* II. Neukirchen-Vluyn: Neukirchener, 1981.

_____. *Genesis 1-11: A Commentary.* Translated by J. J. Scullion, S. J. Minneapolis: Augsburg, 1984.

_____. *Genesis 12-36: A Commentary.* Translated by J. J. Scullion, S. J. Minneapolis: Augsburg, 1984.

_____. *Genesis 37-50: A Commentary.* Translated by J. J. Scullion, S. J. Minneapolis: Augsburg, 1986.

White, K. D. *Roman Farming.* London: Thames & Hudson, 1970.

_____. "Farming and Animal Husbandry." Pages 211-45 in vol. 1 of *Civilizations of the Ancient Mediterranean.* Edited by M. Grant and R. Kitzinger. 3 vols. New York: Scribner's, 1988.

Whitaker, R. E. *A Concordance of the Ugaritic Literature.* Cambridge, Mass.: Harvard University Press, 1972.

Wilkinson, J. G. *The Manners and Customs of the Ancient Egyptians.* 3 vols. London: John Murray, 1842.

Wilkinson, T. "The Structure and Dynamics of Dry-Farming States in Upper Mesopotamia." *Current Anthropology* 35 (1994): 483-520.

Willis, J. T. "The Genre of Isaiah 5:1-7." *JBL* 96 (1977): 337-62.

Wilson, H. *Egyptian Food and Drink.* Aylesbury, Eng.: Shire, 1988.

Wilson, J. A. "The Story of Si-nuhe." In *ANET* [See Pritchard 1969] 18-22.

_____. "The Asiatic Campaigns of Thut-mose III." In *ANET* [See Pritchard, 1969] 234-41.

_____. "Asiatic Campaigns under Pepi I." In *ANET* [See Pritchard, 1969] 227-28.

Wilson, J. V. K. *The Nimrud Wine Lists: A Study of Men and Administration at the Assyrian Capital in the Eighth Century B.C.* Cuneiform Texts from Nimrud, vol.1. London: British School of Archaeology in Iraq, 1972.

Wilson, R. R. *Genealogy and History in the Biblical World.* Yale Near Eastern Researches. New Haven, Conn.: Yale University Press, 1977.

Winkler, A. J., J. A. Cook, M. M. Kliewer, and L. A. Lider. *General Viticulture.* 2d ed. Berkeley: University of California Press, 1974.

Wiseman, D. J. "Mesopotamian Gardens." *Anatolian Studies* 33 (1983): 137-44.

Wright, G. E. *Shechem: The Biography of a Biblical City.* New York: McGraw-Hill, 1965.

Yadin, Y. "Recipients or Owners, A Note on the Samaria Ostraca." *IEJ* 9 (1959): 184-87.

_____. *Hazor II.* Jerusalem: Magnes, 1960.

Yee, G. "The Form Critical Study of Isaiah 5:1-7 as a Song and as a Juridical Parable." *CBQ* 43 (1981): 30-40.

Young, I. "The Style of the Gezer Calendar and Some 'Archaic Biblical Hebrew Passages.'" *VT* 42 (1992): 365-75.

Young, J. H. "Studies in South Attica: Country Estates at Sounion." *Hesperia* 25 (1956): 122-46.

Zaccagnini, C., ed. *Production and Consumption in the Ancient Near East.* Budapest: University of Budapest Press, 1989.

Zapletal, V. *Der Wein in der Bibel.* Freiburg: Herder, 1920.

Zemer, A. *Storage Jars in Ancient Sea Trade.* Haifa: National Maritime Museum, 1978.

Zertal, A. *The Israelite Settlement in the Hill Country of Manasseh.* (Hebrew) Haifa: Haifa University Press, 1988.

Zettler, R., and N. Miller. "Searching for Wine in the Archaeological Record of Ancient Mesopotamia of the Third and Second Millennia B. C." Pages 123-31 in *The Origins and Ancient History of Wine.* Edited by P. McGovern, S. Fleming, and S. Katz. Luxemburg: Gordon & Breach, 1995.

Zimmerli, W. *Ezekiel 1: A Commentary on the Book of the Prophet Ezekiel, Chapters 1-24.* Hermeneia. Philadelphia: Fortress, 1979.

Zohary, D. "The Domestication of the Grapevine *Vitis Vinifera* L. in the Near East." Pages 23-30 in *The Origins and Ancient History of Wine.* Edited by P. McGovern, S. Fleming, and S. Katz. Luxemburg: Gordon & Breach, 1995.

Zohary, D., and P. Spiegel-Roy. "Beginning of Fruit Growing in the Old World." *Science* 187 (1975): 319-27.

Zohary, M. *Plant Life of Palestine.* New York: Ronald, 1962.

_____. *Plants of the Bible.* Cambridge, Eng.: Cambridge University Press, 1982.

Zohary, M., and M. Hopf. *Domestication of Plants in the Old World: The Origin and Spread of Cultivated Plants in West Asia, Europe, and the Nile Valley.* Oxford: Clarendon, 1988.

Zorn, J. R. "Tell en-Nasbeh." Pages 1098-1102 in vol. 3 of *The New Encyclopedia of Archaeological Excavations in the Holy Land.* Edited by E. Stern. 4 vols. Jerusalem: Israel Exploration Society and Carta, 1993.

_____. "Tell en-Nasbeh: A Re-evaluation of the Architecture and Stratigraphy of the Early Bronze Age, Iron Age and Later Periods." Ph. D. diss., University of California, Berkeley. 4 Vols. Ann Arbor, Mich.: UMI, 1993.

Scripture Index

Subject Index

intoxication, 6, 107-8, 227, 242,
253-57, 257n11

Iron Age, 12, 15, 16, 18n35, 20, 45,
59, 92, 107, 128-29, 129n7,
132, 134, 134n35, 135-36,
142-44, 144n75, 146, 150-51,
151n114, 152-53, 155,
157-58, 162-63, 164n196,
172, 174, 194, 201, 206,
212-13, 213n21, 214, 214n31,
217, 245, 249

lmlk, 7, 44, 50-51, 131, 219

Lot, 1, 5, 5n7, 6, 228, 234, 254,
256-57, 257n11

Mesopotamia(n), 10, 14, 19, 19n45,
21, 23-28, 113-14, 188,
200-201, 220nn74, 76,
222n85

Noah, 1, 5, 6, 6n12, 11, 13, 14,
14n14, 43, 53, 65n124, 220,
228, 254-57, 257n11

Pottery, 17, 18, 18n35, 44, 51-52, 57,
95, 135, 140, 150-51,
151n114, 152-53, 154n132,
155, 158, 213, 217-18,
218n63

Samaria Ostraca, 7, 16n27, 44, 46,
51-59, 65, 194, 206, 215, 218

terrace(s), 21, 70n148, 94-95, 95n27,
96, 98, 106, 144, 153

vine(s), 2, 3, 4, 5, 6n12, 8-11, 13,
13n10, 14-16, 19-20, 20n47,
21-27, 27n87, 29, 30, 30n106,
31-33, 36, 38-41, 45-46, 50,
61-63, 66, 69, 75-76, 83,
87-88, 88n2, 89n5, 90-91,
94-96, 98-99, 99n51, 100,
100n53, 101n58,

102-5, 105nn76, 78,
106-7, 107n84, 109-11,
112-19, 119n145, 120,
122-23, 123n163, 124-25,
127-28, 134, 134n38, 136,
140, 142-43, 150, 154,
171-74, 176, 178-79, 184,
202, 205nn192-93,
205n195, 211-13, 224,
231-32, 245n165, 249-52,
255-57

vineyard(s), 2-4, 7, 9-10, 14, 14n14,
15n22, 19-22, 24-25, 28n92,
31-33, 37, 40, 43-44, 44n3,
46, 51, 56, 58, 60-63, 66-85,
87-91, 93, 93n17, 94, 96-98,
98n47, 100-101, 101n57,
102-3, 105-6, 108, 110,
110n105, 111, 111n115, 112,
112n121, 113, 113n124,
114-16, 118-20, 122-24,
124n165, 125, 125n169,
127-29, 131-43, 145, 147,
150, 160-61, 163n192, 165,
167, 170, 172, 173n31,
174-83, 185, 187, 194, 197,
200, 205, 205n196, 209, 211,
221, 224, 229-30, 237, 246,
249-51, 255-57

vintner(s), 2, 7, 10, 20, 21, 31, 43-45,
59-60, 62-63, 67-68, 73,
76-77, 81, 83, 85, 88-89,
89n2, 90-92, 97-98, 100,
101n57, 103, 105-10,
111n115, 112, 114, 119-20,
122-23, 127-28, 130-31, 135,
137, 140-41, 143, 146-47,
154, 160-61, 161n187, 165,
167-70, 172n26, 173, 177,
179, 181, 215n33, 221-22,
224, 229-31, 245, 250, 252,
255, 258

Printed in the United States
By Bookmasters